CONSTRUCTION PROJECT SCHEDULING

McGraw-Hill Series in Construction Engineering and Project Management

Consulting Editors
Raymond E. Levitt, *Stanford University*
Jerald L. Rounds, *Arizona State University*

Barrie and Paulson: *Professional Construction Management: Including CM, Design-Construct, and General Contracting*
Callahan, Quackenbush, and Rowings: *Construction Project Scheduling*
Jervis and Levin: *Construction Law: Principles and Practice*
Koerner: *Construction and Geotechnical Methods in Foundation Engineering*
Levitt and Samelson: *Construction Safety Management*
Oglesby, Parker, and Howell: *Productivity Improvement in Construction*
Peurifoy and Ledbetter: *Construction Planning, Equipment, and Methods*
Peurifoy and Oberlender: *Estimating Construction Costs*
Shuttleworth: *Mechanical and Electrical Systems for Construction*
Stevens: *Techniques for Construction Network Scheduling*

Also Available from McGraw-Hill

Schaum's Outline Series in Civil Engineering

Most outlines include basic theory, definitions, and hundreds of solved problems and supplementary problems with answers.

Titles on the Current List Include:
Advanced Structural Analysis
Basic Equations of Engineering
Descriptive Geometry
Dynamic Structural Analysis
Engineering Mechanics, 4th edition
Fluid Dynamics
Fluid Mechanics & Hydraulics
Introduction to Engineering Calculations
Introductory Surveying
Mathematical Handbook of Formulas & Tables
Mechanical Vibrations
Reinforced Concrete Design, 2d edition
Space Structural Analysis
State Space & Linear Systems
Statics and Strength of Materials
Strength of Materials, 2d edition
Structural Analysis
Structural Steel Design, LFRD Method
Theoretical Mechanics

Schaum's Solved Problems Books

Each title in this series is a complete and expert source of solved problems containing thousands of problems with worked out solutions.

Related Titles on the Current List Include:
3000 Solved Problems in Calculus
2500 Solved Problems in Differential Equations
2500 Solved Problems in Fluid Mechanics & Hydraulics
3000 Solved Problems in Linear Algebra
2000 Solved Problems in Numerical Analysis
800 Solved Problems in Vector Mechanics for Engineers: Dynamics
700 Solved Problems in Vector Mechanics for Engineers: Statics

Available at your College Bookstore. A complete list of Schaum titles
may be obtained by writing to: Schaum Division
                                  McGraw-Hill, Inc.
                                  Princeton Road, S-1
                                  Hightstown, NJ 08520

CONSTRUCTION PROJECT SCHEDULING

Michael T. Callahan
President, CCL Construction Consultants

Daniel G. Quackenbush, AIA
Vice President, CCL Construction Consultants

James E. Rowings, P.E.
Iowa State University

McGraw-Hill, Inc.
New York   St. Louis   San Francisco   Auckland   Bogotá
Caracas   Lisbon   London   Madrid   Mexico   Milan   Montreal
New Delhi   Paris   San Juan   Singapore   Sydney   Tokyo   Toronto

CONSTRUCTION PROJECT SCHEDULING

1 2 3 4 5 6 7 8 9 0 DOH DOH 9 0 9 8 7 6 5 4 3 2 1

ISBN 0-07-009701-1

This book was set in Times Roman by General Graphic Services, Inc.
The editors were B. J. Clark and James R. Belser;
the production supervisor was Denise L. Puryear.
The cover was designed by Carla Bauer.
R. R. Donnelley & Sons Company was printer and binder.

Library of Congress Cataloging-in-Publication Data

Callahan, Michael T.
    Construction project scheduling / Michael T. Callahan, Daniel G.
Quackenbush, James E. Rowings.
        p.   cm.
    Includes index.
    ISBN 0-07-009701-1
    1. Construction industry—Management.   2. Scheduling (Management)
3. Critical path analysis.   I. Quackenbush, Daniel G.   II. Rowings.
James E.   III. Title.
TH438.C26   1992
624'.068'5—dc20   91-36138

ABOUT THE AUTHORS

Michael T. Callahan is President of CCL Construction Consultants, Inc., a firm specializing in construction project management and the resolution of construction and design claims for the domestic and international construction industry. Mr. Callahan holds a BA from the University of Kansas, and both a J.D. and LL.M. from the University of Missouri—Kansas City, School of Law. Mr. Callahan has coauthored six other books on the construction process, including *Construction Delay Claims* and *Arbitration of Construction Disputes*. He is also a visiting professor at the University of Kansas, teaching courses in construction scheduling and construction claims. He has worked on projects and lectured throughout the U.S., Europe, and the Middle East.

Mr. Callahan is a member of the Kansas, New Jersey, and Missouri Bar associations. He is also a frequent arbitrator, negotiator, and mediator, a regional advisor to the American Arbitration Association, in addition to being on its national panel, and an associate of the Chartered Institute of Arbitrators. He is listed in *Who's Who in American Law* and *Who's Who in the Midwest*.

Daniel G. Quackenbush, AIA, is Vice President of CCL Construction Consultants and CPM Scheduling. Mr. Quackenbush has a Bachelor's Degree from the University of Kansas, School of Architecture, and he is a registered architect and a member of the AIA. Prior to joining CCL, Quackenbush worked as a contractor, designer, and consultant for other firms. He has served as an arbitrator for the American Arbitration Association and is widely sought after for expert schedule analysis in construction delay claims.

Mr. Quackenbush has taught construction scheduling at the graduate level at the University of Kansas and has lectured in seminars throughout the United States.

Mr. Quackenbush has personally prepared hundreds of CPM schedules for contractors, owners, and engineers on projects across the United States. He has also been involved in a large number of delay claims from both a preparation and analysis and a review standpoint.

James E. Rowings, Jr. is Professor-in-Charge of Construction Engineering at Iowa State University, the largest construction engineering program in the United States. Dr. Rowings received a B.S., M.S., and Ph.D. from Purdue University. His varied career includes a stint as a boilermaker for CBI, project engineer for multibillion dollar projects in Saudi Arabia, and numerous scheduling assignments for all types of projects. He has lectured throughout the United States and Europe. Dr. Rowings is a registered professional engineer in Indiana and Iowa and a certified cost engineer. Dr. Rowings teaches the fundamentals of cost engineering review course both in the United States and around the world. He is actively engaged in research in project control systems and specifically in the area of scheduling methods and systems.

CONTENTS

FOREWORD

As we grow older, we all learn that while some things in life are ever changing or growing, other things stay the same. I can remember what construction scheduling was like back in the early days of the computer, when critical path scheduling was being introduced to the construction industry and its clients.

I remember when there was no precedence diagram, when the IBM 1620 produced punched output cards that had to be arranged on manually sorting machines at 120 cards per minute per column. We had a "bread board" which was especially wired to make the printer perform and space our output. When I first started, it took a long time just to print the schedule.

Back in the 1960s, my firm was one of just a few that served the industry as a CPM consultant. We had to spend several nights a week at the computer-service bureau to produce our work. There was no such thing as a minicomputer, lap top, plotter, let alone a TV screen to see what we were doing, and so on. As I look back on what we did and the tiny prices we charged, I wonder how we ever did it.

Indeed, it's amazing what advances have taken place. Today's scheduling engineer typically owns a computer, a high-speed printer, a screen and a high-speed plotter that produce time scaled logic in a few moments compared to the days it took us to draw the same chart by hand. This new book covers the state of the art in scheduling today, 30 years later than when I was first learning about scheduling. As I read through it, I could not help but marvel that so many things that never crossed my mind as possible when I started have become ordinary staples in the scheduling engineer's life.

Yet, there are some things that have not changed at all. They even preceded the advent of critical path scheduling and will be ever with us. They are the real standards of who we are and what we are about. I remember in the early marketing efforts by IBM that part of its sales approach was to give away millions of little signs which said but one word: THINK.

Mike Callahan and I have been colleagues and friends for 20 years. Dan Quackenbush and Jim Rowings don't go back with me quite as far as Mike, but they all share the understanding that a construction schedule has no value unless the person who put it together knows first how things go together and then thinks it all through to produce a document that draws respect because its craftmanship must be recognized.

In the sixties, as I tried to sell CPM schedules to the contractors and owners across our country, I often heard that they had tried CPM and it didn't work. Some of the logic produced back then was so wrong, it would have been laughable if it wasn't so sad. Lots of schedules were and still are drawn by people who can't read plans.

Thirty years and a lot of construction jobs have gone by since then, and I still hear that CPM doesn't work, even with all the technology that has developed. I have always been convinced that CPM is a valuable tool that when done artfully, works well. I early recognized that it wasn't the schedules, it was the people that didn't work the way they should have.

This new scheduling book impresses me as a balanced text that deals with all the aspects of the schedule, including the people who prepare and follow or try to follow that which seems so easy on paper. This book deserves a welcome place in the serious practioner's library. I trust you will appreciate what the authors have tried to say to you. Their words are wise.

H. Murray Hohns
Honolulu, Hawaii

PREFACE

We have drawn upon our diverse knowledge of law, construction, architecture, engineering, and education to develop a comprehensive book, which covers both theory and practical application, on construction project scheduling. *Construction Project Scheduling* is written and organized to be valuable to students of construction science, architecture, and engineering as well as to the many participants in the construction process including project managers, project engineers, owners representatives, architects, engineers and, of course, construction schedulers. In addition, this text provides numerous guidelines for preparing construction schedules that, along with the scheduler's calendar included in the Appendix, give the student or the practicing professional a reference for difficult scheduling problems. Consequently, this book shall become the construction scheduler's handbook of the future.

Construction Project Scheduling is written to be understandable even by those not involved in the construction industry allowing laymen from all areas of business to use it for scheduling construction and other industrial scheduling applications. We do not overcomplicate the subject with formulas and rules. Mastering the principles of construction scheduling through the use of this book requires only an understanding of simple mathematics. Knowledge of the construction industry practices is helpful but is not a mandatory prerequisite.

This is the first book of its kind to explain the developmental process of a large construction project schedule. Throughout the text, we discuss scheduling problems and solutions using traditional scheduling techniques. The recommendations and guidelines provided in *Construction Project Scheduling* have been field-tested on hundreds of construction projects. The scheduling calendar provides easy calculation of scheduling start and finish dates for projects through the year 2000.

James E. Rowings, P.E., provided the theory and conceptual basis for scheduling as well as in-depth explanations of the more advanced scheduling applications such as PERT and time-cost trade-offs. He also made sure that the text explained each scheduling concept so that any individual, even without construction experience, could comprehend those concepts readily.

Daniel G. Quackenbush, AIA, provided the hands-on level of information on how construction schedules should be created and the realities of using construction schedules in the field. His extensive experience in the creation of hundreds of CPM schedules and through thousands of updates provides tested real-world solutions to complex scheduling problems.

Michael T. Callahan, Esq., an attorney with extensive experience in the legal interpretation of schedules, scheduling clauses in contracts, and the use of schedules in the presentation and the refuting of delay claims provided the experience to explain how schedules are used to prepare or refute delay claims. Recommendations in *Construction Project Scheduling* comply with current legal trends for the interpretation of CPM schedules by the courts.

While no technical publication can make recommendations that are always applicable to the situations the reader may encounter, we believe *Construction Project Scheduling* provides the concepts and procedures that will guide you in the development and use of quality CPM construction schedules.

McGraw-Hill and the authors would like to thank the following reviewers for their many helpful comments and suggestions: Don Hancher, Texas A&M University; David Haviland, Rensselaer Polytechnic Institute; Ray Levitt, Stanford University; and Jerald Rounds, Arizona State University.

Michael T. Callahan
Daniel G. Quackenbush
James E. Rowings

CONSTRUCTION PROJECT SCHEDULING

CHAPTER
1

REQUIREMENTS FOR CONSTRUCTION SCHEDULING

PLANNING

The planning process is of paramount importance in the success of construction projects. Several studies during the 1980s showed the impact of adequate planning on the eventual outcomes of construction projects. One such study[1] showed that an exceptional planning effort can yield as much as a 40 percent cost savings over reasonable planning. In this same study it was shown that poor planning could create overruns of as much as 400 percent and, on average, projects with poor planning would cost as much as 50 percent more than projects with a reasonable planning effort. A coordinated planning effort is imperative for a cost-effective project.

There seems to be a great deal of confusion in the construction industry between the terms "planning" and "scheduling." These terms are often used synonymously, in a "planning and scheduling engineer." The terms are not synonymous, however; they are very different, though related. The scheduling process for a construction project is just one part of the planning effort. This

[1] The Business Roundtable, *Planning and Scheduling—Report A-6.1,* The Business Roundtable, New York, NY, 1982.

1

section will define planning and explain its importance. Later sections will discuss the specific topic of scheduling.

Project planning serves as a foundation for several related functions. These functions include estimating, scheduling, and project control. Planning involves the process of selecting the one method and order of work to be used on a project from among all the various methods and sequences possible. This process provides detailed information for estimating and scheduling as well as a baseline for project control. Scheduling, in comparison, is the determination of the timing and sequence of operations in the project and their assembly to give the overall completion time. The schedule is obviously a reflection of the plan, but the plan must come first. The process of scheduling may uncover flaws in the plan, leading to revisions, but the plan is still first in the hierarchy. There are many approaches to planning, but the process as a whole is not very scientific or even systematic as it is practiced. Planning generally involves approximately 80 percent memory of historic procedures and 20 percent synthesis or creative thought.

We tend to follow work patterns used previously and try new approaches only when the old way will not work or a traditional method is not apparent. This tendency to follow known methods is consistent with our historic resistance to change. Construction is a high-risk enterprise, and the taking of additional risks by using unproven methods is not well accepted. Modifying or using totally new approaches may also take additional training and supervision. These extra costs add to the risk of untried methods.

Scheduling, on the other hand, is fairly systematic and scientific. There are several well-documented methods, and it is possible to become very competent in performing scheduling computations. Rigid sets of rules can be applied through programming to perform complex schedule computations. The conversion of planning concepts to a structured schedule requires the skills of an experienced and well-trained scheduler.

The planning process is a hierarchical one. Planning involves making decisions about what tasks will be performed, how the tasks will be performed, who will perform the tasks, and when and in what sequence the tasks will be performed. Necessarily, the planning process involves anticipating actions and anticipating performance. Fundamental to the planning process is a chain of operations involving:

Information search and analysis
Development of alternatives
Analysis and evaluation of alternatives
Selection of alternatives
Execution and feedback

Exceptional planning usually involves documentation of the assumptions and analysis made in arriving at a final ordering and identification of work activities. This documentation includes the assumptions made about alternatives, resources, and cost, and the decision criteria used in selecting the plan.

Planning is performed at various levels at various times during the execution of a project. At the onset of a project, early planning deals with conceptual issues and identifies broad areas of responsibility for the various participants (owners, designers, construction managers, and consultants) in the project to undertake. In turn, each of these participants provides additional planning at a more detailed level to further anticipate and make the decisions concerning alternatives related to his or her specific area of responsibility. This hierarchy continues on down through the project organization to the detailed craft level, where day-to-day and hour-to-hour decisions are made. As mentioned earlier, the significant part of the planning process involves learning from history. Documentation of the plan provides a benefit to the project participants in providing a formalized, systematic learning process.

NEED FOR SCHEDULING

Scheduling is the determination of the timing of activities and follows logically from the planning process. Many forms of schedules exist, from bar charts to critical path method schedules. Each has its advantages and disadvantages. Each has its appropriate applications. As the plans for a project become more complex, the need for critical path method schedules to represent the plan increases.

Few people involved in the construction industry will dispute the need for construction project schedules. A 1983 survey of 448 owners involved in building construction projects in excess of $10 million reported that a third of their projects are sometimes completed behind schedule.[2] Only half of the owners reported that their contractors usually use critical path method schedules on their projects. However, owners whose contractors usually use critical path method schedules appear to be less susceptible to delay. Only 27 percent of the owners whose contractors usually use critical path method scheduling said that their construction projects are usually or sometimes behind schedule, while 44 percent of the owners whose contractors sometimes or hardly ever use it experience delays. Among the owners who sometimes or usually experience cost overruns on their projects, 73 percent said that poor scheduling by their contractors was very or somewhat significant in causing the cost overruns. Poor scheduling can be the result of poor planning or poor scheduling techniques. The 1983 survey also found that delayed projects and disputes were closely related. Half of the owners who sometimes or usually had projects behind schedule had been involved in litigation or arbitration.

A 1985 survey[3] of 493 owners involved in building construction reported similar results. Almost 53 percent felt that poor scheduling is a very or somewhat significant cause of cost overruns, and 43 percent said that their projects

---

[2] Opinions of Building Owners on the Construction Industry, A Report to Wagner-Hohns-Inglis, Inc., Opinion Research Division, Fleishman Hillard, Inc., September 1983, St. Louis, MO.

[3] Opinions of Building Owners on the Construction Industry, A Report to Wagner-Hohns-Inglis, Inc., Opinion Research Division, Fleishman Hillard, Inc., April 1985, St. Louis, MO.

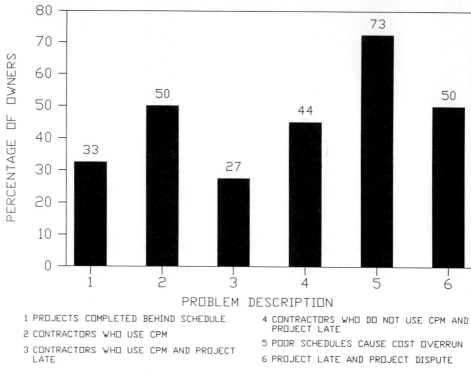

1 PROJECTS COMPLETED BEHIND SCHEDULE
2 CONTRACTORS WHO USE CPM
3 CONTRACTORS WHO USE CPM AND PROJECT LATE
4 CONTRACTORS WHO DO NOT USE CPM AND PROJECT LATE
5 POOR SCHEDULES CAUSE COST OVERRUN
6 PROJECT LATE AND PROJECT DISPUTE

FIGURE 1-1   CPM use versus on-time project completion. (*Source:* Opinions of Building Owners on the Construction Industry, A Report to Wagner-Hohns-Inglis, Inc., Opinion Research Division, Fleishman Hillard, Inc., September 1983.)

were usually or sometimes completed behind schedule. Less than half, 47 percent of the owners, reported that their contractors usually use critical path method scheduling.

Another 1983 survey[4] of 800 construction industry participants reported that over half of the respondents believed poor contract administration to be the most threatening factor to contract relations. Participating in the survey were contractors, designers, subcontractors, construction managers, and owners involved in industrial, commercial, and residential construction. To improve contract administration, 83 percent of the respondents stated they had or intended to install computer-based cost and schedule control systems. This again supports the important role of both planning and scheduling.

These three surveys clearly show both the significance of planning in construction and the importance of proper construction schedules to control and manage time. Construction schedules can reduce delays, cost overruns, and disputes. Without a schedule, a project has a higher probability of delays and the cost overruns and disputes which seem to follow. With a schedule, however, the probabilities of on-time, on-budget, dispute-free completion is improved.

[4] A Report on Current Construction Contract Practices in the United States, Arthur Andersen & Co., Washington, D.C. 1984.

WHAT IS A SCHEDULE?

A construction project "schedule" may mean different things to the designers, contractors, suppliers, subcontractors, and owners involved in the construction process. The schedule may mean the completion date stated in the contract, or interim completion dates required for phases of the work. The schedule may mean the schedule of values the contractor submits against which monthly progress payments will be calculated, or any of the many other lists itemized or required by the contract. The schedule may also refer to the process of sequencing and phasing individual activities required to complete the project.

In this book a construction schedule will mean a tool to determine the activities necessary to complete a project and the sequence and the time frame within which the activities must be completed in order to obtain timely and economical project completion. Construction schedules described here provide time-phased plans that permit portions of the work to be organized, sequenced, and controlled so that the entire project is completed in an organized and efficient manner. Most schedules in this book are presented as graphs that show the relationship between planned starting and finishing dates for the elements of work that comprise the project. The same organization and control may also be achieved (although perhaps not be so readily understandable) by lists of dates and activities without a graph.

Used as a tool to manage time, the primary advantage of a schedule is to produce the necessary planning which all successful projects require. Formal scheduling systems such as the critical path method (CPM) achieve this by forcing the scheduler, superintendent and project manager to think through the entire project in detail at the onset. This detailed thought process avoids inefficient and poor sequencing of the project. Understanding how the entire project will be completed also permits project personnel to recognize the effects that unexpected events or alternative actions have on progress. Sequences can be changed to overcome or reduce the impact. As an example, unexpected delays in structural steel delivery can be overcome by adding additional steel erection crews, additional cranes, or accelerating follow-on work.

As will be described in this book, there are many types of schedules which may be used to manage a construction project. The choice of which type of schedule will be used depends on the characteristics of each project. For a project with either few activities or activities which may proceed without much interaction, a bar chart or linear schedule may be appropriate. For projects with many activities that must continuously interact, a CPM schedule may be more appropriate. A bar chart or linear schedule cannot depict the intricacies of multiple-activity interaction.

Schedules can also be used to control time usage. The schedules prepared in this book are intended for use as both planning and control tools. For use as control tools, however, greater care must be taken to reduce the uncertainty and margin of error which schedules as planning tools accept. As will be described in subsequent chapters, the acceptable margin of error included in schedules as planning tools can be reduced when using schedules to control time by including as-built dates and as-built sequences, and updating the schedule regularly to reflect the contractor's current execution plan and strategy.

SCHEDULING AS AN ART

Project duration and completion dates are commonly determined by the owner's need for the project. The date the owner needs to occupy, use, or rent the completed space is used to establish the construction project completion date. The contractor, however, is traditionally responsible for the detailed planning, and scheduling, necessary to ensure completion of the project within the owner's time frame.

The contractor may prepare the detailed schedule in a number of ways. On projects with which the contractor has had extensive experience, the schedule may be only intuitive, with few, if any, target dates recorded or presented. Large, complex, or unusual projects may require more formal scheduling efforts. On larger, more complex projects the contractor may have to analyze the project scientifically and systematically, determining the various subnetworks, relating each subnetwork to the entire project, and defining activity durations, and then organize the analysis in a formal presentation such as the critical path method.[5]

Whether the schedule is prepared intuitively or scientifically, the scheduler must understand how the various materials, trades, equipment, and subcontractors combine to produce a completed project. As stated by the U.S. Army Corps of Engineers, the biggest builder in the United States:

> Developing this schedule is not merely a mechanical exercise. Many judgments must be made, and the schedule's reliability depends largely on their quality. Personnel making such judgments must have thorough knowledge of the job site conditions, the contractor's capabilities and method of operations, . . . the CPM scheduling system, and the Corps of Engineer's contractual liabilities. . . .[6]

Understanding how the various materials, trades, and subcontractors combine to produce a completed project is an ability acquired only after years of study, observation, and experience. Scheduling is truly an output of the planning process, which is an art. The quality of the exercise is thus a function of the ability of the scheduler to combine creativity with experience in order to develop a product which meets the needs of the project and thus the client's objectives.

[5] Edward M. Willis, *Scheduling Construction Projects,* New York: Wiley, 1986, p. 7.
[6] Modification Impact Evaluation Guide, EP415-1-3, Department of the Army, Office of the Chief of Engineers, Washington, D.C., July 1979.

THE SCHEDULER

Scheduling is done by people, not computers. Computers may assist the planner and the scheduler by sorting, storing, performing math, and matching data, but they do not provide the intellectual direction and creative thought required to produce a schedule for a unique, complex construction project. The scheduler must pull together data from the various individuals and organizations which form the project team, refine these data, remove bias, and develop a comprehensive and coordinated plan represented as a schedule.

The scheduler must, as a minimum, have the mechanistic/scheduling skills which will be described in this book. Beyond this, the scheduler must also know how to query and listen to collect the needed data from other team members in order to produce a comprehensive schedule. The scheduler must have broad and deep experience to judge and probe bias from the input received and to identify the most efficient linkages to coordinate the work activities of the various team members. Finally, the scheduler must be persuasive, so that the schedule is perceived as owned by all team members and not as being developed and owned by the scheduler. This ''ownership'' issue is crucial to the development of commitment to the schedule.

PURPOSES OF CONSTRUCTION SCHEDULES

Construction schedules can be used for a number of different purposes. The purpose to which a schedule is put often is determined by the individual using the schedule.

Among the ways a construction schedule can be used is to predict project completion. Timely completion of the project is particularly important when failure to complete within the time required by contract carries a financial penalty or liquidated damages. By using construction schedules to predict project completion, contractors can adjust crew sizes, shifts, or equipment to speed or slow progress.

As part of the design effort, architects or engineers may prepare a construction schedule to determine how long design and construction will take in order to complete the project when needed by the owner. The schedule's ability to predict performance time is particularly valuable when innovative or state-of-the-art projects are to be built. By sequencing a unique set of detailed activities it is possible to prepare an estimate of time with a higher probability of achievement than by examining the overall project scope.

Interestingly, the designer's or owner's requirement that contractor performance be completed within a certain time has not been held a warranty that the project can be completed within that time by courts in the United States.[7]

---

[7] *American Shipbuilding Co. v. United States,* 654 F.2d 75 (Ct. Cl. 1981); Appeal of D. L. Muns Engineering & Building Contractors, ASBCA No. 30104, 87-2 BCA (CCH) P 19709 (1987).

Despite these court cases, some U.S. contracts (specifically, AIA Document A201-1987, p. 8.2.1) attempt specifically to avoid a warranty of completion within a certain time by providing that the contractor agrees that the contract time is a reasonable period for performing the work.

Construction schedules can also be used to predict when specific activities will start and finish. Subcontractors can use this information to predict when they are needed at the site. General contractors can also use activity start and finish dates to arrange for needed material, tools, and equipment. Owners can use activity completion dates to determine when to deliver owner-furnished equipment or to plan partial occupancy.

Contractors can use construction schedules to expose and adjust conflicts between trades or subcontractors. After the schedule has been completed, the schedule may reveal conflicts between trades required to occupy the same space or expose work required to be completed in adverse weather. Once identified by the schedule, potential conflicts may be eliminated and special protection may be arranged. The ability to anticipate and correct potential conflicts is particularly useful in fast-track projects.

A construction schedule can control a variety of resources. Both owners and contractors can use schedules to plan cash flow. By assigning a monetary value to activities on the schedule, the value of the anticipated work for each month can be predicted. The projected monthly values can be used to plan borrowing by the owner. The contractor can use the anticipated monthly values to plan cash flow. The contractor's cash flow for a project is the difference between the funds received and the expenses paid out. It is common for the contractor to pay out more than is received during the first part of the project. The contractor must anticipate the shortfall and provide additional funds to make up the difference. Dollar-loaded schedules can help contractors plan cash flow and arrange for necessary borrowing. Actual progress can be used to calculate monthly contractor progress payments.

Both contractors and owners can also use schedules to evaluate the effect of changes on project completion and cost. Most contracts permit contractors to collect any additional costs which change orders or variations impose on the contractor's performance. By projecting anticipated changes on the schedule, owners can prospectively evaluate the potential additional costs of the change to determine if the change should be ordered. In the same way, contractors can use schedules to model the effect of the change on performance and estimate the value of the change order. Schedules can also be used to measure delay and time extensions. When used to measure delay time and extensions, however, great care must be taken to make the schedule conform to actual progress and sequences.

A schedule may also serve as a record of the project's progress. If the schedule has been regularly updated to reflect changes in work sequences, unanticipated delays, actual activity completion dates, and change orders, the schedule can be used as a historical record of the project. When used in this manner, both the owner and the contractor can use the updated schedule to retrospectively measure the impact of additional work and unanticipated delays, or to measure percentage completion for retainage reduction. By providing a standardized yardstick by which delays and time extensions can be measured, extended disputes can be avoided and differences settled quickly.

An updated schedule can also be used as notice of claims or extensions of time. Most contracts require contractors to request time extensions or extra payments within a reasonable time after the event which causes the additional cost or performance time. An updated schedule which is regularly exchanged between owner and contractor can be used as notice of delay or extra work if the updated schedule shows that the schedule was affected by the unanticipated events.

The project schedule has a great many uses and provides a service to the project which alternate methods cannot. The effort required to produce a quality project schedule and maintain its accuracy throughout the project is a most worthwhile investment.

EXERCISES

1 Define the terms "planning" and "scheduling." Describe the difference between the terms.
2 Develop and document a plan for changing a flat tire on your car. Document your assumptions concerning resources, sequence, and timing.
3 Develop and document a plan to build a new home after your firm has decided to relocate you to a different city.
4 Describe the skills needed by a scheduler. How would one develop these skills through experience?
5 Describe the ways that schedules would be used for the development of a new superhighway from Chicago to Kansas City.
6 Describe the ways that schedules would be used for the construction of a 140-story building in a major city.
7 Who should be involved in the planning and scheduling of the project described in Exercise 5?
8 Who should be involved in the planning and scheduling of the project described in Exercise 6?
9 Describe in your own words the significance of the data presented in Figure 1-1.

CHAPTER
2

GANTT
CHARTS

DESCRIPTION

The most frequently used construction schedule is the bar or Gantt chart. The bar chart was developed by Henry L. Gantt during World War I. The bar chart is widely used as a construction schedule because of its simplicity, ease of preparation, and easily understandable format.

A bar chart is a collection of activities listed in a vertical column with time represented on a horizontal scale. The projected start and finish are shown for each activity, and the duration is indicated by the placement of a horizontal bar to the right of the description. Approximate dates of start and finish for each activity can be determined from the horizontal time scale at the top of the chart. The length of the bar represents the duration of the activity. Normally the activities are listed in chronological order according to their start date.

On most bar charts, each bar indicates when a particular activity will begin and when work will be completed. The bar does not necessarily mean that work will be performed continuously. Rather, the bar usually represents the first day the activity's work is to be performed and the last day on which the work must be completed. The number of days of actual work between these two dates is not specified, although a continuous line may lead one to assume that work will be performed continuously. Figure 2-1 shows a bar chart schedule using continuous bars. A bar chart can be prepared to indicate when actual

Bar charts provide a simple scheduling technique for small projects.

FIGURE 2-1 Simple bar chart.

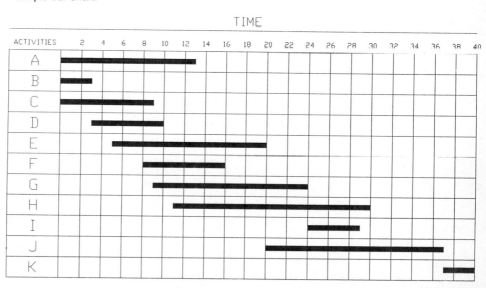

work is planned to be interrupted by using a dashed (noncontinuous) bar. Use of dashed bars may present a clearer picture of the intent of the scheduler. Figure 2-2 shows a bar chart schedule using dashed bars to indicate planned interruptions in actual work. It is usually preferable to separate activities which must be broken into separate bars with unique descriptions. This prevents user confusion over the meaning of the dashed lines. Figure 2-3 shows a comparison of the two cases.

A bar chart is prepared by following a few simple steps. The first step is to determine which activities will be listed on the schedule. This can be done by breaking the work into smaller, finite construction activities and listing these work-related activities on the bar chart. The process is identical to the process described in Chapter 3 for the critical path method. Often, however, bar chart activities are selected on the basis of physical location or responsibility rather than a work relationship. Bar chart activities tend to be selected on the basis of cost accounts or pay estimate item, by subcontractor, or by specification sections. Bar charts based on work-related activities are more valuable as planning and managing tools. Selection, however, depends on what the scheduler wishes to portray for planning or control.

After activities have been selected and listed, durations are estimated. Again, this process is described in detail in Chapter 3. Finally, the scheduler determines a sequence and plots the bars on the schedule. The sequence is usually not explicit but can be inferred from the end point for one activity and the beginning point for another.

Traditionally, bar chart activities are of longer duration than those on network schedules. Bar chart activities may be several months and in some cases

FIGURE 2-2   Bar chart with noncontinucus lines.

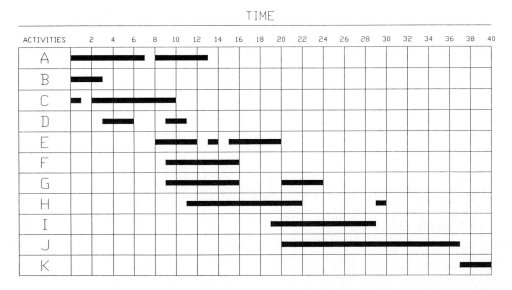

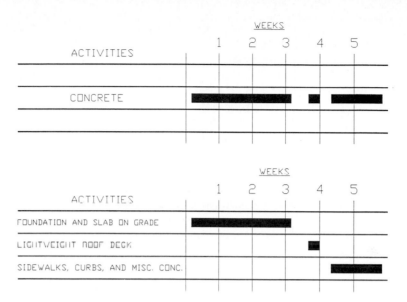

FIGURE 2-3   Alternative bar chart activity representation.

several years long. Any bar chart activity with a duration of more than 3 months probably should be subdivided into greater detail.

The determination of what data belong on a bar chart is a matter of judgment and preference. As a minimum, a bar chart should contain a heading with the project title and location, a brief description of each activity, and bars representing the activities drawn to indicate the period of time during which the activities will be performed. Traditionally, the dates shown on the bar chart for a specific activity represent the earliest start time and the earliest finish time for each activity. The latest starting time and the latest finish time can be added by using dashed lines or icons to mark these points. The date of preparation and update should be shown on the bar chart. As a general rule, a bar chart will contain no more than 100 activities. Bar charts with more than 100 activities become difficult to read and use. Selection of the activities and the objective for use of the bar chart determines the ideal number of bars.

The bar chart can also contain additional data. Additional data may make the bar chart more useful to some users but may also increase the required size of the paper on which the bar chart is to be printed or drawn, making the bar chart difficult to use, copy, or exchange. The addition of too many data may also make the chart less understandable.[1]

Information in the heading may include the project location, project owner, project designer, project number, contract amount, award date, and date of current status update, and all previous updates. Information provided with the activity descriptions may include an activity duration, activity code, dollar

[1] Edward M. Willis, *Scheduling Construction Projects,* New York: Wiley, 1986, p. 59.

value, earned value, cost codes status (revised planned duration and actual duration), I node/J nodes (terms to be explained in Chapter 3), and resource types and quantities. The current status of partially completed activities, an indication of how actual status differs from scheduled status, and the cause of any variation can all be shown.

A bar chart is relatively easy to prepare and can be quickly understood even by lay people. A bar chart may be the only scheduling tool used on a project. Bar charts may also be used to reduce complex scheduling methods to a more clearly understood form.

Figure 2-4 is a portion of a bar chart for the completion of a 42-story building. This schedule is easily understandable. The task titled "granite panels" is scheduled to be performed during July, August, and September 1988. Installation of granite panels is scheduled to be complete the first week of September. The task titled "roofing and sheet metal" is scheduled to start during the last week of August. Note that this bar chart does not identify a specific day for the start or finish of any activity. Most bar charts do not identify specific start and finish dates; however, approximation of dates is possible from the horizontal time scale.

Many of the bars in Figure 2-4 overlap. For example, completion of granite panels overlaps the start of roofing and sheet metal. This indicates that the roofing and sheet metal can begin shortly before completion of granite panels.

Most of the activities on this bar chart do not have a specific start date. Rather, at the time this schedule was prepared, most of the activities had already started. The schedule shows granite panels, glass and aluminum, and skylights being completed at approximately the same time. Similarly, the 43rd-floor mechanical and electrical, the 5th-floor mechanical and electrical, and the service elevators and escalators are scheduled to be complete at approximately the same time. Low-, medium-, and high-rise elevators are not scheduled to be complete until February 1989, or some $3\frac{1}{2}$ months later.

The specific reasons why the scheduler chose the sequence which Figure 2-4 illustrates are not indicated on the schedule. However, logic and an understanding of the construction process can be combined to interpret the intent of the scheduler. Rough-in and finish work at each floor has been scheduled in groups of three. Floors 8, 9, and 10 are scheduled to be completed the first week of November; floors 11, 12, and 13 approximately a week later; and completion of three-floors units continues up the building until floors 41 and 42 are completed in the middle of March 1989. Although we can presume that rough-in crews will work on several floors at any one time, the precise number of crews and the specific floors on which they will work are not identified. Rough-in on the lobby through 7th floors is shown completing after floors 8 through 34 have been roughed in and after completion of the food service equipment. Presumably, food service equipment will be installed on one or more of the first seven floors, which prevents earlier completion of rough-in.

The bar chart also indicates that ceiling tile work will begin the first week in October and complete in the middle of March of the following year. Although

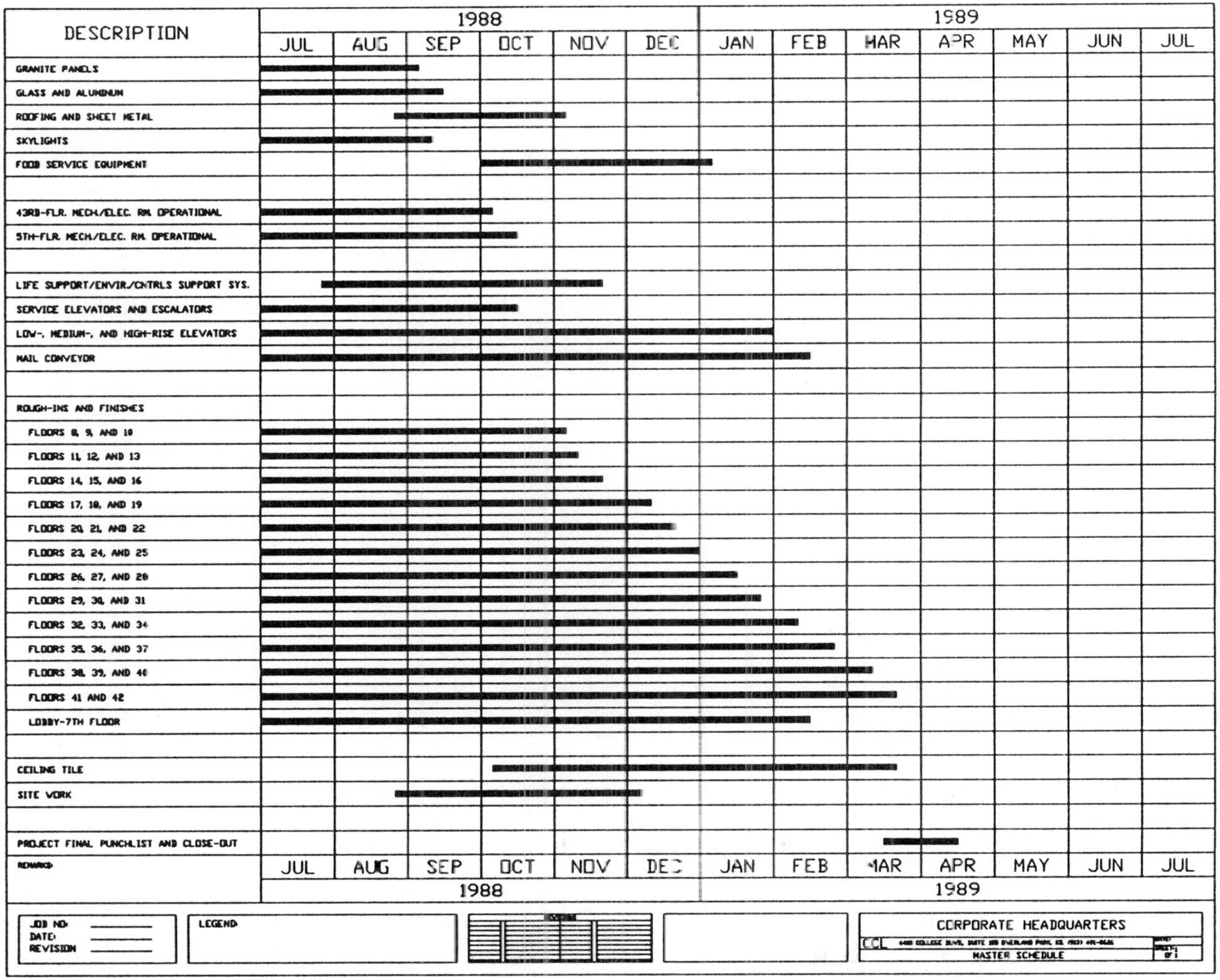

FIGURE 2-4 Bar chart schedule.

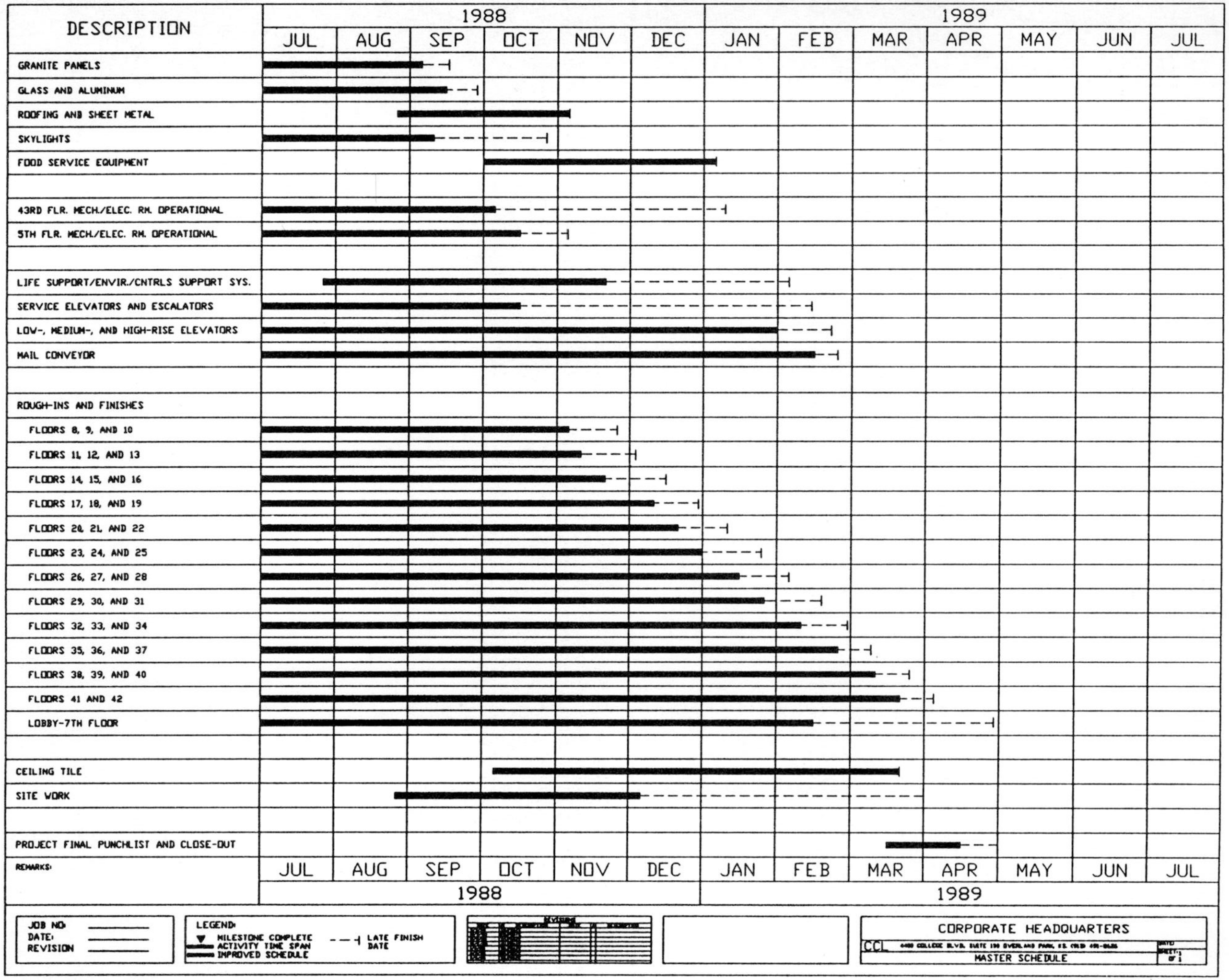

FIGURE 2-5 Bar chart with late dates.

the bar chart does not indicate on which floor ceiling tile work is to begin, it can be presumed to begin on the 8th floor as rough-in and finish are completed there.

Site work begins when granite panels, glass and aluminum, and skylights are almost completed. Presumably, these activities require exterior material-handling equipment which prevents site work from making major progress until the equipment has been removed. Project punchlist and close-out is to begin after all other activities have ended and complete in the middle of April 1989.

In addition to illustrating the presumptions necessary when reading a bar chart, Figure 2-4 shows how a bar chart schedule can be used to schedule a large, complex building project.

Figure 2-5 shows the same bar chart as Figure 2-4 with the addition of dashed lines and late-completion markers. This figure uses vertical lines to indicate the activity's late finish date (term to be explained in Chapter 4). This additional information is valuable for determining work priorities on the job site. It is difficult to determine late finish dates accurately without the help of a CPM schedule.

SIZE

The bar chart in Figure 2-4 has been reduced to fit in this book. Just as an illustration of a bar chart was selected that would fit into this book, schedulers need to be concerned with the size of the schedule when preparing bar charts. By combining several sheets of paper, bar charts of virtually any size can be created. Because the information on any schedule must be communicated to many people involved in the construction project, some handling, distributing, or reproducing can be anticipated. Bar charts should be limited in size so that such communication can be easily facilitated.

Standard paper or plan sizes are best suited for bar charts. Adjusting the units of time in which the activities are scheduled to fit on standard paper may be the best way to control the size of a bar chart. For instance, the bar chart may be ruled off in days, weeks, or months as in Figure 2-4. Bar chart sizes may also be controlled by limiting the length of time the schedule attempts to control. If it is necessary to show an entire project scheduled to take many years to complete, durations may be limited to months. Supplemental schedules may be created in work days for a limited "look-ahead" period to provide necessary details in a manageable size. The scheduler or manager can use the supplemental "look-ahead" schedules in conjunction with the overall project schedule to provide necessary perspective. When multiple pages are used, it is important to try to group activities which have sequential relationships on the same page.

Do not make the mistake of scheduling only a portion of the project while assuming that the remainder of the project will fit into the remaining contract time. Interim or partial schedules should be used only while the final schedule is being developed. The complete project schedule should be finalized and distributed for use as soon as reasonably possible.

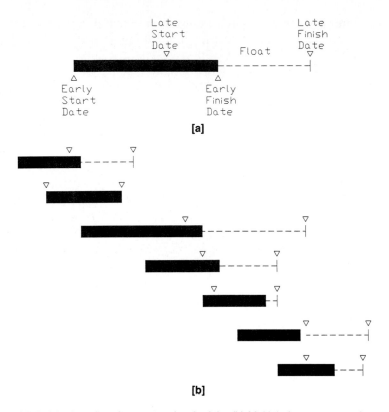

FIGURE 2-6   (a) Activity bar chart from network schedule. (b) Multiple bar summary chart.

COMPUTER-GENERATED BAR CHARTS

Computer systems that include a printer are capable of producing a bar chart schedule. If a plotter is available, a higher-quality and more detailed bar chart can be produced. Several graphics programs produce bar charts by simply specifying start and finish dates for the bars.

Most commercial scheduling software can be used to produce a bar chart summary of the network schedule. Although the symbols used for a summary bar chart vary from one software system to another, the four CPM dates (defined in Chapter 4) are usually identified for each activity's bar instead of the traditional two (early start and early finish). Figure 2-6a shows how to identify the four CPM dates on a summary bar chart. Figure 2-6b shows how a multiple bar chart schedule generated by a scheduling computer software system might look.

PERCENTAGE COMPLETION CURVES

Managers often find a cumulative percentage completion or ''S-curve'' helpful for monitoring and controlling progress. Traditionally, the S-curve is super-

Simple bar charts list many activities but may not clearly show the planned order of building finishes.

imposed on a bar chart to give the manager a cumulative and incremental measure of planned completion. An S-curve reflects the project work anticipated to be completed at any particular time. For comparison, the actual work performed to date is plotted with another line.

To prepare an S-curve, the ratio of each activity's work to the total work of the project is computed. There are several ways of computing the percentage of total project work which each activity represents. Commonly, the proportion is determined as the ratio of the monetary value of each activity to the total contract amount as indicated on the project's schedule of values for progress payments. The proportion may also be calculated by the level of effort (man-hours or work days) to be performed in a period. Once each activity's proportion of the total project has been determined, the proportion is distributed over the performance of the activity. Figure 2-7 shows the distribution of the determined proportion over the performance of the activities.

At the bottom of Figure 2-7 is a line titled "Cumulative %." The information on this line represents the cumulative percentage of the total project work that will be completed at the end of each period during the project's duration. The column called "% Period" represents the period's proportion to the entire work.

Once the determined proportion is distributed over the performance of the activities, and then totaled by time period and by cumulative progress along

TIME

ACTIVITIES	PERCENT PROJECT	2	4	6	8	10	12	14	16	18	20	22	24	26	28	30	32	34	36	38	40
A	14	1	2	3	3	3	1														
B	1	1																			
C	7	1	1	2	2	1															
D	6	1	2	2	1																
E	8			1	1	1	2	2	1												
F	5				1	1	1	2													
G	14				1	2	2	2	2	2	2	1									
H	19					1	2	2	2	2	2	2	2	2	2						
I	6									1	1	1	1	1	1						
J	17										1	1	1	1	2	3	3	3	1		2
K	3																		1		2
PLANNED % PERIOD		3	6	7	8	8	7	8	6	6	6	5	4	5	5	3	3	3	2	2	2
PLANNED CUMULATIVE %	100	3	12	19	27	35	42	50	56	62	68	73	77	82	87	90	93	96	98		100
ACTUAL % PERIOD																					
ACTUAL CUMULATIVE %																					

FIGURE 2-7 Distribution of percentage completion.

20

TIME

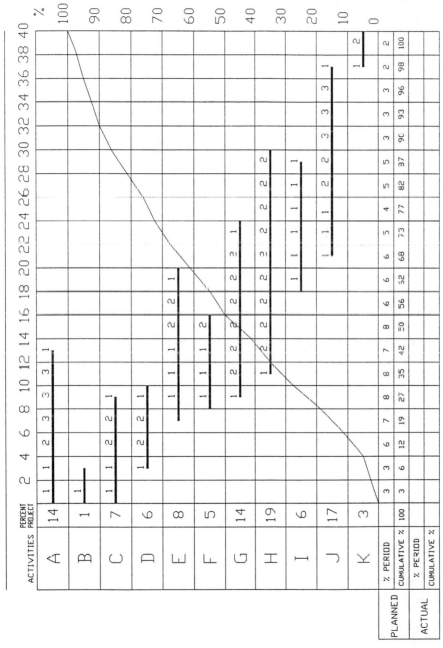

FIGURE 2-8 Plotting an S-curve.

the bottom of the bar chart, a scale for project percentage is added to the bar chart and the S-curve is plotted. Figure 2-8 shows an S-curve plotted from the proportion distributed in Figure 2-7.

The S-curve in Figure 2-8 represents the scheduled cumulative percentage of project completion as of the end of each period during the scheduled project performance. The S-shaped curve is characteristic of construction projects. It indicates that progress is slow when the project begins, accelerates as more activities begin and are completed, and slows again as fewer activities remain to be completed.

Bar charts can be made very complex and can provide a great deal of information. Bar charts can show:

Anticipated cash flow
Actual monthly draws
Anticipated total manpower
Actual total manpower
Anticipated manpower by trade
Actual manpower by trade
Anticipated productivity
Actual productivity

Bar charts have a multitude of uses independently and in conjunction with CPM schedules. The bar chart's ease of readability makes it an excellent communication tool.

Bar chart use in itself could be the subject of an entire book. The authors have, however, elected to keep discussion of bar charts short in order to allow more space for CPM techniques.

ADVANTAGES AND DISADVANTAGES OF BAR CHARTS

The primary advantage of a bar chart is its simplicity. The simplicity of the bar chart has led to its wide acceptance as an effective planning and scheduling device for certain projects. The bar chart is simple to read and interpret. The bar chart can communicate in a straightforward manner the results of a network schedule. However, the bar chart cannot depict the intricacies of multiple-activity interaction. Thus, the inherent limitations and simplicity of a bar chart reduce its effectiveness and accuracy in projects involving a large number of activities.

The limitations and simplicity of a bar chart make it an attractive schedule for some types of construction, such as highway or pipeline construction where the contractor must constantly adjust schedules to overcome weather-related problems. For example, when soil moisture makes compaction inefficient, equipment may be moved to another location where soil conditions may be better. Weather-related shifts cannot be planned and shown on the schedule, but a bar chart schedule does not need to be revised when they occur, because it shows so little information about sequence and relationships anyway. The bar chart gives contractors who perform repetitious work more flexibility and does not require constant revision to match actual sequence.

The primary disadvantage of a bar chart relates to its preparation. A bar chart schedule is difficult to prepare accurately when there are continuous relationships between many activities and if multiple-activity interaction is required to complete the project. Bar charts can also be prepared all too easily, encouraging unrealistic schedules.

There is a tendency to work backward when preparing a bar chart. Knowing a project's contract completion date, the scheduler may plot the bar representing the final activity so that it ends at the time the project must be completed. The scheduler may then spread the other activity bars so they cover the time available, adjusting the starting time for a bar on the basis of start and end times of previous activities. Thus, there is a tendency to produce bar chart schedules that contain arbitrary, if not unrealistic, activity starting and finishing times, rather than anticipate realistic sequences, constraints, and durations. This is why contractors who use bar charts have a higher proportion of delayed completion than those who use CPM.[2]

When a bar chart is cost loaded and used for monthly payment to the contractor, it is easy to "shift" costs to earlier scheduled activities, thus improving the contractor's cash flow. While this is just good business sense for the contractor, it is not in the best interest of the owner, who is paying for work that has not yet been performed. This technique is known as early loading or front-end loading.

The major disadvantage of the bar chart is that the logic the planner used in developing it is not obvious. To understand the impact of changes this logic must be known. This drawback and others have led to the development of a more complex approach—network scheduling using the critical path method—which can incorporate a wide variety of activities and depict their interrelationships graphically.

EXERCISES

1 Identify the primary advantages and disadvantages of a Gantt chart for construction projects.
2 Draw a bar chart for changing a flat tire on your car. Your chart should have sufficient detail to show all activities that require more than 15 seconds to complete.
3 Develop a bar chart for removal of an existing sidewalk and construction of new sidewalk 6 feet wide and 220 feet long with joints every 4 feet along its length.
4 Describe the type of bar chart and the level of detail which would be appropriate for:
   (a) A 3-mile highway reconstruction project
   (b) A four-span, two-lane bridge project
   (c) The design of a strip shopping center
   (d) Construction of an airport terminal building
   (e) Construction of a power plant

[2] Edward M. Willis, *Scheduling Construction Projects,* New York: Wiley, 1986, p. 62.

TABLE 2-1 ACTIVITY LIST OF A SMALL COMMERCIAL PROJECT

Activity	Description	Duration	Cost
1	Foundations and slab utilities	6	12,500
2	Structural steel	4	23,000
3	Roof deck	4	10,000
4	Floor slab	2	5,500
5	Siding	6	19,000
6	Windows and doors	2	4,700
7	Mechanical	14	16,800
8	Electrical	14	15,700
9	Insulation	4	8,800
10	Finishes	5	7,000

TABLE 2-2 ACTIVITY LIST FOR GAS STATION CONSTRUCTION

Activity	Duration	Cost
Order and deliver lights and signs	15	0
Order and deliver fuel tanks	20	0
Order and deliver hydraulic lift	30	0
Order and deliver fuel pumps	25	0
Order and deliver air compressor	20	0
Order and deliver grease equipment	15	0
Layout and excavate	2	2,000
Install fuel tanks	2	12,000
Pipe up fuel tanks	4	7,000
Construct light and sign foundations	1	4,000
Install exterior underground conduit and piping to tanks	5	7,000
Deliver and erect fence	3	6,000
Install lights and signs	2	9,000
Connect and hook up power and lights (meter)	1	5,000
Construct pump islands	2	4,000
Construct building foundations	3	14,000
Construct hydraulic lift foundations	1	3,000
Place underground conduit and piping	3	3,000
Backfill fuel tanks and grade site	2	2,000
Backfill building interior	1	1,000
Form, reinforce, and pour concrete floor	3	10,000
Pave service and parking areas	5	10,000
Erect prefabricated building walls	3	18,000
Erect building roof	10	13,000
Install hydraulic lift	3	7,000
Install greasing equipment	4	6,000
Install interior plumbing and light fixtures	2	5,000
Install interior air compressor	4	7,000
Form sidewalks	1	1,000
Install doors and hardware	2	3,000
Glaze windows	1	1,000
Final test and inspection	3	1,000
Install fuel pumps on islands	2	4,000
Pour and cure concrete sidewalks and strip forms	6	3,000
Paint	2	3,000

5 Develop a bar chart for construction of a free-standing, two-car garage with wood framing and siding and with a shingle roof. List your assumptions.

6 Develop a summary bar chart for the activities listed in Table 2-1. Assign the costs proportionately to the duration for each day of each activity and develop the total cost for each day. Plot the cost-versus-time curve for this project.

7 Figure 2-9 is a section and plan for construction of a gas service station. Table 2-2 is a list of the activities (not in chronological order) required to complete the gas station.

The project shown in Figure 2-9 calls for building walls to be load bearing. Interior below-grade piping should be installed before pouring the concrete floor. Excavation around the building should be backfilled prior to erection of the building walls to provide a safe work surface. Sidewalks run parallel to both streets, and the area is paved to the sidewalks. Since the hydraulic lift is embedded in the concrete floor, it must be installed prior to pouring the floor. Sitework is independent of the building.

The building should be enclosed before installing interior equipment and fixtures in order to prevent vandalism. The roof should be installed before doors and windows.

FIGURE 2-9   Gas station design.

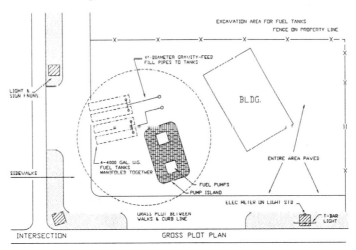

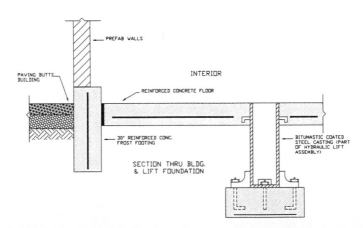

The building should be weathertight before painting. Sidewalks should be installed prior to paving for grade control. The electric meter is located on the light standard. With the exception of underground conduit and interior fixtures, all electrical work should be assumed to be included within the listed activities.

An aluminum cyclone fence separates the station from adjoining property. Assume that the fence lines have been located and fence material is available on 24-hour notice.

Assume that all plans have been approved and building permits obtained, the contract has been signed, and the owner has cleared and leveled the site. However, the owner has not yet issued a notice to proceed, so the contractor does not know the project start date. After receiving the notice, the contractor will have 13 weeks within which to complete the work. Assume that the contractor will work 5 days a week. Thus the contractor will have $13 \times 5 = 65$ work days to complete the project.

Prepare a bar chart schedule in work days showing that the project will be completed within 13 weeks.

8 Distribute the costs from Table 2-2 onto the activities on your bar chart. Tabulate monthly draw predictions and draw a cash flow curve.

9 On tracing paper, overlay your cash flow bar chart from Exercise 8. Now early-load the individual bars (just slightly) and recalculate the monthly draws. Replot the cash flow curve.

DISCUSSION TOPICS

1 Can a substantial-sized project be adequately planned using only bar charts?
2 Can comparisons of actual progress to planned progress curves be misleading? Actual to planned cash flow?
3 Is front-end loading ethical?
4 How can front-end loading be detected?

CHAPTER
3

CRITICAL
PATH
METHOD
SCHEDULES

INTRODUCTION

Network schedules have contributed significantly to the planning, control, and on-time completion of construction projects. Preparation of network schedules forces the manager to define and plan the work in detail from start to finish, thus permitting early identification of potential problems. Network schedules support efficient communication between field and office forces. If management staff changes are made during the project, the schedule can assist in a smooth transition for execution of the plan. Network diagrams make job coordination easier among material suppliers, contractors, subcontractors, owners, and designers. Networks reveal whether the manager understands the project and disclose how much the planner understands about the construction process. Networks can form a sound basis for related capital equipment and human resource planning.

Although there are a variety of network scheduling techniques, the construction industry has traditionally used only two methods to schedule projects.

Cast-in-place reinforced concrete structures require careful sequencing of the shoring and reshoring operations to maintain efficiency, economy, and safety.

Both methods can be termed "the critical path method," CPM. These methods are known as the "I-J" method (also called "activity on arrow," AOA, or "arrow diagramming method," ADM) and the "precedence diagramming method," PDM (also called "activity on node," AON). The primary difference between these two methods is that the I-J method defines the activity as an arrow between two numbered nodes, while in precedence diagramming the activity is represented as a node and the arrows indicate the logical relationships among the activities.

Precedence diagramming is actually a derivation of the original activity-on-node (AON) type of schedule diagramming. Pure AON diagramming does not allow for leads, lags, finish-to-finish, start-to-start, or start-to-finish relationships. (All these terms are defined later, in Chapters 4 and 5.) Pure AON diagramming uses only finish-to-start relationships, with no lags, just as does I-J scheduling.

Since the vast majority of computer software available today which uses the AON technique also allows leads, lags, finish-to-finish, and start-to-start relationships, the authors will not elaborate on pure AON diagramming. The reader will see the difficulties encountered with these additional relationships when reading Chapter 5, and will understand the merits of pure AON scheduling. The scheduler can perform pure AON scheduling techniques using precedence software just by avoiding use of the additional relationships and lags.

Network schedules were first developed by E. I. Du Pont de Nemours Company in conjunction with the UNIVAC Applications Research Center of Remington Rand between 1956 and 1958 while evaluating the potential use of computers to schedule construction projects. The technique was developed by John W. Mauchly of the UNIVAC Applications Research Center, James E. Kelly, Jr., of Remington Rand, and Morgan Walker of Du Pont. Network scheduling was first applied to equipment maintenance on a chemical plant in Louisville, Kentucky, where it was credited with saving Du Pont $1 million during the first complete year's use of the technique. The particular equipment involved could not be maintained while the unit was in production. Maintenance was performed when production permitted the equipment to be shut down. Past maintenance had required a shutdown of 125 hours. Du Pont wanted the maintenance period to be reduced so the equipment could be returned to production faster. Using the new network scheduling method, maintenance was completed in 93 hours. Subsequent application of network scheduling techniques further reduced the shutdown period to 78 hours.[1]

In 1961, network schedules were first applied to construction projects. In that year, Perini-Canada completed the Port-Mann Bridge using network schedules. Network technique applications in construction were not widespread in the 1960s, however. In the late 1960s and early 1970s, as students who had learned of the techniques in engineering curricula entered the construction industry workforce, more applications were attempted, many with success. Several owner agencies began requiring network scheduling, and usage has continued to grow. Still, there are many projects which are not yet benefiting from a network approach to scheduling.

DEFINITIONS

The elements into which a construction project are subdivided for network schedules are called "activities." An activity is a single work step that has a recognizable beginning and end. Activities are time-consuming tasks. No standards exist as to the number of activities any CPM schedule should have, but sufficient activities are needed to control the schedule. Activities are defined within a hierarchical system which subdivides larger elements of the project. This system is commonly referred to as the "work breakdown structure."

In I-J method schedules, the activities are separated by "events." An event is a point in time which recognizes the end of one or more preceding activities and the beginning of one or more subsequent activities. The I-J method tends to focus attention on events. In PDM schedules the activity receives the focus of attention.

The time which each activity requires to be completed is called its "duration." No standards exist as to how long activity durations should be. Most construction activity durations are expressed in work days, but durations can

[1] J. O'Brien, *CPM in Construction Management* 7, 2nd ed., McGraw-Hill, New York, 1971.

also be expressed in calendar days, months, weeks, hours, or minutes, depending on the work to be scheduled. All time-consuming activities are assigned a duration.

The order in which the activities are to be accomplished is called the "logic." The start of some activities depends on the completion of others. For example, the roof cannot be constructed until the roof deck is on and the parapet has been constructed. There are also activities which are independent of each other and can proceed concurrently. The order of activities or logic must be developed with safety, space, and structure in mind. Much of the logic for construction projects originates in well-established work sequences that are customary in construction contracting. Nevertheless, there is always more than one way to complete a project. Except in the rarest of cases, no single unique project logic exists. Usually there are many ways to construct a project, and thus many different schedules are possible. Each set of logic has its own characteristics of resources, risk, and cash flow associated with it.

The "logic diagram" is nothing more than a graphic display of the sequence in which the activities will be completed. The logic diagram may also be called the "network." Network schedules are symbols to indicate the activities necessary to complete the project and the logical relationships among those activities. I-J and PDM use different symbols to represent the activities and their relationships. I-J or "activity-on-arrow" diagrams use an arrow to represent the activity and circles (called "nodes") to represent events. The order in which the activities are arranged indicates relationships. The length of the arrows is not related to durations. In PDM or "activity-on-node" diagrams, the nodes are usually depicted by a box (although a circle can also be used). The nodes represent the activities, and arrows show only the logical relationships among the activities. Figures 3-1 and 3-2 show the differences between I-J and PDM symbols.

A "network analysis" or "printout" is the mathematical calculation and sorting of the logic information. The network analysis computes values for starting times, ending times, the critical path, and any other information desired. The "critical path" is the sequential combination of activities and relationships

FIGURE 3-1   Activity-on-arrow diagrams.

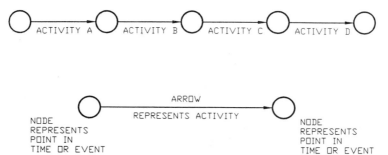

FIGURE 3-2   Activity-on-node diagrams.

from project start to finish that requires the longest time to complete. Activities on the critical path are called "critical activities." A critical activity which is not completed within the scheduled performance time may cause completion of the project to be delayed. In contrast, a noncritical activity's delayed completion may not affect the project completion date.

DETERMINING WORK ACTIVITIES

The first step in the preparation of any type of schedule is to break the total project scope into individual work "activities." The objective for determining the number and size of activities is to divide the project into sufficiently small activities that the work can be controlled. By identifying the logical relationships among the activities the sequence of construction can be identified so that a manager can control progress and project completion can be obtained as required.

The size of each activity may vary depending on the type of work involved and the importance of the activity to project completion. However, no activity should be so small as to unnecessarily complicate or lengthen the schedule or so large that the work cannot be controlled. For example, a construction schedule activity should never be as small as the units of reach, move, turn, apply pressure, grasp, position, release, etc., used in the "methods-time measurement system" (MTM)[2] used in industrial engineering. A construction schedule activity should not be as large as a major phase of a construction project, such as "interior finishes," "electrical work," or "mechanical work." Such large activities are themselves composed of many smaller activities. By neglecting the smaller activities and preparing a schedule using only large descriptions of major phases of a project, the required dependency among the smaller activities is not disclosed. This neglect prevents the manager from anticipating and eliminating problem areas and thus controlling the project's progress.

An activity is a single discrete work step in the total project. The size of the activity is determined by how the scheduler intends to use the schedule. Day-to-day control of field work may require activities of no more than several days duration. Predicting completion time or measuring progress may indicate activities of larger durations. For example, consider work activities that will result

---

[2] See, generally, Harold B. Maynard, G. J. Stegemerten, and John L. Schwab, *Methods-Time Measurement,* McGraw-Hill, New York, 1948; Delmar Karger and Franklin Bayha, *Engineered Work Measurement,* Industrial Press, New York, 1957.

in the erection of an interior concrete block wall. In addition to actually laying the blocks, electricians and plumbers must install conduit and pipe within the wall. If prediction of start and finish dates for the project is all that is necessary, a single activity such as "erect block wall" may be used. The activity "erect block wall" would be understood to include work by masons, electricians, and plumbers. If prediction of start and finish dates for each craft involved in completing the wall is important, three activities can be used: "erect block wall," "install embedded conduit," and "install embedded plumbing." Using three activities to schedule interior block wall construction rather than only one is neither right nor wrong; each is correct depending on the needs of the scheduler.[3]

The size of the activities is also influenced by several other factors. The timing of the activity, its relation to other activities, and when it will be performed must be considered. Erection of structural steel must occur early in the project, while installation of reinforced steel bars in the concrete slab on deck may occur later. Thus the timing of steel erection can influence the activities into which steel-related work may be divided.

The trade, craft, or subcontractor that will perform the work can also influence activity identification. Plumbers installing embedded sleeves in concrete may be assigned a separate activity rather than including the plumbers' work with forming and pouring concrete.

Design can also influence the selection of activities. Major steel columns and beams may be separated from the installation of intermediate steel framing; however, the design may require that all steel be firmly installed before it is safe to continue upward.

Any activity that can or must be interrupted to be completed (noncontinuous work) should also be broken into activities for each required interruption. An example of this would be the major steel columns and beams that must be in place and plumbed up prior to proceeding upward with additional steel while intermediate steel and decking can lag behind. Major equipment may need to be installed in the steel framing, before the intermediate steel, to allow maneuvering room during equipment installation.

Installation of different materials may warrant separate activities. Often this is related to procurement or to craft jurisdictions. Similarly, differences in required equipment may require separate activities. For example, concrete slab placement and finishing requires different equipment than concrete column placement. The two concrete placements should then be assigned as separate activities.

No two schedules for the same project will be exactly the same. This is because no two schedulers will break down a project into the same activities. How any individual scheduler determines the activities for a schedule will depend on that scheduler's background, experience, and expectations of how the schedule will be used. Although the activities in a schedule will not nec-

[3] Edward M. Willis, *Scheduling Construction Projects,* New York: Wiley, 1986, p. 21.

essarily be the same, the major subparts of a schedule will be similar. Five major subparts of the schedule include mobilization activities, engineering and procurement activities, owner activities, construction activities, and completion or startup activities.

"Mobilization" activities may also be necessary. The contractor may be required to obtain permits, payment and performance bonds, liability and builders risk insurance, and easements before work can actually begin. The location or nature of the work may also require a field office, telephones, reproduction equipment, and power sources. Portable toilets and drinking water, and if the project is remote recruitment and training of craftspeople, may be necessary before actual construction can begin. Specialized equipment may need to be fabricated or erected. Finally, temporary roads, housing for employees and support personnel, or other temporary facilities may need to be constructed. A time allowance must be made for required mobilization activities and appropriate activities included in the construction schedule.

In addition to activities for all the work necessary to complete the project, additional activities are required for "engineering" and "procurement." These terms include the engineering activities which must be completed, such as preparation and approval of shop drawings, and the materials management activities, such as ordering and delivering materials and equipment to the site. Shop drawings provide information about the design and how the contractor will construct certain aspects of the project in greater detail than is provided by the design drawings. Shop drawings are commonly required on U.S. projects designed by U.S. architect/engineers. Most contracts require shop drawings to

Sequencing of work activities is often dictated by unique soil conditions and early availability of shoring and forming materials. The sitework and substructure activities typically control the schedule in the early phases of construction.

be reviewed and approved by the designer before equipment can be ordered or fabricated.

Similarly, the contractor may be required to submit procedural requirements for asbestos removal or hazardous waste disposal, or descriptive literature on manufactured materials and fixtures. Procurement activities should include contractor submission of required mix designs, samples, and technical data. Delivery of material from off-shore locations must also be considered. Because timely delivery of equipment and material is necessary for timely project completion, sufficient activities should be included in the construction schedule to understand how procurement of major equipment and supplies will affect construction activities. Procurement of necessary equipment and material can be a major cause of project delay. Although it is often overlooked, procurement is a necessary part of all major construction schedules. "Off-the-shelf" or construction materials in small quantities that are normally kept in stock by suppliers and can be delivered within a short time of the placement of the order are normally not included in the schedule.

Significant owner activities should be included in the schedule if they directly constrain any of the construction, procurement, engineering, and mobilization activities. These might include:

Release of work areas
Inspections
Relocation of existing facilities and equipment
Land acquisition
Government agency approvals
Startup activities

Startup or completion activities also must be included in the schedule. These could include final cleanup and punch list, startup of HVAC or other equipment, the move in process of owner furnishings, and the final adjusting and balancing of various systems.

A complete construction schedule will include each of these five subparts. Although actual construction activities may appropriately be given greater attention and more detail, procurement, engineering, mobilization, and owner activities affect the ability to complete the project on time and should be included in the schedule.

DETERMINING DURATIONS

Activities will have a duration assigned to them. An activity's "duration" is the amount of time estimated to complete that activity. An activity's duration can be expressed in any unit of time: minutes, hours, work days, calendar days, weeks, or months. Construction schedules most commonly express durations in either work days or calendar days. Duration of activities depends on the quantity of work, the type of work, the type and quantity of available resources that may be used to complete the activity, whether the work will be completed using single or multiple shifts or overtime, and the environmental factors that affect the work.

Activity durations are estimated. It is not essential that all durations be exact. Rather, for schedules used to manage time, it is only important that each activity have a reasonable duration. If all durations are reasonable and the critical path (defined later) is made up of many activities, then variations in activity durations will offset each other, resulting in a reasonably accurate project duration. To ensure reasonable durations, the scheduler should work closely with the project's estimator and superintendent. Activity durations are the result of careful consideration of such factors as the following:

Method of construction
Project time limits
Construction sequencing
Weather and site effects on production
Concurrent work activities
Quality of supervision
Labor training and motivation
Complexity of the tasks

For production activities, such as paving, duration is established by examining the inherent sequential tasks which make up the activity and the expected daily production and quantity of work involved in each task. The duration is calculated by dividing the quantity of work by the daily production rate. As an example, assume a daily production rate for a paving crew of 500 cubic yards (cyd) per day, with 8000 cubic yards required for the complete activity:

$$\text{Duration} = \frac{8000 \text{ cyd}}{500 \text{ cyd/day}} = 16 \text{ work days}$$

A standard production rate may not include startup or setup time. It may take one or two days before full production can be met. Therefore, an activity of 17 or 18 working days may be more appropriate.

Schedules used to measure time require greater accuracy in durations than schedules used to manage time. For time measurement schedules, actual durations may be required.

ACTIVITY DESCRIPTIONS

In addition to durations, activities on a construction schedule usually include a description of the activity. The activity descriptions assist in reading the schedule. Most descriptions are abbreviated. Abbreviated descriptions are required for two reasons. In computer data bases, there are a limited number of spaces available for activity descriptions. Second, activity descriptions are either hand-drawn on the logic diagram or hand-entered into the computer.

Abbreviated descriptions minimize the time required to enter data or letter them and speed completion of the schedule. When plotted diagrams are produced, abbreviated descriptions are more likely to be complete, as short activities have very little room for descriptions.

For example, the activity list for construction of a gas station (Table 2-2) contains activity descriptions that are too long for a computer printout or logic diagram. To shorten the activity description, the following list demonstrates how abbreviations may be used:

Order and deliver grease equipment	Ord. & Del Grease Eq.
Construct light and sign foundation	Const. Light & Sign Fdns.
Install exterior underground conduit and piping to tanks	Inst. Ext. U/G Cond. & Pipe to Tks.
Connect and hook up power and lights (meter)	Hook Up Power & Lights
Construct building foundations	Const. Bldg. Fdns.
Construct hydraulic lift foundations	Const. Hydr. Lift Fdns.
Place underground conduit and piping	Place U/G Cond. & Pipe
Backfill fuel tanks and grade site	Bkfl. Fuel Tks. & Grade Site
Backfill building interior	Bkfl. Bldg Int.
Form, reinforce, and pour concrete floor	F R & P Conc. Flr.
Pave service and parking area	Pave Serv. & Park Area
Erect prefabricated building walls	Erect Bldg. Walls
Install greasing equipment	Inst. Grease Equip.
Install interior plumbing and light fixtures	Inst. Int. Fix.
Install interior air compressor	Inst. Int. Air Comp.
Install fuel pumps on islands	Inst. Fuel Pumps
Pour and cure concrete sidewalks and strip forms	Pour & Strip Sdwks.

Even the abbreviated activity descriptions can be further abbreviated if even shorter activity descriptions are necessary. Too much abbreviation, however, will make the schedule difficult to read—if not incomprehensible.

ACTIVITY CODES

Activities may be coded to include attributes of the activity. Many schedules are "cost-loaded." Each activity can have assigned to it the cost of completing that activity's work. Not all activities, however, will have a cost assigned to them. For instance, activities to be completed by the owner, such as furnishing specialized equipment or necessary interim inspections, will have no costs (assuming the schedule is prepared by the contractor) unless the owner furnishes a value. Time for completion of these activities must be included in the

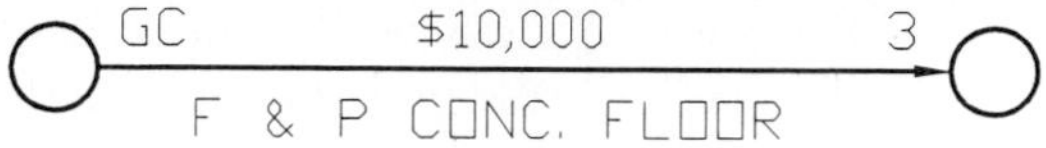

FIGURE 3-3   Activity with codes.

schedule, but the cost is usually not included. Some software programs permit two or more cost code fields. This allows the scheduler to keep track of labor and material costs separately. The scheduler can identify these fields in almost any way desired.

Activities in a schedule may also be identified by which subcontractor or trade group will perform the activity. Activities which indicate the group responsible for completion of that activity are "responsibility-loaded" schedules. By assigning responsibility to each activity, the schedule can be used to predict crew sizes and predict times when specialty contractors will be present at the site. Responsibility-loaded schedules allow the computer to sort information by subcontractor, making the schedule easier to use for the various trades.

Figure 3-3 shows an activity for forming and pouring a concrete floor from an I-J schedule. The activity has a duration, activity description, monetary value, and responsibility code. The duration assigned is located near the head of the arrow. The responsibility is located next to the tail of the arrow. The monetary value appears in the center of the arrow, while the activity description appears below the arrow. In this case, form and pour concrete floor has a duration of 3 work days, will be performed by the general contractor ("GC"), and has a value of $10,000. Location on the arrow of the various codes will vary among software plotting programs. You may locate the codes to your own liking on hand-drafted logic diagrams. The authors have found that on hand-drafted diagrams, placing the description below the line is most efficient and reduces the difficulty of adjusting or lengthening descriptions.

DETERMINING LOGICAL RELATIONSHIPS

After the work activities and durations have been determined, the next step in preparing a CPM schedule is to arrange the activities in the order in which the activities will be performed. The order in which the activities will be performed is called the "logic." How any activity relates to any other activity is called its "logical relationship."

Each work activity is related in some way to every other work activity in the schedule. There are three possible logical relationships among activities: predecessor, successor, and concurrent or independent. Predecessor relationships occur when an activity must be completed before a particular work activity. For example, foundation work usually precedes installation of the roof. The foundation work has a predecessor relationship to the roof work. Successor relationships occur after completion of a particular work activity. For example, interior dry finish work can begin following completion of the roof. Interior dry finish work has a successor relationship from roof work. If an activity is neither a predecessor nor a successor, it is considered to be independent or concurrent. An independent or concurrent relationship between activities does not necessarily mean that the two activities will be performed at the same time. Independence or concurrence means that completion of an activity is not dependent on the start or finish of another activity.[4] For example, installation of elevators can be completed any time after the roof has been installed. Since interior dry finishes can also be applied after the roof has been completed, elevator installation and dry interior finishes can be considered independent or concurrent relationships. However, elevator installation and dry interior finishes do not necessarily have to be completed at the same time. Predecessors, successors, and independent or concurrent relationships are shown in Figure 3-4.

All activities are related to each other. However, not all activity relationships are required to be specifically drawn on the logic diagram. Many of the relationships are implied. When sequencing work activities, one builds a chain, linking activities that are immediately related to each other. The chain of activities may become quite large. For large construction projects it is not uncommon for several thousand activities to comprise a schedule. Although the first activity and the last activity are related, the relationship is shown through the many other activities which must be completed after the first and before the last activity. No individual arrow is drawn between the first and last activities by the scheduler. Nevertheless, the first and last activities are "implicitly" related to each other.

Experienced schedulers develop the activities, durations, and logical relationships at the same time. For those just learning the art of scheduling, it may be advantageous first to develop a complete list of all project activities. The scheduler can then determine accurate durations for each activity and finally organize the activities through logical relationships.

In I-J schedules, the logical relationship is shown by the way the arrows are arranged. The manner of arrangement and presentation of the arrows influences the ease with which the activities may be identified on the logic diagram.

There are two ways in which the arrows may be drawn: sloping format, and staff format.

The sloping format combines straight lines at different angles. At least one

[4] Edward M. Willis, *Scheduling Construction Projects,* New York: Wiley, 1986, p. 74.

FIGURE 3-4   Logical relationships.

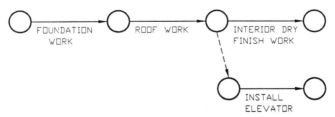

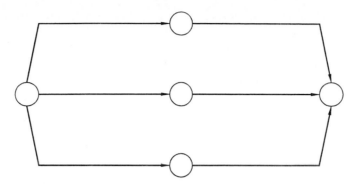

FIGURE 3-5   Sloping format.

straight line is horizontal to permit labeling the activity. Each arrow is separate; no arrow shares part of another arrow. Most schedulers use the sloping format when preparing the logic diagram, because each arrow is drawn separately without concern for arrangement or possible sharing of parts of other arrows. It is also easier to read a schedule prepared in the sloping format. Particularly when many other activities start or finish at a single activity, it is often difficult to distinguish arrows that are shared by many activities. Figure 3-5 shows a logic diagram drawn in the sloping format.

The staff format is drawn using only horizontal and vertical lines. Part of one arrow may be shared with another arrow. Logic diagrams produced on a computer-driven plotter or printer will usually be in the staff format. The staff format can be difficult to read. Because the staff format shares parts of arrows with other activities, when many activities start or finish at the same point, the reader may have difficulty determining where arrows begin or end. Figure 3-6 shows a logic diagram drawn in the staff format.

FIGURE 3-6   Staff format.

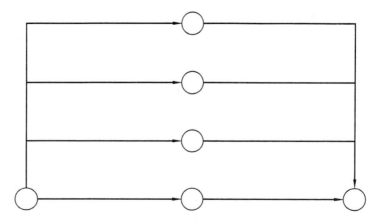

I-J critical path method schedules are composed of circles and arrows. The circles are usually drawn so that they are located in equally spaced vertical columns, and the arrows are aligned horizontally. By "structuring" a logic diagram in this manner, both the numbering of the nodes is facilitated and the likelihood of computational errors is reduced when performing the network analysis.[5] Figures 3-5 and 3-6 are structured logic diagrams. Figure 3-7 shows an unstructured logic diagram.

The length of the arrow for any activity carries no significance. The arrow is intended only to identify logical relationships. As long as the arrow is sufficiently long to allow the activity description to be included, arrow length is usually unimportant. However, as will be discussed later, some logic diagrams are "time-scaled." Time-scaled logic diagrams distribute the length of all arrows proportionately to the scale of the diagram. The length of an arrow is important in time-scaled logic diagrams. Although they are easier to understand because of their similarities to a bar chart, time-scaled logic diagrams can be time-consuming and difficult to prepare, even with the assistance of a computer-driven plotter.

I-J critical path method schedules use a connecting activity called a "dummy" to help indicate logic correctly and to ensure that each activity has a unique set of node numbers. This connecting activity has no duration or responsibility and is sometimes referred to as a "restraint" or a "tie." Figure 3-8 illustrates the use of dummy activities.

Figures 3-8a and 3-8b represent two different logics. In Figure 3-8a, both activities Y and Z require completion of both activities W and X. In Figure 3-8b, only activity Y requires completion of activities W and X. Activity Z can begin as soon as activity X is complete. The proper use of dummy activities assists the scheduler in correctly portraying the construction sequence and dependencies.

[5] Edward M. Willis, *Scheduling Construction Projects,* New York: Wiley, 1986, p. 81.

FIGURE 3-7   Unstructured logic.

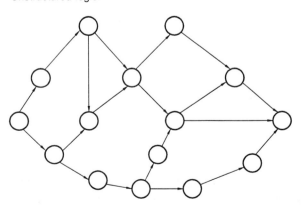

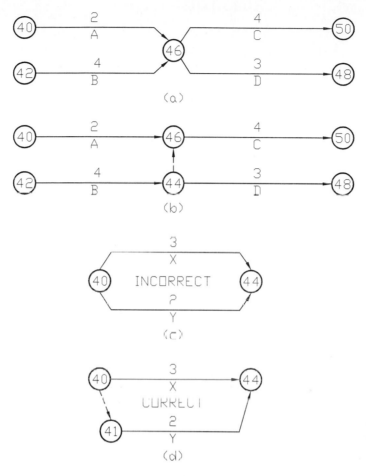

FIGURE 3-8   Use of "dummy" activities.

A second use of dummy activities is to ensure that each activity has a unique set of node numbers for computation and sorting by the computer. Figures 3-8c and 3-8d illustrate the incorrect and correct ways to represent two activities with the same starting and ending events.

Critical path method schedules permit the scheduler to determine how long project completion will require. By adding the durations of activities in the order in which the activities have been arranged, the scheduler can determine the chain of activities through the project that requires the longest period to perform. That longest chain is called "the critical path." The process of adding the durations of activities to determine the amount of time necessary to complete a chain is called "network analysis." By arranging the activities in different ways, the scheduler can vary the amount of time necessary to complete the project. The critical path concept will be discussed in much greater detail in subsequent chapters.

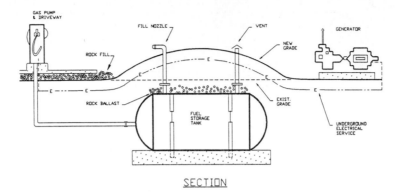

FIGURE 3-9 Remote fuel station.

EXERCISES

1 Draw a logic diagram that conforms to the following statements:

X follows A.
B precedes I and follows W.
G starts when A is completed.
M follows H and must be done before L can start.
W is the first activity on the job.
A and H can start concurrently with B.
E is the last activity, but it cannot start until G and L are completed.
L starts when X and I are completed.

2 Figure 3-9 shows a section for a remote fuel pumping station. Prepare a list of construction activities necessary for completion of the project, and assign a duration to each activity. Include procurement and mobilization activities. Anticipate that 20-25 activities should be necessary. The activities do not have to be listed in the sequence that they will be completed.

3 Identify the list of predecessors for the following activity list for a small bridge project.

Activity code	Activity description	Predecessors
A	Award contract	
B	Order and deliver piling	
C	Order and deliver beams	
D	Drive pile east support	
E	Construct pile cap east support	
F	Construct east abutment	
G	Construct pile cap center support	
H	Drive pile center support	
I	Erect center columns	
J	Construct west abutment	
K	Drive pile west support	
L	Construct pile cap west support	

Activity code	Activity description	Predecessors
M	Set beams east half	
N	Construct deck east half	
O	Construct approach east side	
P	Seed and sod	
Q	Construct guard rail	
R	Erect signs	
S	Paint road markings	
T	Set beams west half	
U	Construct deck west half	
V	Construct approach west side	

4 Draw the logic diagram for the project involving the activities listed in Exercise 3.

5 Draw the logic diagram for the following set of activities:

F follows A and E.
G follows E.
E follows B.
D precedes K.
C precedes H.
F precedes M and L.
L follows F, G, and H.
A, B, C, and D start the project.

The project is complete when M, L, and K are finished.

6 Draw the logic diagrams for the following set of activities:

A, B, and C start the project.
D follows A.
B precedes E.
C follows F.
K follows E and F.
G and H follow D and E.
G precedes L.
M follows H and K.

7 Sketch the logic and draw the logic diagram for the following set of activities. Try to sketch so that no lines cross.

A and B precede E.
B precedes F.
C precedes G.
D precedes H and K.
N follows E, H, and L.
M follows E.
F and G precede L.
O follows M.
P, Q, and R follow O and N.
N precedes S.
T follows K and S.

S and K precede Y.
Z follows Y.
U follows P.
V follows Q and precedes X.
W follows R.
U precedes X.
When X, W, T, and Z are complete, the project is done.

CHAPTER
4

ACTIVITY-
ON-ARROW
SCHEDULES

ACTIVITY-ON-ARROW DIAGRAMS

Understanding an I-J Diagram

An activity-on-arrow diagram is composed of arrows and circles. The arrows represent activities. The circles or "nodes" represent events. The node at the beginning of the arrow (the tail) is referred to as the "I" node. The node at the head of the arrow is referred to as the "J" node.[1]

Because activity-on-arrow diagrams link the nodes of all activities together, the J node of the preceding activity is also the I node of succeeding activities. Activity-on-arrow diagrams are also referred to as I-J diagrams because of the I or J designation of the nodes.

Each activity (arrow) carries a brief description. The activity's description is usually printed on the logic diagram and the printout. The activity description can be printed either above or below the arrow. In addition to the activity description, each activity is assigned a duration. In this text, most activity descriptions appear below the arrow, and durations and responsibility codes are shown above the arrow. Descriptions placed below the line give the sched-

[1] Edward M. Willis, *Scheduling Construction Projects*, New York: Wiley, 1986, p. 72.

Earthwork activities require careful selection and utilization of equipment resources. These operations are usually very sensitive to weather. (Photo courtesy of John Deere.)

uler space in which to add or to write a more complete description of the activity.

Each node represents a point in time. The I node is the point in time at which the activity may begin. The J node is the point in time at which the activity may end. Thus the point in time represents the "event" of either starting or finishing an activity. Completion of the preceding activity also represents the point in time when the succeeding activity may begin. Figure 4-1 shows a portion of an I-J network.

Each node is assigned an identifying number. Each arrow or activity can be identified by the number located in the I node and J node. For instance, the activity "lay out and excavate" in Figure 4-1 is designated activity 12-18. The activity "lay out and excavate" has an I node of 12 and a J node of 18. Some

FIGURE 4-1 Partial I-J diagram.

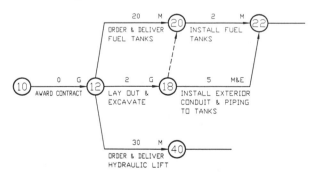

I-J diagrams may initially have the same I-J nodes for different but concurrent activities. Each activity should have a unique I-J number, primarily in order to avoid communication problems. The same I-J number should not be used for different activities, even if the distinction to be made is only between different trades scheduled for the same activity.

A node represents an event or point in time. In Figure 4-1, node 12 represents the point in time when it is possible to commence the activity "lay out and excavate" and the activity "order and deliver fuel tanks." Node 18 is the point in time at which activity "lay out and excavate" can be completed and when activity "install exterior conduit and piping to tanks" can begin. Node 22 is the point in time at which activities "install fuel tanks" and "install exterior and piping to tanks" are to be completed.

There are three paths in Figure 4-1 between nodes 12 and 22. The first path comprises nodes 12, 20, and 22: activity 12-20, "order and deliver fuel tanks," and activity 20-22, "install fuel tanks." The length of the first path is 22 days. Activity 12-20 has a duration of 20 days. The duration is shown centered on the activity's arrow, above the arrow. Activity 20-22 has a duration of 2 days. A second path comprises nodes 12, 18, and 22: activity 12-18, "lay out and excavate," and activity 18-22, "install exterior conduit and piping to tanks." The length of the second path is 7 days. Activity 12-18 has a duration of 2 days, and activity 18-22 has a duration of 5 days. A third path comprises nodes 12, 18, 20, and 22, the length being 4 days total. Activity 18-20 has no work or time associated with it. This activity is also known as a "dummy" or "restraint." The first path is the "critical path" because it is the longest path. Activities 12-20 and 20-22 are "critical activities" because they are on the critical path. As far as this partial schedule is concerned, the first path, the critical path, will control when the activities identified in Figure 4-1 will be completed.

Since the second path has a combined duration of 7 days, and the third path a duration of 4 days, they are noncritical paths. Thus, even if the activities which comprise the second path are not completed within their scheduled durations and are delayed by a few days, the total time necessary for completion of the activities in Figure 4-1 will not be extended. Unless completion of the activities on the second path is delayed longer than the activities on the first path, the total time necessary to complete all the activities in Figure 4-1 will not be affected. If, however, activities 12-18 and 18-22 along the second path *are* delayed longer than the first path, the second path will become the critical path and control completion of the activities in Figure 4-1. This is an important concept to understand because it will be applied to calculating delay and time extensions, to be discussed in Chapter 12.

When drawing a logic diagram it may not be possible to avoid having arrows cross. When this occurs one of several types of "crossover" symbols should be used. Figure 4-2 shows two kinds of crossover symbols.

Without crossover symbols, the schedule reader may misinterpret the relationship the arrow is intended to depict. Crossover symbols minimize errors

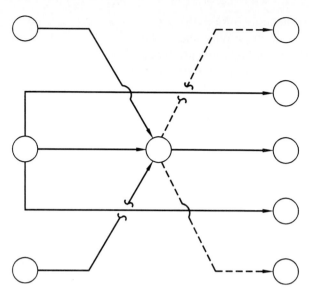

FIGURE 4-2   Use of crossover symbols.

in reading the logic diagram. On plotted schedules, the computer usually just leaves a gap in the activity or restraint line. This can be very confusing, especially on large, detailed schedules.

I-J Logic Assumption

The I-J method of critical path scheduling is based on the premise that a given activity cannot start until all the activities that immediately precede it have been completed. In Figure 4-1, "lay out and excavate" cannot start until "award contract" has been completed. "Install fuel tanks" cannot begin until "order and deliver fuel tanks" has been completed. The I-J premise is not always a true representation of the field, but it is rigid in the mathematical calculation of times for the scheduling process. Reality in the field suggests that perhaps 2 percent or 5 percent of the preceding activity may not be complete but still permit following work to begin.

All activities in the I-J network must have definite events that define their performance. In I-J scheduling, the end of one activity cannot overlap the start of the following activity. If the beginning of one activity overlaps the completion of another (by any appreciable amount), the activities should be subdivided further.

Numbering the Nodes

Each node is assigned a unique identifying number. The same number should not be used more than once. This way, each activity will have its unique set of I-J numbers and cannot be mistaken for another activity.

To help the scheduler number the nodes so that each node will have its own number, the logic diagram is completed before any node numbers are assigned. In other words, the scheduler should arrange the activities which comprise the project in the order in which they will be completed in the field before numbering any nodes.

Schedulers also observe other rules when numbering the nodes. It is customary, but not mandatory, to give the I node at the tail of the arrow a lower number than the node at its head, the J node. J greater than I for all activities is known as "forward numbering." This rule was initially imposed because of the limitations of early CPM computer programs. Although it is no longer necessary for most programs, the practice persists because it makes it easier to locate events and activities on the diagram. Knowing that the J node always has a higher number than the I node, the reader can search chronologically/numerically for the location of any known activity in the logic diagram and/or the printout.

Forward numbering also helps prevent "loops." A loop occurs when activities are arranged in such a way that an activity returns to an earlier activity, forming a circle. By strictly following the arrows, a loop creates a continuous circle which prevents consideration of all remaining activities. Circling in a logic loop, the project schedule will never complete. Figure 4-3 shows a logic loop. Although most software includes an antiloop routine, loops slow calculations and prolong completion of the schedule. Loops must be removed before the scheduler can calculate the schedule.

A careful scheduler is unlikely to create a loop while preparing a schedule. However, lack of concentration can lead the scheduler to create a loop. A construction conflict may also cause a loop. For example, an existing clarifier pipeline on a wastewater-treatment plant renovation was to remain in operation throughout construction. Although the clarifier pipeline was shown to be operating in the old plant, it was also shown to be shut down and replaced in the renovated plant. Attempting to schedule this conflict could easily cause a loop. A change in the operating procedure of the wastewater treatment plant during construction was required to resolve this "loop."

Loops are more likely to be created by a draftsperson who unintentionally misnumbers a node (say, numbering node "5225" as "5252" by mistake) or a data processor making a keypunch error. If the drafting or keypunch error results in a duplication of node numbers, the computer will combine logic at

FIGURE 4-3   Logic loop.

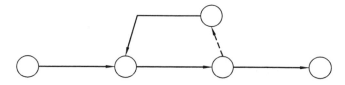

the duplicated node number, sometimes creating a loop. Sometimes data key-punch errors do not create loops. Without a careful data check, the error may go undetected—resulting in field use of an erroneous schedule.

Even a careful scheduler, however, can create a loop when revising or resequencing a schedule. For example, when changing the interior finishing of an 11-story office building to top-to-bottom sequence from bottom-to-top sequence, it is easy to miss a change and create a loop.

By forward numbering, a scheduler can discover most if not all loops which may have been inadvertently included in the schedule. Careful study of the occasions when logic indicates that activities return to earlier numbered activities will prevent most loops. This reverse numbering is known as "back-branching." Although forward numbering will not eliminate all loops (loops may still occur during entry of the data into the computer), forward numbering can significantly reduce the potential for loops.

Schedulers also group numbers when assigning numbers to nodes. When arranging activities in the order in which they will be completed, schedulers often use "subnetworks" or groups of related activities which are later tied together to produce a completed project schedule. Common groups include "foundation, structural frame, enclosure," "interior rough-in," "interior finishing," and "utilities," for example. Subnetworks may comprise any number of individual activities. Subnetworks and the scheduling of large projects is discussed in detail in Chapter 6. Experienced schedulers number the nodes so that all activities in a related group are in the same number group. All foundation activities, for instance, may carry 2000-series numbers, and only foundation numbers will carry 2000-series numbers. All frame and enclosure activities may be numbered in the 3000 group. Forward numbering is not possible in all situations; backbranches are sometimes inevitable. With careful planning and proper sequencing of subnetworks on very large schedules, however, forward numbering is usually possible.

When numbering the nodes, it is also a good idea not to use consecutive numbers. Nonconsecutive numbers, in multiples of 2, 3, 5, or 10, are used so that node numbers will be available for adding activities later. Thus the nodes on an initial logic diagram may be numbered as shown in Figure 4-4.

Figure 4-5 shows how a new activity which succeeds activity 10-20 but precedes activity 20-30 should be added. Method 1 in Figure 4-5 uses "double noding" and dummy activities. Nodes 22 and 25 are the revised logic. Nodes 20 and 25 are double nodes. Implicit in a double node is a dummy or restraint activity linking the new activity to the existing logic. The dashed line between nodes 20 and 22 is a dummy. When adding activities which succeed and precede existing activities, double nodes and dummies are often used to permit the

FIGURE 4-4   Numbering nodes.

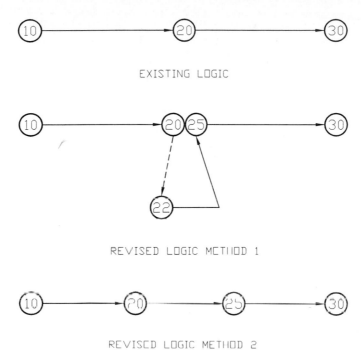

EXISTING LOGIC

REVISED LOGIC METHOD 1

REVISED LOGIC METHOD 2

FIGURE 4-5   Adding new activities.

scheduler to evaluate the effect of the new activity on the original logic. The double nodes and dummy permit the scheduler to compare the duration of the schedule both with and without the new activities and to determine if the critical path of the schedule has changed. Changes in duration and critical path of a schedule as a result of new activities are termed the "impact" which the new activities have on the schedule.

If the scheduler does not wish to compare before and after effects of the change, then method 2 in Figure 4-5 provides the quickest and easiest way to modify the schedule.

Figure 4-6 shows how a new activity which can be performed concurrently with existing activities may be added to the schedule. As with new activities that succeed and precede existing activities, double nodes and restraints, as shown in method 1, can be used to enable the scheduler to evaluate the impact of the new activity on the duration and sequence of planned project completion.

"Double noding" can also be used in complex logic diagrams to maintain correct relationships among activities. Figure 4-7 shows a more complex logic arrangement which requires double noding. In rare instances triple noding may be required, although this usually can be avoided.

The double noding shown in Figure 4-7 is necessary to indicate that activities F and G cannot proceed until activities A, B, C, and D have been completed, but activity E can proceed when activities B and C have been completed.

EXISTING LOGIC

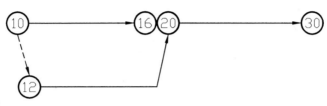

REVISED LOGIC METHOD 1

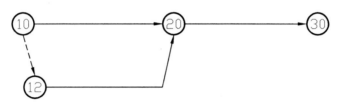

REVISED LOGIC METHOD 2

FIGURE 4-6   Adding concurrent activities.

Activity E is not dependent on activities A and D. For clarity, the scheduler shows a dummy between nodes X and Y to maintain the relationships indicated. This logic could not be represented without the use of double nodes.

Many schedulers like to minimize restraints and extra nodes. These schedulers will avoid double noding for many good reasons. Double noding increases

FIGURE 4-7   Double noding.

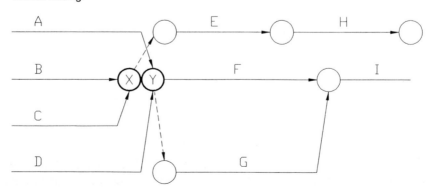

the amount of data that must be input into the computer and slows the time required for analysis. Double noding also clutters the logic diagram, making it more confusing. Programs which plot logic diagrams have difficulty clearly drawing double nodes, especially in time-scale format. Double noding, in short, is not right for all schedules.

While the concept of avoiding double nodes is good, sometimes those extra restraints serve an important function. Figure 4-8a could be drawn correctly without restraints 2-4 and 6-10, as shown in Figure 4-8b. Figure 4-8b follows the theory of scheduling more closely, will produce a shorter printout, and will reduce calculation time. The printout made from Figure 4-8a, however, will be easier to read. No two work activities in Figure 4-8a have the same I node. When reviewing or updating a schedule, the scheduler needs to refer only to the I node, and not to both the I and J nodes as would be required in Figure 4-8b. This distinction is known as the "independent I-node rule" and is often followed for any large I-J schedule that will be computerized. Although double noding and independent I nodes may help the scheduler, it is not necessary for project control.

Some scheduling software programs include an automatic numbering feature. Instead of numbering nodes manually, the computer may do the node num-

FIGURE 4-8   (a) Proper numbering sequence. (b) Minimum restraint technique.

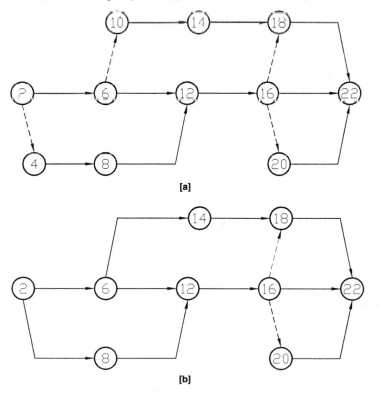

[a]

[b]

Careful placement of large equipment often requires the contractor to mobilize special heavy-lift equipment such as this "ringer crane." (Photo courtesy of Manitowoc Company, Inc.)

bering automatically. For example, one program assigns node numbers automatically as activities are entered into the schedule. Experienced schedulers ignore automatic numbering routines in scheduling software, because they eliminate control of numbering. The ability to distinguish subnetworks according to certain number groups, to locate activities chronologically in the schedule, and to permit the addition of new activities without renumbering are important to good schedules. Automatic numbering often precludes use of these techniques and contributes to schedules which are harder to read and understand.

Finally, when numbering nodes, the scheduler should work one subnetwork at a time, numbering the full subnetwork from left to right. The scheduler should pick up all nodes on all paths within the subnet in the single left-to-right numbering process.

THE LOGIC DIAGRAM

Determining the Sequence

The most important part of schedule preparation is determining the sequence in which the activities will be completed. The sequence depicted on the logic diagram must be logical, reasonable, and possible. To the extent that the logic diagram portrays an unreasonable, illogical, or impossible sequence, the schedule is wrong. The schedule must reflect the plan of the contractor and in particular the team who will be executing the project.

Two things must be understood about determining the sequence on the logic diagram. First, proper sequencing requires an understanding of the construction industry, not just scheduling. Second, there is always more than one correct sequence for completing a construction project.

A properly sequenced construction schedule requires an understanding of the construction industry. A scheduler must understand how the various materials, trades, subcontractors, and suppliers interact to complete a construction project. Technical reasons and traditional work order also influence sequence. Sometimes, technical considerations require accomplishing interior finish work in a multiple-story building from the bottom up. Other times technical reasons will require completion of interior finish work from the top down. Tradition also plays an important role in sequencing activities. All sorts of nontechnical customs influence how construction projects are completed. Different areas observe different work rules and trade restrictions. Religions may influence work seasons or customary work days. For example, in predominately Christian areas, work is accomplished Monday through Friday, while in predominately Islamic areas, work is accomplished Sunday through Thursday. Different nations observe different holidays. Even within a single nation, work rules may vary from one section to another.

How these technical and traditional rules influence the manner in which a construction project can be completed is not part of learning construction project scheduling. Scheduling is a discipline that can and is applied to any process or industry. Understanding scheduling permits the preparation of a schedule for any number of events. A scheduler who has mastered the principles of scheduling should be able to prepare a schedule for the preparation of a meal, assembly of an automobile, or servicing of an airplane as long as technical assistance in establishing the proper sequence is available.

However, proper sequencing, as opposed to scheduling techniques, does play a major role in schedule preparation. A construction scheduler learns the technical and traditional restrictions on sequence from experience in the field, education in construction science and design, and actual preparation of construction schedules. An understanding of the construction industry is acquired after years of participation, observation, and instruction. Schedulers who do not have such experience or understanding must review their proposed sequence with someone who does.

The second concept that must be understood when sequencing construction activities is that a construction schedule may be prepared with many different sequences. There is no one correct way to sequence any construction project. Different sequences may be equally possible, logical, and reasonable. There are many ways to complete any project. Personal choices of the various managers play an important role in sequencing. Differences between two schedules thus do not necessarily make one incorrect.

Consultation and understanding of how the project manager and craft superintendents plan to execute the project is essential to developing an accurate schedule. A qualified scheduler can develop a reasonable and correct schedule for a project, but without input from the actual managers it is almost impossible for the scheduler to come up with the contractors' plans for executing the project.

An important corollary to the concept that there is more than one way to schedule construction projects is that the sequence shown on the schedule, although reasonable, possible, and representative of the contractor's original intent, is not necessarily what will be executed in the field. This does not mean that the schedule is incorrect. The scheduler necessarily makes assumptions about how work in the field will be sequenced. Most buildings can be completed by pouring the concrete floor slab either before or after the roof is completed. Both ways are correct. Good reasons may support either sequence. Completing the roof first permits the concrete to be poured with the roof providing protection from rain. Pouring the concrete first permits greater efficiency and productivity in placing the concrete, since no roof or columns interfere with the pour. Even though the scheduler has assumed one way, it does not make the schedule wrong if the work is actually completed in another way in the field. When different sequences are chosen in the field, however, the schedule should be changed to match the actual sequence. Failure to revise the schedule as field conditions change planned sequences can render the schedule ineffective as a management tool (although arguably still able to be used to measure time). The process of revising the schedule to match the sequence actually followed in the field is known as "updating" and will be discussed in greater detail in Chapter 8.

Developing the sequence also requires an intent or plan. An experienced scheduler will subconsciously know the project time frame while developing activities and sequencing. An experienced scheduler will consider ways to "tighten" the project schedule while preparing the logic if the scheduler knows the project has a short time frame. Individual activity durations are estimated considering the most efficient use of labor, materials, and equipment.

Time constraints are not generally considered during development of the schedule's sequence. The logic diagram is prepared assuming reasonable times and efficient resources. Activity durations are determined after the sequence and are based on limitations of equipment, materials, and labor.

Developing the sequence also includes required revisions. Only in extremely rare cases does the "first run" of a schedule finish on the exact project completion date. Considerable discussion with the managers of the project is usually required to bring the project schedule in line with the contract completion date.

Restraints

To be realistic, a CPM schedule must reflect the practical restraints of safety, available space, and structure that apply to the job activities. Completion of the supporting structure restrains the installation of the roofing membrane.

Structural steel cannot be erected until it is available at the site. Steel availability depends on preparation, transmittal, approval of erection drawings, shop drawings, mill orders, rolling, delivery, steel fabrication, and delivery to the job site. Thus the start of steel erection is restrained by the necessary preliminary actions of detailing, engineering approvals, steel fabrication, and delivery. These and similar restraints arise from the necessary order of construction operations. Most often, restraints are indicated by the manner in which the work activities are arranged in the logic diagram.

Restraints that are not reflected in the arrangement of activities but are still necessary to reflect the reality of construction are represented on the network as dashed lines. These restraints are shown as ''dummies'' to explain certain relationships among activities that the arrangement of activities on the logic diagram does not indicate. Regardless of how a logic diagram is sequenced, dummies are required to show correct job logic. The use of dummies is the most difficult aspect of I-J diagramming.

In Figure 4-1, the dashed arrow 18-20 is an example of a dummy. This is not a time-consuming activity, but dummy 18-20 is necessary if job logic is to be portrayed correctly. A correct sequential relationship stipulates that the start of ''install fuel tanks'' must await completion of ''order and deliver fuel tanks.'' The tanks must be ordered and delivered before they can be installed. But ''install fuel tanks'' must also await completion of ''lay out and excavate.'' Dummy 18-20 shows that the start of ''install fuel tanks'' depends on the completion of not only ''order and deliver fuel tanks'' but also the restraint of ''lay out and excavate.''

The direction of the dummy designates flow of activity dependencies. Dummy 18-20 illustrates that completion of ''lay out and excavate'' permits installation of fuel tanks to begin.

Not only can proper use of dummies indicate correct job logic, it can also correct improperly drawn logic. Unnecessary logic constraints commonly occur when more than one arrow enters a node or when more than one arrow exits a node. For example, Figures 4-9a and 4-9b show two logic diagrams for the forming, placing of concrete, setting, and hookup of a piece of equipment. Figure 4-9a is drawn without dummies. The manner in which the activities are arranged imposes incorrect constraints on the final mechanical and electrical hookup of the equipment. Also, two activities have the same I and J nodes.

The logic diagrams in Figures 4-9a and 4-9b consist of eight activities.

Form foundation
Place concrete
Set equipment
Mobilize electrical contractor
Mobilize mechanical contractor
Electrical hookup
Mechanical hookup
Start up equipment

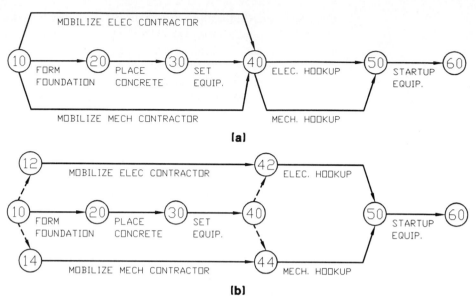

FIGURE 4-9   (a) Incorrect constraints. (b) Correct constraints.

The scheduler has arranged them logically as shown in Figure 4-9a. Without dummies, however, the logic imposes improper constraints on the schedule. Improper constraints occur where more than one arrow from node 10 enters node 40 and more than one arrow from node 40 enters node 50. As drawn, "mobilize mechanical contractor" constrains "electrical hookup" of the machine. This is not a true constraint, as the mobilization of the mechanical contractor can occur independently of the electrical hookup. Similarly, the logic shows "mobilize electrical contractor" constraining "mechanical hookup." This is also not a true constraint. The mobilization of the electrical contractor can occur independently of mechanical hookup. Properly included in the logic diagram, dummies can be used to correct these logic constraints. Figure 4-9b shows the corrected diagram. Note that dummies 10-12 and 10-14 have been added to follow the independent I-node rule.

Too many dummies on the logic diagram can cause as many problems as too few. Numerous dummies can make the diagram hard to read and difficult to analyze. As will be discussed in the next section, dummy activities impose additional computations on the calculation of the critical path and activity event times. Too many additional computations imposed by dummies will slow the calculation and delay network analysis. To avoid difficult-to-read diagrams and lengthy network analysis, dummy activities should be kept to a minimum.

A restraint can represent any of many different types of limitations. Restraints can reflect physical, crew, weather, crowding, preferred sequence, access, equipment, facility, contractual, phasing, or other requirements:

Physical restraints are used to show physically dependent activities, such as that a foundation must be completed before structural steel can be erected upon it.

Crew restraints limit construction by the number of crews available to work on a given type of activity at the same time. A crew restraint prevents a subcontractor from painting all seven floors of an office building simultaneously when only two painting crews are available.

Weather restraints can prevent concrete work when subzero weather is anticipated or limit the amount of concrete work which can be executed because of limitations in the availability of weather-protection equipment.

Crowding restraints prevent too many different trades from working in the same area simultaneously even though it may be physically possible.

Preferred sequence restraints may be used to control the way certain work proceeds, such as installing carpet in an office building at a time that will avoid the need for extra cleaning.

An access restraint is used to show a limitation or control of a work area by the owner. Renovation of a hospital could have an access restraint limiting the contractor to 12 patient rooms at a time, thereby allowing the owner to "juggle" patients around the renovation.

Equipment restraints control the availability of major equipment on a project. A tower crane which is being used 100 percent by the curtain wall erector may not be available to other contractors until after certain parts of the curtain wall have been erected. Another example is a project with only one backhoe: The scheduler cannot expect to trench for utilities and at the same time excavate for a retaining wall. The scheduler uses an equipment restraint to show this relationship even though no physical relationship may exist.

Facility restraints limit the amount of construction that can go on simultaneously. A project with multiple structures may have restraints limiting the total amount of construction to a manageable amount for any one contractor.

Contractual restraints are usually dictated by the owner, who may require a certain sequence of construction to permit continuous operations. For example, an owner may need the loading dock completed early.

Phasing restraints may require certain elements of the work to be completed prior to the start of other work items. Typically, wastewater-treatment plant expansions and renovations require phasing restraints to keep the plant operational throughout construction.

These types of restraints are all reasonable ways to limit and control a project schedule. However, excessive use of restraints can make a schedule difficult to read and use. Adequate restraints must be included to show all the limitations and constraints on the construction.

Restraints may also be shown on the diagram and in the computer database by the use of an imposed start or finish date (often termed a "plug date"). This type of feature adds a benchmark or fixed time point to the schedule. Figure

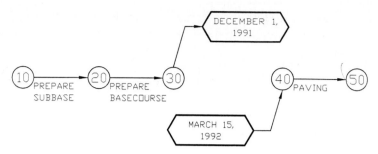

FIGURE 4-10 Imposed start/finish dates.

4-10 illustrates the use of this type of restraint and shows one method of showing this on the diagram.

The authors recommend against using plug dates in a schedule unless they reflect a contractual requirement. This issue is discussed in detail in Chapter 9.

Restraints may also be arranged to make critical a particular chain of activities that are anticipated to be delayed in the future. Sometimes referred to as "spring-loaded," these schedules have been manipulated to reroute the calculated critical path to a path on which delays are anticipated. One can identify a "spring-loaded" schedule in one of several ways. First, durations may be in even multiples of a 5-day or 7-day week, indicating that no real effort has been made to estimate activity durations. Instead a particular chain of activities has been "set" to be critical by a series of even multiple durations. Second, a higher proportion of activities will be critical than is usual. If restraints have been manipulated to make otherwise noncritical activities critical, more and different activities will be critical. When reviewing a schedule that might be spring-loaded, the scheduler must ask: "Are normally flexible activities critical?" Usually 5 percent to 15 percent of the schedule's activities are critical or near-critical. More than that should raise suspicions of manipulation or unreasonable project duration.

Finally, one can identify a "spring-loaded" schedule by a lower-than-usual average total float per activity. Usually, activities will average 15 to 35 days average float per work activity for most construction projects. Note, however, that very long or very short projects or those with unusual procedures may have a different average. Float analysis is discussed in detail in Chapter 6.

ANALYZING THE NETWORK

Time Versus Date

Nodes represent points in time. The initial node represents that point in time when the project begins. All subsequent nodes also represent points in time. Any node's point in time can be calculated by adding all preceding activity durations in the same chain to the time that the project started. Calculated in this way, any node's point in time will be a number that represents the value of all preceding activity durations in the same chain.

Points in time represented by nodes can also be expressed as a calendar date. By determining the date that a project begins, the initial node and all subsequent nodes' points in time can be translated to a calendar date by identifying the date that corresponds to the point-in-time value of any event.

Most CPM printouts express the time value of nodes as calendar dates. A table is available in this book (see Appendix) that assists in translating from points in time expressed as numbers to points in time expressed as calendar dates. The table is based on calendar and work days. Many different work day tables are available, reflecting the variety of different days that may constitute work days. Which days are nonworking and thus which table is to be used must be carefully considered. There are considerable differences among nations and cultures as to work days, holidays, and weekends. Most software programs allow the scheduler either to select the correct table or to revise available tables to reflect actual work schedules.

For weather-sensitive work it is also necessary to convert the schedule's critical activity work days to an anticipated calendar-date target. Figures 4-11, 4-12, 4-13, and 4-14 show the expected distribution or work days by month for various activities for a given location in the United States. Those tables are based on historical data over a 10-year period. Owners and contract administrators use these types of figures to establish calendar-date milestones based on the work day analysis.

The date for starting an activity, whether based on calendar dates or work days, traditionally refers to the beginning of the day. The date for completing an activity refers to the end of the day. This issue is clarified in Figure 4-19, later in this chapter.

Durations

Durations for activities can be assigned to individual activities either while or after the logic diagram has been completed. Time constraints are generally not considered when the novice scheduler first prepares the logic diagram. Proper CPM technique calls for the logic to be developed under the assumption that reasonable time and resources are available. When actually assigning durations, however, the most cost-effective mix of resources which comply with the project plan should be used.

Equipment, material, and labor limitations should be considered when activity durations are assigned. Ideally, durations will be determined with the assistance of the estimator who prepared the contractor's bid. More often, however, durations are determined by "informed" judgments of the scheduler. Many experienced schedulers are better at estimating durations of work than are estimators, who are mostly concerned with manhour totals.

Some schedulers do not have the construction experience necessary to es-

Distribution of Working Days
by Months, Through 110 W.D. Year
Cumulative Totals Per Month

Light Grading and Urban

WD	Mon	JAN	FEB	MAR	APR	MAY	JUN	JUL	AUG	SEP	OCT	NOV	DEC
		JAN											
0	JAN		FEB										
0	FEB			MAR									
0	MAR				APR								
4	APR	4	4	4	4	MAY							
3	MAY	17	17	17	17	13	JUN						
7	JUN	34	34	34	34	30	17	JUL					
8	JUL	52	52	52	52	48	35	18	AUG				
8	AUG	70	70	70	70	56	53	36	18	SEP			
8	SEP	88	88	88	88	84	71	54	36	16	OCT		
7	OCT	105	105	105	105	101	88	71	53	35	17	NOV	
5	NOV	110	110	110	110	106	93	76	58	40	22	5	DEC
0	DEC												0
0	JAN												
0	FEB												
0	MAR												
4	APR	114	114	114	114	110	97	80	62	44	26	9	4
3	MAY	127	127	127	127	123	110	93	75	57	39	22	17
7	JUN	144	144	144	144	140	127	110	92	74	56	39	34
8	JUL	162	162	162	162	158	145	128	110	92	74	57	52
8	AUG	180	180	180	180	176	163	145	128	110	92	78	70
8	SEP	198	198	198	198	194	181	164	146	128	110	93	88
7	OCT	215	215	215	215	211	195	181	163	145	126	110	105
5	NOV	220	220	220	220	216	203	185	168	150	132	115	110
0	DEC												
0	JAN												
0	FEB												
0	MAR												
4	APR	224	224	224	224	220	207	190	172	154	136	119	114
3	MAY	237	237	237	237	233	220	203	185	167	149	132	127
7	JUN	254	254	254	254	250	237	220	202	184	166	149	144
8	JUL	272	272	272	272	268	255	238	220	202	184	167	162
8	AUG	290	290	290	290	286	273	256	234	220	202	185	180
8	SEP	303	303	303	303	304	291	274	256	235	220	203	198
7	OCT	325	325	325	325	321	303	291	273	255	237	220	215
5	NOV	330	330	330	330	326	313	296	278	242	225	220	
0	DEC												
0	JAN												
0	FEB												
0	MAR												
4	APR	334	334	334	334	330	317	300	282	264	246	229	224
3	MAY	347	347	347	347	343	330	313	295	277	259	242	237
7	JUN	364	364	364	364	360	347	330	312	294	276	259	252
8	JUL	382	382	382	364	378	365	348	330	312	294	277	272
8	AUG	400	400	400	400	396	383	356	343	330	312	295	290
8	SEP	418	418	418	418	414	401	394	365	348	333	313	309
7	OCT	435	435	435	435	431	318	401	383	365	347	330	325
5	NOV	400	400	400	400	436	423	406	393	370	352	335	330
0	DEC												

FIGURE 4-11   Distribution of work days—light grading and urban.

timate certain durations. In these cases the scheduler should accurately determine the quantity of work to be performed, as in cubic yards, square feet, etc. Then the scheduler should determine the available or estimated number of workers who will perform that part of the work. Consultation with an estimating manual will provide the scheduler with an estimate of production for a crew

Distribution of Working Days by Months, Through 100 W.D. Year
Cumulative Totals Per Month
Medium and Heavy Grading

Month	JAN	FEB	MAR	APR	MAY	JUN	JUL	AUG	SEP	OCT	NOV	DEC
JAN												
FEB												
MAR												
APR	4	4	4	4								
MAY	17	17	17	17	13							
JUN	32	32	32	32	28	15						
JUL	48	48	48	48	44	31	15					
AUG	65	65	65	65	61	48	33	17				
SEP	80	80	80	80	76	63	48	32	15			
OCT	95	95	95	95	91	78	63	47	30	15		
NOV	100	100	100	100	100	96	68	52	35	20	5	
DEC												0
JAN												
FEB												
MAR												
APR	104	104	104	104	100	87	72	56	39	24	9	4
MAY	117	117	117	117	113	100	85	69	52	37	22	17
JUN	132	132	132	132	128	115	100	84	67	52	37	32
JUL	148	148	148	148	144	131	116	100	83	68	53	48
AUG	165	165	165	165	161	148	133	117	100	85	70	65
SEP	180	180	180	180	176	163	148	132	115	100	85	80
OCT	195	195	195	195	191	178	163	147	130	115	100	95
NOV	200	200	200	200	196	183	168	152	135	120	105	100
DEC												
JAN												
FEB												
MAR												
APR	204	204	204	204	200	187	172	156	139	124	109	104
MAY	217	217	217	217	213	200	185	169	152	137	122	117
JUN	232	232	232	232	228	215	200	184	167	152	137	132
JUL	248	248	248	248	244	231	216	200	183	168	153	148
AUG	265	265	265	265	261	248	233	217	200	185	170	165
SEP	280	280	280	280	276	263	248	232	215	200	185	180
OCT	295	295	295	295	291	278	263	247	230	215	200	195
NOV	300	300	300	300	296	283	268	252	233	220	205	200
DEC												
JAN												
FEB												
MAR												
APR	304	304	304	304	300	287	272	256	239	224	209	204
MAY	317	317	317	317	313	300	285	269	252	237	222	217
JUN	332	332	332	332	328	315	300	284	267	252	237	232
JUL	348	348	348	348	344	331	316	300	283	268	252	248
AUG	365	365	365	365	361	348	333	317	300	285	270	265
SEP	380	380	380	380	376	363	348	333	315	300	285	280
OCT	395	395	395	395	391	378	363	347	330	315	300	295
NOV	400	400	400	400	396	383	368	352	335	320	305	300
DEC												

FIGURE 4-12 Distribution of work days—medium and heavy grading.

of workers for the specific type of work being considered. Where possible, the scheduler should consult with the project team members who will ultimately be responsible for constructing the work. Factoring all of this information together will give the scheduler reasonable durations.

Durations should be reasonable. Allowance should be made for contingen-

Distribution of Working Days
by Months, Through 135 W.D. Year
Cumulative Totals Per Month

	JAN	FEB	MAR	APR	MAY	JUN	JUL	AUG	SEP	OCT	NOV	DEC
JAN												
FEB												
MAR												
APR	14	14	14	14								
MAY	30	30	30	30	16							
JUN	48	48	48	48	34	18						
JUL	67	67	67	67	53	37	19					
AUG	85	85	85	85	71	55	37	18				
SEP	103	103	103	103	80	73	55	36	18			
OCT	120	120	120	120	106	90	72	53	35	17		
NOV	135	135	135	135	121	105	87	68	50	32	15	
DEC												0
JAN												
FEB												
MAR												
APR	149	149	149	149	135	119	101	82	64	46	29	14
MAY	165	165	165	165	151	135	117	98	80	62	45	30
JUN	183	183	183	183	169	153	135	116	93	80	63	48
JUL	202	202	202	202	188	172	154	135	117	109	82	67
AUG	220	220	220	220	206	190	172	153	135	117	100	85
SEP	238	238	238	238	224	203	190	171	153	135	118	103
OCT	255	255	255	255	241	225	207	188	170	152	135	120
NOV	270	270	270	270	255	240	222	203	185	167	150	135
DEC												
JAN												
FEB												
MAR												
APR	284	284	284	284	270	254	236	217	199	181	165	149
MAY	300	300	300	300	286	270	252	233	215	197	180	165
JUN	318	318	318	318	304	288	270	251	233	215	198	183
JUL	337	337	337	337	323	307	289	270	252	234	217	202
AUG	355	355	355	355	341	325	307	288	270	252	235	229
SEP	373	373	373	373	359	343	325	306	283	270	253	238
OCT	390	390	390	390	376	359	342	323	305	287	270	265
NOV	405	405	405	405	391	375	357	328	320	302	285	270
DEC												
JAN												
FEB												
MAR												
APR	419	419	419	419	405	389	371	352	334	316	299	284
MAY	435	435	435	435	421	405	387	368	350	332	315	300
JUN	453	453	453	453	439	423	405	386	368	350	333	315
JUL	472	472	472	472	458	442	424	405	387	369	352	337
AUG	490	490	490	490	476	460	442	423	405	387	370	355
SEP	508	508	508	508	494	478	460	441	423	405	385	373
OCT	525	525	525	525	511	495	477	458	440	422	405	399
NOV	540	540	540	540	526	510	492	473	455	437	420	405
DEC												

FIGURE 4-13　Distribution of work days—bridges.

cies when assigning durations. Normal weather, a reasonable labor supply, normal durations for receipt of materials or approvals, and some defective work can be anticipated. However, the scheduler should not assume that either excessive or minimal delays may occur. Rather, the scheduler should allow for the normally predictable unfavorable conditions that may cause delay.

Distribution of Working Days by Months, Through 150 W.D. Year — Cumulative Totals Per Month — Traffic W. Day Schedule

WD	Month	JAN	FEB	MAR	APR	MAY	JUN	JUL	AUG	SEP	OCT	NOV	DEC
0	JAN	0											
0	FEB	0	0										
0	MAR	0	0	0									
10	APR	10	10	10	10								
15	MAY	25	25	25	25	15							
22	JUN	47	47	47	47	37	23						
22	JUL	69	69	69	69	59	44	22					
23	AUG	92	92	92	92	82	67	45	23				
23	SEP	115	115	115	115	105	90	68	46	23			
18	OCT	133	133	133	133	123	108	86	64	41	18		
15	NOV	148	148	148	148	133	123	101	79	56	33	15	
2	DEC	150	150	150	150	140	125	103	81	59	35	17	2
0	JAN	·	·	·	·	·	·	·	·	·	·	·	·
0	FEB	·	·	·	·	·	·	·	·	·	·	·	·
0	MAR	·	·	·	·	·	·	·	·	·	·	·	·
10	APR	160	160	160	160	150	135	113	91	68	45	27	12
15	MAY	175	175	175	175	165	150	128	106	83	60	42	27
22	JUN	179	179	179	179	187	172	150	128	105	82	64	40
22	JUL	219	219	219	219	209	194	172	150	127	104	86	71
23	AUG	242	242	242	242	232	217	195	173	150	127	109	94
23	SEP	265	265	265	265	255	240	218	195	173	150	132	117
18	OCT	283	283	283	283	273	258	236	214	191	173	150	135
15	NOV	298	298	298	298	288	273	251	229	205	183	165	150
2	DEC	300	300	300	300	290	275	253	231	203	185	167	152
0	JAN	·	·	·	·	·	·	·	·	·	·	·	·
0	FEB	·	·	·	·	·	·	·	·	·	·	·	·
0	MAR	·	·	·	·	·	·	·	·	·	·	·	·
10	APR	310	310	310	310	300	285	263	241	218	195	177	162
15	MAY	325	325	325	325	315	300	278	256	233	210	192	177
22	JUN	347	347	347	347	337	322	300	278	255	232	214	193
22	JUL	369	369	369	369	359	344	322	300	277	254	236	221
23	AUG	392	392	392	392	382	367	345	323	300	277	259	244
23	SEP	415	415	415	415	405	390	360	346	323	300	282	267
18	OCT	433	433	433	433	423	408	385	364	341	318	300	285
15	NOV	448	448	448	448	439	423	401	379	356	333	315	300
2	DEC	450	450	450	450	440	425	403	381	358	335	317	302
0	JAN	·	·	·	·	·	·	·	·	·	·	·	·
0	FEB	·	·	·	·	·	·	·	·	·	·	·	·
0	MAR	·	·	·	·	·	·	·	·	·	·	·	·
10	APR	460	460	460	460	435	435	413	391	368	345	327	312
15	MAY	475	475	475	475	465	450	428	406	383	360	342	327
22	JUN	497	497	497	497	487	472	450	428	405	382	364	349
22	JUL	519	519	519	519	509	494	472	450	427	404	386	371
23	AUG	542	542	542	542	532	517	495	473	450	427	409	394
23	SEP	565	565	565	565	555	540	518	495	473	450	432	417
18	OCT	583	583	583	583	573	553	536	514	491	468	450	435
15	NOV	598	598	598	598	588	573	551	529	506	483	465	450
2	DEC	600	600	600	600	590	575	553	531	508	483	457	452

FIGURE 4-14 Distribution of work days—traffic.

As time is introduced into the schedule, the scheduler may find it necessary to redefine or subdivide certain activities, to condense others, or to expand some into additional ones. The need to revise the logic when durations are added can be anticipated. Adding time to a schedule is only the next step in the schedule preparation process. Experienced schedulers who add durations

Steel erection is often one of the controlling activities in a construction schedule.

as the logic is developed make these adjustments as they put the logic on paper, avoiding the adjustments required in a two-step process.

Estimated activity durations are *estimates*. Actual durations will rarely match the estimated durations. Anticipated variation in individual activity duration, however, should have little effect on estimated project duration. Variations which increase estimated durations should offset variations that decrease estimated durations, if all durations are estimated reasonably. Where a single repetitive or only a few activities make up the critical path, the scheduler should carefully evaluate the accuracy of those activity durations.

Durations are normally expressed in whole days. Most construction durations do not include parts of days.

The Forward Pass

After a time duration has been assigned to each time-consuming activity on the logic diagram, the "critical" or longest path through the logic is calculated. The critical path determines the period in which the project may be completed and the period of time within which each activity on the critical path must be accomplished if the predicted completion time is to be met.

Calculation of the critical path involves determining four event times for each activity: early start (ES); late start (LS); early finish (EF); and late finish (LF).

The "early start" is the earliest time an activity can start after completion of the preceding activities. The early start is calculated by taking the project start date and adding each activity's duration following the logic diagram up to the activity in question.

Note that the four times calculated for each activity assume that actual time required to complete the work will match the durations placed on the activities.

The "early finish" time is the earliest time an activity can be completed if it is started at its early start time and is completed using its estimated duration. Early finish time is calculated by adding the activity's duration to its early start time.

The "late finish" is the latest time an activity can be completed without delaying the scheduled project completion time. Late finish is calculated by working backward through the logic, starting at the schedule's end date and subtracting in turn each activity's duration to determine the latest completion time which will not cause a delay in the project.

The "late start" is the latest time an activity can be started without delaying the project. Late start is calculated by subtracting the activity's duration from the late finish time.

When the early start and late start are identical and the early finish and the late finish are identical, that activity is on a critical path. Critical paths through the logic are determined by comparing early and late times for each activity. Once early and late times have been calculated, each time can be translated into a date. The process of calculating early and late times, and translating those times into dates, is referred to as "network analysis."

Activity time calculations can be accomplished manually or by computer. Manually computing early and late times enables the scheduler to understand and, if necessary, to revise the schedule. The network analysis may result in the critical path going through activities which those experienced in construction recognize will not be critical despite the indication of the schedule. For example, experience in the field teaches that completion of the flagpole or toilet accessories will almost never affect substantial completion of the project. A critical path which includes these activities should be studied by the scheduler and revised as required. Manual computation also helps the scheduler identify computer input errors.

Activity time calculations are begun by computing each activity's early start and early finish times. The basic assumption in calculating early activity times is that every activity will start as early as possible. Early time calculations begin at the first node (or event) and work forward. The first node has an early start time of zero. The value of all subsequent early event times is the sum of the event time and the activity's duration. For each event there is only one early and one late event time. Each activity will have its own early and late activity times. This process is known as the "forward pass." Where two or more activities merge into a node, the value is calculated for each path and the largest value is used.

Calculation of early times indicates the earliest time at which the last activity in the logic diagram will finish. If the project's activities have been arranged properly, *if activity durations have been estimated accurately,* and if everything goes as anticipated, project completion can be expected within the early event times.

When the event times are calculated manually, it is convenient to write the results of the computations next to the nodes. In this text, early event times are recorded in small boxes above the nodes. Figure 4-15 shows a logic diagram with small boxes added in which to record early event times.

In order to compute the early event time for an event, it will be necessary to compute a separate time for each activity or restraint that terminates at a node. The largest of these times is the early event time. That value is printed in the box on the logic diagram. The early event times are computed systematically by examining the nodes one at a time in the order of the activity arrows. The forward pass is made from left to right following the logic arrows until all nodes have been assigned an early event time. An early event time can be recorded when the early event times have been computed for every node that precedes it and the preceding activity's durations have been added. If more than one activity precedes it, several trial values must be calculated. The activity's duration must be added to each preceding node's early event time. The largest trial value is selected as the early event time to be recorded.

The first node in a logic diagram always has an early event time of zero. The value of all remaining nodes' early event times is the sum of the preceding event's early event time and the activity's duration. The last node's early event time is equivalent to the project duration. In Figure 4-15 the last node is number 28.

In Figure 4-16, early event times have been calculated for the diagram in Figure 4-15. Notice that several trial early start times have been written next to the box, but only the largest value is recorded in the box. The early event time for the last node also represents the project's scheduled duration. In this

FIGURE 4-15 Early event time diagram.

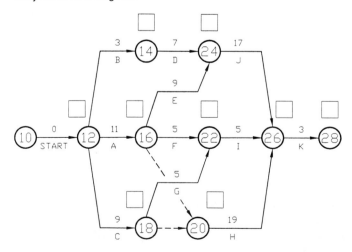

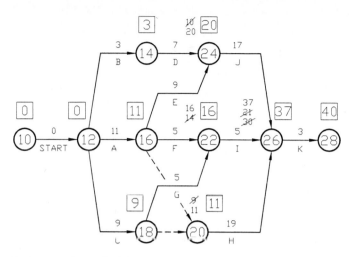

FIGURE 4-16   Early event time calculations.

schedule the early event time for node 28 is 40. The scheduled duration for the project shown in Figure 4-16 is 40 days.

The Backward Pass

The "backward pass" calculates late start and late finish times by starting at the project end (terminal node) and working backward through the network, generally from right to left. The backward-pass calculation is based on the early completion time for the last activity. During the backward pass, the presumption is that each activity finishes as late as possible without delaying scheduled completion.

As in the forward pass, one or more trial values will be calculated for the late event times of each node. The trial values are calculated by taking the late event time at the head of the arrow and subtracting the duration. A unique calculation is made for each arrow that originates at that node. The smallest of all calculated values for a particular node is the late event time. The late event time for the last node is equal to the scheduled duration for the project determined in the forward pass.

Late event times are traditionally recorded above the early event times. Figure 4-17 shows diamonds above the boxes used to record early event times.

Because one or more trial values may be calculated before the smallest late event time can be determined, trial late times are also recorded on the diagram. There will be as many trial values recorded for a node as there are immediately subsequent arrows. Only the smallest late event time, however, is recorded in the diamond. Trial values considered in the backward pass are recorded next to the diamond.

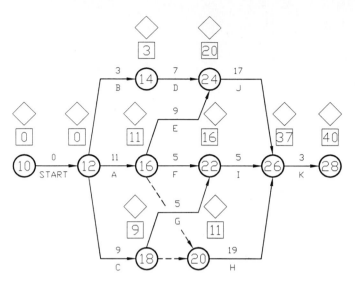

FIGURE 4-17   Late event time diagram.

Calculation begins with the early event time of the terminal node. The late value of subsequent nodes is determined by subtracting the activity's duration from the late event time. Figure 4-18 shows the late event time and trial values for the diagram in Figure 4-17.

With the completion of the forward and backward pass, the four times necessary for each activity can be calculated as follows:

Early start time = early event time

Early finish time = early start time + activity duration

Late finish time = late event time

Late start time = late finish time − activity duration

The four activity times are normally listed on the computer printout that accompanies the logic diagram, along with the activity's I and J nodes, duration, and description. From the event times shown on Figure 4-18, the four activity times can be calculated as shown in Table 4-1. Note that the list of four activity times also includes dummy activities.

The four activity times for activity 16-22, F, can be read as follows: The earliest that activity F can start is on day 11; the earliest that activity F can finish is on day 16; the latest that activity F must start in order not to extend completion is on day 27; the latest that activity F must finish in order not to extend scheduled completion is on day 32.

Although the four times which result from the forward and backward passes seem very specific, the calculation is based only on assumptions the scheduler has made concerning activity durations and the sequence in which the activities

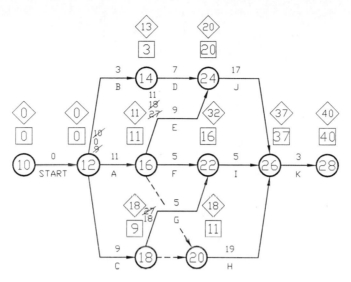

FIGURE 4-18   Late event time calculations.

will be performed. It is unlikely that, when the activity is actually performed, the estimated durations will match the actual durations exactly. Actual crew sizes may be larger or smaller than were assumed by the scheduler when the durations were estimated. Delays caused by weather may be longer or shorter than anticipated. Similarly, it is unlikely that, when the work is actually performed, scheduled sequences will exactly match actual sequences. Field personnel may perform activities in a different sequence than was visualized by the scheduler. Also, some of the restraints shown on the diagram may not have

TABLE 4-1   EVENT TIMES FOR FIGURE 4-18

I node	J node	Duration	Activity description	ES	EF	LS	LF
10	12	0	Start	0	0	0	0
12	14	3	B	0	3	10	13
12	16	11	A	0	11	0	11
12	18	9	C	0	9	9	18
14	24	7	D	3	10	13	20
16	20	0	Dummy	11	11	18	18
16	24	9	E	11	20	11	20
16	22	5	F	11	16	27	32
18	20	0	Dummy	9	9	18	18
18	22	5	G	9	14	27	32
20	26	19	H	11	30	18	37
22	26	5	I	16	21	32	37
24	26	17	J	20	37	20	37
26	28	3	K	37	40	37	40

FIGURE 4-19 Manual calculation of dates.

been essential. Regardless of the assumptions and guesses which were used to produce specific event times, however, the initial CPM schedule is used regularly as a tool with which to manage completion and to measure progress and delay. It is considered the most precise calculation of time and date for construction projects available today.[2]

Note that when one is calculating ES, EF, LS, and LF times manually, it is possible that the first activity in a chain will have the same EF date as the following activity's ES date. Does this mean that the two activities overlap by one day? Not really. When calculating a CPM schedule manually, it may be helpful to assume that the date calculated is at 12 noon. Therefore, if a scheduler is calculating two consecutive 2-day activities, the dates will be as shown in Figure 4-19. Activity A EF is the same date as activity B ES, but they do not overlap.

Software programs calculate activity dates slightly differently. The software uses "inclusive calculation of dates." Therefore, an activity of 2 days' duration which starts on Monday will end on Tuesday. The following 2-day activity will show its ES on Wednesday and EF on Thursday. Thus the software program's dates on the printout are inclusive and avoid any confusion of identical dates used on sequential activities.

The Critical Path

Early and late start and early and late finish times will be the same for some activities. When this occurs, the activities are "critical." Critical activities are those that must start and complete on their early start and finish times for the project to be completed within the scheduled time.

Critical activities form a continuous chain through the network known as the "critical path." Critical activities are not necessarily the most difficult or those that are the most important job elements. Critical activities merely form the longest path through the job logic from the beginning of the project to the end.

The scheduler needs, however, to be able to distinguish the real critical path from idiosyncrasies of the logic. The critical path will sometimes indicate that

---

[2] Edward M. Willis, *Scheduling Construction Projects,* New York: Wiley, 1986, p. 130.

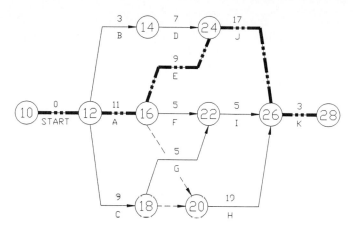

FIGURE 4-20   Critical path of Figure 4-19.

activities are critical which experience shows are never critical to substantial completion. For example, the real critical path will probably never run through the activity "install toilet accessories," because such accessories are not necessary for substantial completion of the building. The flagpole may be important to the owner but inconsequential to completion and should never be critical. Careful study and adjustments to logic, durations, or restraints may be necessary to allow the critical path to indicate substantive activities important to project completion to be critical. The scheduler must not make arbitrary adjustments to the logic just to move the critical path to where he or she believes it should be. The only adjustments that should be made are when the logic does not represent the current plan for completing the project.

The event times shown in Table 4-1 indicate the critical path through the logic diagram in Figure 4-18. Activities 10-12, 12-16, 16-24, 24-26, and 26-28 are critical. In each case, the early and late start and the early and late finish times are the same.

The critical activities in Figure 4-18 form a continuous path through the logic. Figure 4-20 shows the critical path through the network.

FLOAT

Definition

Activities with early and late times which do not match are flexible. Such activities do not have to begin or end with the early start and finish times to permit completion within the scheduled completion date. This flexibility is a measure of the ability of a given activity to have its peformance time extended and is called "float" or "slack." Float measures the amount of time a particular activity's performance can be delayed while still permitting the project to be completed within the scheduled times. Another way to look at float is as a measure of "criticality." The more float an activity has, the less critical the activity is. The less float an activity has, the more critical it is. Activities which have no float are critical and cannot be delayed without delaying the project's scheduled completion date.

The typical construction schedule has many possible logic chains or paths. Not all paths will be critical, but there may be more than one critical path. Even those paths that are composed of activities which contain float may become critical paths if float is eliminated.

Two classifications of float are generally recognized: total float and free float. The "total float" of an activity is the difference between its earliest and latest start time or its earliest and latest finish time.

Total float of an activity indicates the amount of time by which an activity may be delayed without affecting the project's scheduled completion date. Activities share total float with other activities on the same path. Using some total float in the performance of one activity will reduce it in all the remaining activities which share it. Individual activities in the path with total float do not own it.

For example, consider Figure 4-21, which describes the installation of fuel tanks at a garage. The fuel tanks are scheduled to be installed in 8 days. The mechanical contractor is permitted 2 days to install the fuel tanks and 4 days to "pipe up" the tanks. After the fuel tanks have been piped up, the general contractor is given 2 days to backfill and grade site. The installation of fuel tanks is part of a chain of activities. Whenever the chain is started, it is scheduled to be completed 8 days later.

If the mechanical contractor starts to install fuel tanks a week late, the general contractor's backfilling and site grading will similarly be delayed a week (assuming that the durations of activities 20-30 and 30-40 cannot be reduced).

Float is shared among activities in a chain in a similar manner. If the mechanical contractor's installation of fuel tanks has 10 days float, piping up the fuel tanks will also have 10 days float. Each activity does not have 10 separate days of float: The float is shared. If the mechanical contractor uses 5 days of the chain's float by starting fuel tank installation 5 days after its scheduled early start date, the float available for pipe up and backfill activities will be reduced by 5 days. By delaying its start 5 days, the mechanical contractor will reduce the general contractor's available float because both contractors share the float in the chain of activities. Sharing float is an important concept to understand, because it influences the "ownership" of float and allocation of

FIGURE 4-21   Installation of fuel tanks.

delay, which will be discussed later. How activities share float will also be demonstrated when methods for calculating float values are explained.

The second type of float, "free float," is the difference between an activity's early finish time and the earliest start time for any succeeding activity. The free float of an activity is the amount by which that activity can be delayed without delaying the early start of any following activity or affecting any other activity in the network.[3]

In contrast to total float, which is shared by all activities in a path, free float occurs only in certain circumstances. Free float can occur only when more than one arrow goes into a node. One or more of the multiple arrows entering the node can possess free float. Free float is not shared. Free float is generally not a concept that has proved useful to the construction industry. As a practical matter, contractors are not concerned with how much delay they can enjoy without affecting another contractor's work as much as with what other contractors do with the total float that is shared. The limited value of free float is for busy subcontractors juggling efforts at multiple job sites. Free float is not an important number in the field. However, free float can be useful when selecting activities for resource leveling, as will be shown in Chapter 11.

Calculating Float

Total float is calculated either by subtracting the early finish time from the late finish time or by subtracting the early start time from the late start time.

Table 4-1 lists the early start, late start, early finish, and late finish times for the project diagrammed in Figure 4-18. The total float of noncritical activities can be calculated from Table 4-1. For example, activity 12-14 has an early start time of 0 and a late start time of 10, so $10 - 0 = 10$. Thus, activity 12-14 has 10 days of total float. Similarly, activity 12-18 has an early start time of 0 and a late start time of 9, for $9 - 0 = 9$. Activity 12-18 thus has 9 days of total float. By subtracting early start time from late start time, total float for all the activities listed in Table 4-1 can be determined. Table 4-2 lists the total float for the activities in Table 4-1.

Free float is calculated by subtracting an activity's early finish time from the early start time of the next activity. Activities 14-24, 18-20, 18-22, 20-26, and 22-26 have free float.

Figure 4-22 shows another network with activities that have free float. In essence, free float is the flexibility a contractor has in start and finish dates which will not affect any other activity. Let us assume that activity E-F has 10 days of total float. The critical path is elsewhere in the logic and is not shown in Figure 4-22. Within the subnet shown, activities A-B and B-E also have $+10$ days of total float. Activities A-B, B-E, and E-F have 0 free float, because any change in these activities' start or finish dates will affect another activity. Activity C-E has 15 days total float because its duration is 5 days less

[3] Edward M. Willis, *Scheduling Construction Projects,* New York: Wiley, 1986, p. 116.

TABLE 4-2   FLOAT CALCULATIONS

I node	J node	Duration	Activity description	ES	EF	LS	LF	FF	TF
10	12	0	Start	0	0	0	0	0	0
12	14	3	B	0	3	10	13	0	10
12	16	11	A	0	11	0	11	0	0
12	18	9	C	0	9	9	18	0	9
14	24	7	D	3	10	13	20	10	10
16	20	0	Dummy	11	11	18	18	0	7
16	24	9	E	11	20	11	20	0	0
16	22	5	F	11	16	27	32	0	16
18	20	0	Dummy	9	9	18	18	2	9
18	22	5	G	9	14	27	32	2	18
20	26	19	H	11	30	18	37	7	7
22	26	5	I	16	21	32	37	16	16
24	26	17	J	20	37	20	37	0	0
26	28	3	K	37	40	37	40	0	0

than the concurrent path of $+10$ $(10 + 5 = 15)$. Activity C-E has 5 days of free float: The start of activity C-E can be delayed 5 days without affecting any other activity. Activity D-E has 13 days of total float and 3 days of free float. If the start of activity D-E is delayed 2 days, the total float will be 11 days and the free float will be 1 day.

Note that Figure 4-22 follows the independent I node rule. If, for example, the additional restraint was placed at the end of the activity instead of the beginning, the free float would occur on the dummy or restraint. The reason this occurs is that CPM treats dummies and restraints just like other activities except that they have no durations.

Figure 4-23 shows a diagram with different free float. Activities C-E and D-F have no free float because if they are delayed, activities E-G and F-G will be delayed. Activities E-G and F-G, however, each have 2 days of free float. If activities C-E and D-F are completed by their respective early finish dates, then E-G and F-G will each have 2 days of free float in which to adjust their work without affecting the start of G-H.

FIGURE 4-22   Free float.

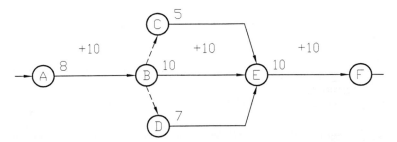

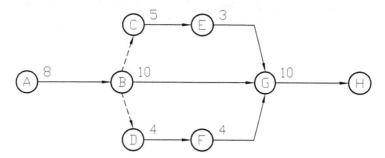

FIGURE 4-23   Another example of free float.

The Computer Printout

The time values for early start, late start, early finish, late finish, and float have so far been expressed as number of days since the project began. In Table 4-2, for instance, the time value for the early start of activity 16-22 was determined to be 11, for the late start it was 27, for the early finish, it was 16, for the late finish it was 32, and the total float was 16. Expressing time in this way, however, is not very useful for managers who need to check progress or plan future work. To make the time values more useful and understandable, most network analyses translate the time values into calendar dates. The translation is easily accomplished once the project start date is known and the work calendar has been established.

A table is provided in the Appendix to facilitate the manual translation of time values to dates. Available tables take many forms, and care must be exercised in their selection. Some tables list only working days and omit weekends and holidays. When these tables are used, the scheduler must make sure that holidays and weekends are identified. Holidays vary among companies and among nations. Weekends may also vary from country to country.

Computers are commonly used to translate times into calendar dates and perform the forward and backward passes necessary to produce the event times. Figure 4-24 is a computer printout showing one page of a network analysis of a schedule for renovation and addition to a hospital. The computer has determined the critical path and calculated float values. Time values have been expressed as dates to facilitate the use and understanding of the schedule. Critical activities are those activities which have no total float.

Very often the scheduler prepares the list of activities and logic diagram but relies on data processors to input the individual activity data and operate the computer. The scheduler who uses data processing support should check the network analysis once the computer printout is available. This is known as "proofing the schedule." Proofing is performed by comparing the activity's description, I and J nodes, durations, and any other information contained in the logic diagram to the computer printout. Proofing the schedule ensures that the printout accurately reproduces the data in the logic diagram.

CCL CONSTRUCTION CONSULTANTS PROJECT ID: UPDATE # 0 RUN # REPORT DATE: 13APR90 PAGE 38

I NODE	J NODE	ACTIVITY DESCRIPTION	DUR	ORGZNL CODE	EARLY START	EARLY FINISH	LATE START	LATE FINISH	FLOAT FREE	FLOAT TOTAL
2822	2830	RESTRAINT	0		23AUG90	23AUG90	20SEP90	20SEP90	1	19
2824	2828	RESTRAINT	0		24AUG90	24AUG90	24AUG90	24AUG90	0	0
2828	2834	SET INTERMEDIATE STL & BRACING 3RD ETSS	3	SS	24AUG90	28AUG90	24AUG90	28AUG90	0	0
2830	2832	LAY DECKING & SET SLAB ANGLES 2ND L ETSS	4	DK	24AUG90	29AUG90	20SEP90	25SEP90	0	18
2832	2838	RESTRAINT	0		30AUG90	30AUG90	26SEP90	26SEP90	0	18
2834	2838	RESTRAINT	0		29AUG90	29AUG90	26SEP90	26SEP90	1	19
2834	2844	TORQUE UP 1ST SECTION ETSS	3	SS	29AUG90	31AUG90	29AUG90	31AUG90	0	0
2838	2840	LAY DECKING & SET SLAB ANGLES 3RD L ETSS	4	DK	30AUG90	05SEP90	26SEP90	01OCT90	0	18
2840	2842	RESTRAINT	0		06SEP90	06SEP90	02OCT90	02OCT90	0	18
2840	2848	ELEC SLEEVES 1 2 3RD LEVELS ELEV T ETSS	2	EL	06SEP90	07SEP90	02OCT90	03OCT90	0	18
2842	2848	MECH SLEEVES 1 2 3RD LEVELS ELEV T ETSS	2	ME	06SEP90	07SEP90	02OCT90	03OCT90	0	18
2844	2845	RESTRAINT	0		04SEP90	04SEP90	04SEP90	04SEP90	0	0
2844	2848	RESTRAINT	0		04SEP90	04SEP90	04OCT90	04OCT90	4	22
2844	2854	SET 2ND SEC COLS @ ELEV TOWER ETSS	3	SS	04SEP90	06SEP90	04SEP90	06SEP90	0	0
2845	2854	SET MANLIFT ETSS	3	EQ	04SEP90	06SEP90	04SEP90	06SEP90	0	0
2848	2849	POUR SLABS 1,2,3, LVL ELEV TWR & CR ETSS	3	GC	10SEP90	12SEP90	04OCT90	08OCT90	0	18
2849	2850	CURE SLABS 1,2,3 LVL ELEV TWR & CR ETSS	5	GC	13SEP90	19SEP90	09OCT90	15OCT90	0	16
2850	2852	CORRIDOR STRUCTURE TOPPED OUT ETSS	0	XX	20SEP90	20SEP90	31OCT90	31OCT90	0	29
2850	2888	RESTRAINT	0		20SEP90	20SEP90	16OCT90	16OCT90	3	18
2852	2954	RESTRAINT	0		20SEP90	20SEP90	31OCT90	31OCT90	29	29
2854	2858	SET 4TH LEVEL BEAMS ETSS	2	SS	07SEP90	10SEP90	07SEP90	10SEP90	0	0
2858	2860	RESTRAINT	0		11SEP90	11SEP90	02OCT90	02OCT90	0	15
2858	2864	SET 5TH LEVEL BEAMS ETSS	2	SS	11SEP90	12SEP90	11SEP90	12SEP90	0	0
2860	2862	SET INTERMEDIATE STL & BRACING 4 LV ETSS	2	SS	11SEP90	12SEP90	02OCT90	03OCT90	0	15
2862	2868	RESTRAINT	0		13SEP90	13SEP90	04OCT90	04OCT90	0	15
2862	2870	RESTRAINT	0		13SEP90	13SEP90	05OCT90	05OCT90	0	16
2864	2870	RESTRAINT	0		13SEP90	13SEP90	05OCT90	05OCT90	0	16
2864	2872	TORQUE UP 2ND SEC. ETSS	3	SS	13SEP90	17SEP90	13SEP90	17SEP90	0	0
2868	2874	LAY DECKING & SET SLAB ANGLES 4 LVL ETSS	3	DK	13SEP90	17SEP90	04OCT90	08OCT90	0	15
2870	2878	SET INTERMEDIATE STL & BRACING 5 FL ETSS	2	SS	13SEP90	14SEP90	05OCT90	08OCT90	0	16
2872	2875	RESTRAINT	0		18SEP90	18SEP90	19SEP90	19SEP90	0	1

FIGURE 4-24 Network analysis printout Aldergraf.

USE OF THE I-J METHOD

Advantages

I-J is a well-established and easily understood scheduling method. Techniques for drawing and calculating project durations do not vary greatly from scheduler to scheduler. Most software packages calculate I-J identically, avoiding the problems of varying interpretations of I-J schedules. Thus I-J is readily communicated throughout the construction world.

Preparation of an I-J network requires careful analysis of the project and results in a well-founded understanding of how the project will be completed. I-J format is very structured and requires the scheduler to "follow the rules." This minimizes the production of erroneous or misleading schedules.

Because activity sequences and durations are only estimated, it is anticipated that the actual sequences and times necessary to perform the activities will vary from those shown in the schedule. However, if the sequences and durations have been reasonably determined, the I-J method provides an accurate schedule for project management purposes. Use of a schedule to determine length of delay or additional time due to changes is discussed in Chapter 12.

I-J is a straightforward technique for representing the complete job logic in an efficient and accurate way. I-J networks can also be converted to other types of scheduling formats.

Disadvantages

The primary disadvantage of I-J is that it cannot easily model the relationships among activities in the field. I-J's one logical relationship, that activity B cannot start until activity A is completed, requires many subdivided activities linked as a "stair step" to resemble construction in the field, where activities may start or finish together or the start of one activity may overlap the completion of its predecessor. The logic sometimes appears rigid when it is not really so rigid in actual application.

I-J networks tend to include more activities than precedence networks. It can be very confusing to follow the logic unless extreme care has been given to their development.

EXERCISES

1 Use the section and plan for construction of a fuel station in Figure 2-9 and list of activities and durations shown in Table 2-2 to prepare a logic diagram in I-J format. Make a forward pass and a backward pass. Present the time values as dates assuming contract start of September 1, 1993. A chart for use in converting time values to dates is shown in Figure 4-25.

PRIMAVERA PROJECT PLANNER

REPORT DATE 4MAR91
9:20

PROJECT CALENDAR

DATA DATE 1SEP93   START DATE 1SEP93

PAGE NO.   1

SUNDAY	MONDAY	TUESDAY	WEDNESDAY	THURSDAY	FRIDAY	SATURDAY
			1SEP93 1	2SEP93 2	3SEP93 3	4SEP93 NO WORK
5SEP93 NO WORK	6SEP93 NO WORK	7SEP93 4	8SEP93 5	9SEP93 6	10SEP93 7	11SEP93 NO WORK
12SEP93 NO WORK	13SEP93 8	14SEP93 9	15SEP93 10	16SEP93 11	17SEP93 12	18SEP93 NO WORK
19SEP93 NO WORK	20SEP93 13	21SEP93 14	22SEP93 15	23SEP93 16	24SEP93 17	25SEP93 NO WORK
26SEP93 NO WORK	27SEP93 18	28SEP93 19	29SEP93 20	30SEP93 21		
3OCT93 NO WORK	4OCT93 23	5OCT93 24	6OCT93 25	7OCT93 26	1OCT93 22	2OCT93 NO WORK
10OCT93 NO WORK	11OCT93 28	12OCT93 29	13OCT93 30	14OCT93 31	8OCT93 27	9OCT93 NO WORK
17OCT93 NO WORK	18OCT93 33	19OCT93 34	20OCT93 35	21OCT93 36	15OCT93 32	16OCT93 NO WORK
24OCT93 NO WORK	25OCT93 38	26OCT93 39	27OCT93 40	28OCT93 41	22OCT93 37	23OCT93 NO WORK
31OCT93 NO WORK					29OCT93 42	30OCT93 NO WORK
	1NOV93 43	2NOV93 44	3NOV93 45	4NOV93 46	5NOV93 47	6NOV93 NO WORK

FIGURE 4-25   Primavera Project Planner/Project Calendar.

PROJECT CALENDAR

REPORT DATE 4MAR91 9:20   DATA DATE 1SEP93   START DATE 1SEP93   PAGE NO. 2

SUNDAY	MONDAY	TUESDAY	WEDNESDAY	THURSDAY	FRIDAY	SATURDAY
7NOV93 NO WORK	8NOV93 48	9NOV93 49	10NOV93 50	11NOV93 51	12NOV93 52	13NOV93 NO WORK
14NOV93 NO WORK	15NOV93 53	16NOV93 54	17NOV93 55	18NOV93 56	19NOV93 57	20NOV93 NO WORK
21NOV93 NO WORK	22NOV93 58	23NOV93 59	24NOV93 60	25NOV93 NO WORK	26NOV93 61	27NOV93 NO WORK
28NOV93 NO WORK	29NOV93 62	30NOV93 63	1DEC93 64	2DEC93 65	3DEC93 66	4DEC93 NO WORK
5DEC93 NO WORK	6DEC93 67	7DEC93 68	8DEC93 69	9DEC93 70	10DEC93 71	11DEC93 NO WORK
12DEC93 NO WORK	13DEC93 72	14DEC93 73	15DEC93 74	16DEC93 75	17DEC93 76	18DEC93 NO WORK
19DEC93 NO WORK	20DEC93 77	21DEC93 78	22DEC93 79	23DEC93 80	24DEC93 NO WORK	25DEC93 NO WORK
26DEC93 NO WORK	27DEC93 81	28DEC93 82	29DEC93 83	30DEC93 84	31DEC93 NO WORK	
						1JAN94 NO WORK

81

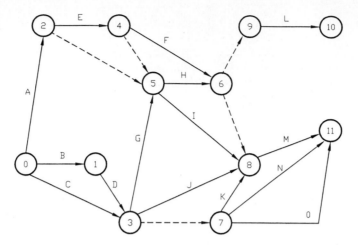

FIGURE 4-26 An incorrect logic diagram.

2 Figure 4-26 is an incorrect logic diagram: It has unnecessary restraints and gaps in the logic. Redraw and renumber the figure so as to correct the errors.

3 Complete the activity-time computations for the network shown in Figure 4-27. Mark the critical activities with asterisks.

4 Prepare a properly drawn I-J diagram which meets the requirements of the following activity list. Identify ES, EF, LS, LF, and float times. Determine the critical path

FIGURE 4-27 Activity-time calculations.

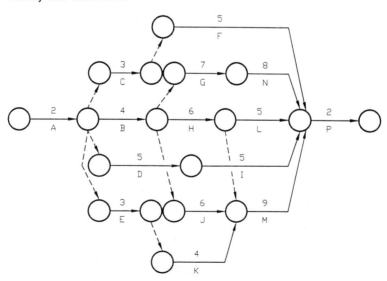

and the number of days required to complete the project. Highlight the critical path. The critical path should form the center of the network.

Activity	Duration	Depends on:
A	3	—
B	9	—
C	4	A, B
E	7	B
F	4	B
G	4	B
H	7	C, E
J	4	F, and lags the start of H by 2 days
K	5	F, H
L	4	F, H
M	3	F, H (lags finish of H by 1 day)
N	5	K
P	5	Q
Q	5	J, M
R	4	K, L

5 Draw the I-J network for the following activity list. Number all nodes, using the even numbering system, working from top to bottom, left to right. Label each activity. Minimize dummies and crossovers. Calculate the ES, LS, EF, LF, total float (TF), and free float (FF). Label the critical path.

Activity	Duration	Depends on (must follow)
A	6	—
B	3	A
C	4	A
D	9	A
E	12	—
F	8	—
G	3	B, D
H	6	C, D
I	4	C
J	4	E
K	3	E
L	1	E
M	6	G, H
N	7	H
O	4	N, I, J
P	2	K, L

6 Draw the I-J arrow diagram for the following activity list and compute the ES, LS, EF, LF, total float (TF), and free float (FF).

Activity	Duration	Succeeding activities
A	22	D, J
B	10	C, F
C	13	D, J
D	8	—
E	15	C, F, G
F	17	H, I, K
G	15	H, I, K
H	6	D, J
I	11	J
J	12	—
K	20	—

7 For the data below, draw an I-J network diagram and determine the critical path. Calculate the event times, free float, and total float.

Activity	Preceded by	Duration	ES	EF	LS	LF	FF	TF
A	C, H, N	4						
B	C, H, N	6						
C	—	3						
D	E, G, T, U	5						
E	A	6						
F	A, B	3						
G	A, B	7						
H	—	1						
I	C, H, N	5						
J	I	3						
K	J	4						
L	J	5						
M	D, L	3						
N	—	2						
O	P, Q, R	3						
P	G, U	4						
Q	F	2						
R	E, G, T, U	5						
S	I	6						
T	I	8						
U	A	8						

8 Create activities, relationships, durations, and responsibilities and draw a complete I-J logic diagram for the construction of a two-story residence. The home will have a full concrete basement plus two wood-framed floors complete with fireplace, brick veneer, and all other typical amenities. Include activities for delivery of long-lead-time items. Perform a forward pass and a backward pass, and highlight the critical path.

9 In your own words, describe what is meant by free float and total float. What is the significance of total float to the contractor?

10 Describe the steps you would take to develop an I-J schedule for:

(*a*) A high-rise building project

(*b*) A cable-stayed bridge project

DISCUSSION TOPICS

1 Is it reasonable to place a duration on a restraint or dummy? To hit a desired start date for an activity? To reduce float? To show a waiting time delay before starting the next activity?

2 Do you agree with the independent I node rule? Why or why not?

CHAPTER
5

PRECEDENCE DIAGRAMMING METHOD

INTRODUCTION

Professor John W. Fondahl of Stanford University presented the fundamental concepts for the precedence diagramming technique in 1961. Fondahl placed the activity on the node rather than on the arrow, as in the I-J method. The arrows connecting the nodes of the network define the relationships between the activities. Fondahl called the new technique "circle and connecting line." Later the term "activity on node" (AON) was applied.[1]

The term "precedence diagramming" first appeared around 1964 in the User's Manual for an IBM 1440 computer program. One of the principal authors of the manual was J. David Craig of the IBM Corporation. Craig was also apparently responsible for naming the technique the "precedence diagramming method" (PDM). PDM is a more complex version of activity-on-node diagramming.[2]

[1]James J. O'Brien, *CPM in Construction Management,* 3rd ed., New York: McGraw-Hill, 1984, p. 129.
[2]James J. O'Brien, *CPM in Construction Management,* 3rd ed., New York, McGraw-Hill, 1984, p. 129.

Concrete conveyors are sometimes used to place concrete. (Photo courtesy of Morgan Manufacturing Co.)

There are many differences between I-J, AON, and PDM. In I-J, arrows represent activities and nodes represent events. In AON and precedence diagramming, the activities are represented by nodes rather than arrows. Arrows represent the logical relationships among the activities. In I-J the nodes are most commonly represented by circles sufficiently large for a node number to be recorded in them, while in AON and PDM the nodes are most commonly represented by even larger boxes. The boxes for AON and PDM nodes are larger than the I-J nodes because within the node is printed the activity duration, activity number, activity description, ES, EF, LS, LF, and available float. Further, the method of calculating activity times in activity-on-node and precedence diagramming differs somewhat from that used in I-J.

Figure 5-1a shows four common methods of diagramming activities in AON and PDM. Which symbol is selected is unimportant as long as the scheduler is consistent.

Other arrangements can be created and additional divisions within the activity box can be made to include various other activity codes. Each software program has its own method of indicating the activities when a plot is made. Some software programs allow the scheduler to select what information will be shown on the plot.

Figure 5-1b shows the difference between I-J and AON/PDM diagrams. Both I-J and AON/PDM describe the same work sequence. Note that activity numbers are shown above the activity description, while activity durations are shown below the descriptions in the AON/PDM network shown in Figure 5-1. A single number is assigned to each activity. In contrast to I-J scheduling,

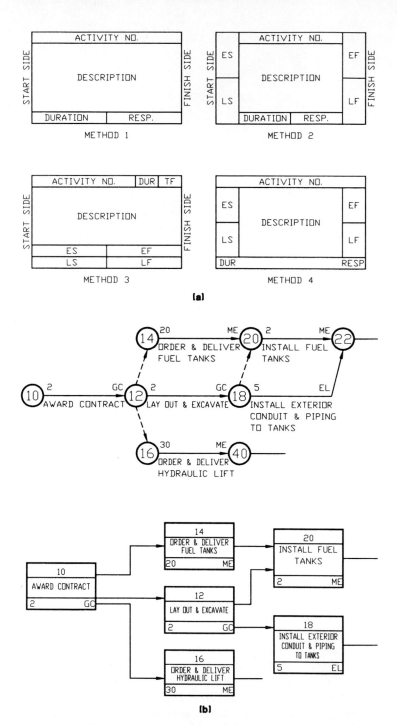

FIGURE 5-1 (a) AON and PDM activity diagramming methods. (b) Precedence diagramming method. (I-J diagram provided for comparison)

where activities are identified by I and J node numbers, only one number is necessary to identify an activity in AON/PDM schedules. References to the logical relationships are more complex, since both predecessors and successors must be identified, which is not necessary in I-J.

PDM LOGICAL RELATIONSHIPS

Four Logical Relationships

Figure 5-1a indicates that activities in AON and PDM have a start side and a finish side. The left side of the AON/PDM node is designated the "start side" and the right side is designated the "finish side." In AON/PDM, whether an arrow originates or terminates at the start side or the finish side of a node is significant. Because of the importance of the sides, PDM uses arrowheads to identify the direction of relationships as does I-J. Relationship arrows in precedence diagramming can be drawn in either the sloping or staff format.

Figure 5-1b indicates some similarity between activity-on-node diagramming and I-J. Precedence diagramming allows more flexibility in modeling relationships than AON or I-J diagramming. I-J and AON allow only one kind of logical relationship between activities: A preceding activity must be complete before any succeeding activity can begin. PDM, in contrast, employs four logical relationships between activities. The PDM method can also use the concept of lag (days between) activities to further create a flexible scheduling tool.

The four logical relationships used by PDM are:

1 Finish-to-start
2 Start-to-start
3 Finish-to-finish
4 Start-to-finish

The four relationships are commonly abbreviated FS, SS, FF, and SF. The four kinds of relationships are shown in Figure 5-2. Note that PDM's finish-to-start relationship is the same as the one logical relationship that I-J and AON use. If only finish-to-start relationships are used in a precedence diagram, PDM is similar to an I-J diagram and identical to an AON diagram.

The three forms that each of the four relationships may take concern the use of "lag." An activity with a lag relationship must wait until the period of lag has expired before beginning. Thus lag is the condition of waiting for a prescribed period before action can start.

Most relationships have either no lag or positive lag. Although used infrequently, relationships may also use negative lag. Figure 5-3 shows a relationship with a 1-day lag. Figure 5-3 indicates that 1 day after the start of activity 20, Fuel Tank Installation, activity 18, Install Exterior Conduit & Piping to Tanks, can start. Lag is commonly shown below a horizontal portion of the relationship arrow near the arrowhead.

Because of precedence diagramming's use of four relationships instead of

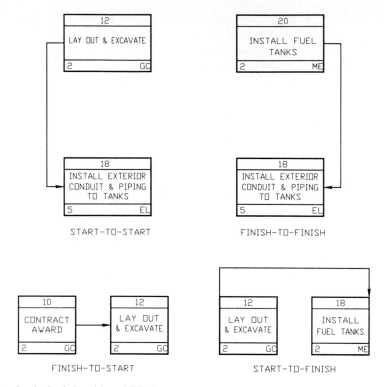

FIGURE 5-2   Logical relationships of PDM.

FIGURE 5-3   Relationship with lag.

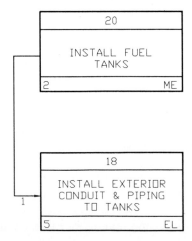

I-J's one, the calculation of event times differs from the calculation method for I-J and activity-on-arrow event times.

Conceptually PDM's use of the variety of logical relationships between activities is different than I-J. In I-J, one complete activity is related to another, but in PDM the start or finish of an activity can relate to the start or finish of another activity. In I-J diagramming, if only parts of activities are related, the activities must then be further divided. Figure 5-3 shows how relationships in PDM can be between the starts of two activities. Only part of activity 20, Install Fuel Tanks, relates to activity 18, Install Exterior Conduit & Piping to Tanks. Only the start of activity 20 relates to activity 18. Similarly, only the start of activity 18 relates to activity 20. One day after the start of activity 20, activity 18 can start. The relationship shown in Figure 5-3 exists between a part of one activity and a part of another.

The ability to depict relationships between the start or finish of activities is the primary attraction of PDM over I-J. Some construction activities relate to only the start or finish of another activity. The four logical relationships possible in PDM facilitate depicting realistic relationships between the starts and finishes of activities.

PDM's four logical relationships, along with lag, allow some complex CPM schedules to be created with fewer activities and thus more quickly than when using AON or I-J techniques. The creation of PDM schedules is also a natural progression from the creation of bar charts. Activities can be conceptualized as they are for a bar-chart schedule. The relationships are then tied from the start of one bar or activity to the next bar or activity.

A problem with this PDM complex relationship is that it is not always clear what part of the activity is related to another activity. For example, in Figure 5-3 it is possible that the majority of the piping and conduit could be installed prior to, or unrelated to, the fuel tanks. It is also not clear how much of the tank installation process must be completed prior to the piping and conduit work. The start-to-start lag indicates 1 day but fails to indicate what work must be completed. While the scheduler who developed the PDM network may know exactly what work is represented by the lag, other schedule readers may not be able to determine the exact work represented by the lag.

Finish-to-Start Relationships

A finish-to-start relationship is the most common precedence diagramming relationship. The finish-to-start relationship in precedence diagramming is the traditional relationship used in I-J. The succeeding activity cannot start until the preceding activity has been completed. Finish-to-start relationships can be expressed in three forms if lag is used. Finish-to-start relationships may have zero lag, positive lag, or negative lag. Figure 5-4 shows activity relationships with zero lag and positive lag.

Positive lag is most commonly used in situations that require material to cure or strengthen before additional work can be performed. For example,

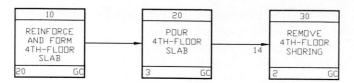

FIGURE 5-4 Finish-to-start relationships with zero lag and with positive lag.

forms for concrete cannot be removed until the poured concrete has been given sufficient time to cure. This relationship is shown in precedence diagramming by using positive lag. Figure 5-4 shows a typical relationship for forming and pouring a suspended slab. Zero lag is shown between the end of forming and the start of pouring because concrete can be poured as soon as the reinforcing and formwork has been completed. A lag of 14 days is shown between completion of the concrete pouring and removal of the shoring. The relationship indicates that one must wait 14 days after the concrete has been poured before removing the shoring. This delay allows time for the concrete to gain sufficient strength to support its own weight. The waiting period between concrete pouring and removal of the shoring is shown by adding positive lag on the logic diagram.

Negative lag, sometimes called ''lead,'' is used in situations which permit succeeding activities to begin before preceding activities have been completed. Figure 5-5 shows the relationship between layout and excavating for a gasoline service station and the installation of fuel tanks. Layout and excavating is scheduled to require 3 days to complete. However, not all of the 3 days' work must necessarily be completed before installation of the fuel tanks can begin. It is possible that excavation for the fuel tanks can be completed, permitting installation to begin, before excavation for the entire site has been completed. It may be possible that fuel tanks can be installed after 2 of the 3 days' work for layout and excavating has been completed. This relationship can be indicated in PDM by the use of negative lag. By including negative lag of 1 day on the diagram, the scheduler has indicated that installation of fuel tanks can begin 1 day before the completion of layout and excavating, or that installation of fuel tanks can begin after 2 days' excavation work has been completed.

As will be discussed later, this same relationship can be shown by using start-to-start relationships with positive lag. The selection is up to the individual scheduler. Although the critical path will be the same length in either case, it

FIGURE 5-5 Finish-to-start relationship with negative lag.

will not pass through the layout and excavating activity when a start-to-start relationship with positive lag is used. As will be discussed in detail later, there is another difference between the two alternatives. The actual duration of layout and excavating may exceed the 3 days allowed in the PDM schedule. Should this occur, the forward-pass calculations could vary substantially between a finish-to-start relationship with -1 lag and a start-to-start relationship with $+2$ lag.

Start-to-Start Relationships

As described in the discussion of finish-to-start relationships with negative lag, some activities do not require completion of preceding activities before they are able to start. Figure 5-5 shows that fuel tanks can be installed 2 days after the start of layout and excavating. This relationship can also be shown using a start-to-start relationship with positive lag as shown in Figure 5-6 below. Figure 5-6 indicates that fuel tank installation can start 2 days after layout and excavating has started.

Note that the start-to-start relationship has been drawn with the activities one above the other. By showing start-to-start relationships in this manner, the diagram is easier to read than if the activities had been drawn side by side and the relationship arrow shown parallel to the top or bottom of the preceding activity. Note also that a horizontal portion of the relationship arrow has been provided for the lag and that the lag has been placed near the arrowhead.

Start-to-start relationships with zero lag are used to show the relationship between two activities which should be started concurrently. For example, forming a slab on grade and installing reinforcing bars in the slab are activities which could be performed nearly concurrently. Figure 5-7 shows how PDM can express this relationship.

FIGURE 5-6   Start-to-start relationship with positive lag.

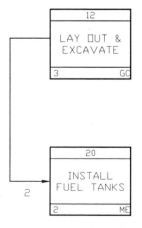

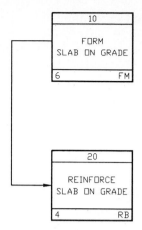

FIGURE 5-7   Start-to-start relationship with zero lag.

Start-to-start relationships with zero lag are frequently used to show relationships between concurrent work performed by two different subcontractors or concurrent work performed by the same contractor which use different equipment, material, or craftspeople.

Start-to-start relationships with negative lag are rarely used. Figure 5-8 shows this relationship. The same relationship can be expressed, however, by reversing the arrow and using positive lag.

Start-to-start relationships with negative lag are difficult to understand and complicate the logic diagram. Start-to-start relationships with negative lag are best avoided.

Finish-to-Finish Relationships

As with start-to-start relationships, finish-to-finish relationships are used to show the relationship between the finishes (or completion) of two activities. The same examples as were used to show start-to-start relationships can be used to show finish-to-finish relationships.

Figure 5-9 shows a finish-to-finish relationship with zero lag. When forming of the slab on grade has been completed, the reinforcing steel placement can also be finished.

FIGURE 5-8   Start-to-start relationship with negative lag.

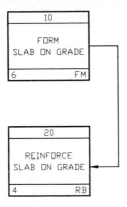

FIGURE 5-9   Finish-to-finish relationship with zero lag.

Figure 5-10 shows a finish-to-finish relationship with positive lag. Fuel tank installation cannot be completed until 1 day after the completion of layout and excavation.

Figure 5-11 on page 96 shows a finish-to-finish relationship with negative lag. Forming of the slab on grade cannot be completed until 1 day after the completion of reinforcing steel. As with start-to-start relationships with negative lag, finish-to-finish relationships with negative lag complicate a precedence diagram. Negative lag is difficult to understand and increases the complexity of the forward- and backward-pass calculations. Use of negative lag should be avoided if possible.

Start-to-Finish Relationships

Precedence diagramming allows for the use of start-to-finish relationships. An example of a start-to-finish relationship follows: a major office building is to

FIGURE 5-10   Finish-to-finish relationship with positive lag.

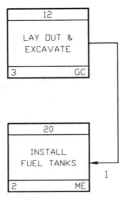

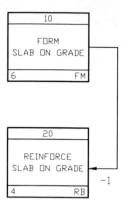

FIGURE 5-11 Finish-to-finish relationship with negative lag.

be constructed with carpet and wood millwork base throughout. The wood base could be installed before, after, or concurrently with the carpet everywhere except in the CEO's office which has wood paneling which must be installed prior to carpeting. A relationship from the start of activity "wood paneling and base" to the end of "install carpet" (possibly with a positive lag) would be appropriate.

Use of start-to-finish relationships are generally avoided as less confusing methods for showing interdependencies are available. Figure 5-12 indicates a start-to-finish relationship. Start-to-finish relationships may also include lag.

Combinations of Relationships

Activity relationships can be better clarified by using two or more of the four possible PDM relationships than by using only one relationship. For example, consider the excavation and pouring of concrete footings. The bar chart in

FIGURE 5-12 Start-to-finish relationship.

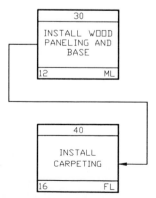

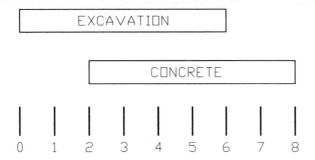

FIGURE 5-13   Relationship of excavation and concrete work.

Figure 5-13 shows the relationship of both 6-day activities and the overall time necessary to complete the footings. Both activities are concurrent for part of the time, but concrete work cannot start until after 2 days of excavation has been completed. A two-activity I-J diagram, as shown in Figure 5-14, cannot represent the same relationship between excavation and concrete work as is shown in the bar chart.

Although an I-J diagram with four activities can be used to represent correct relationships by breaking each bar into two activities, the resulting "fragnet" requires four activities and one dummy. A "fragnet" is the term used to indicate a small part of a larger network diagram. Figure 5-15 depicts an I-J fragnet showing the correct relationship between excavation and concrete work.

Using PDM, however, with two of the four possible relationships and lag, the same correct relationship can be shown with only two activities. Figure 5-16 shows the PDM alternative.

Because precedence diagramming can express relationships between the start and finish of activities, calculation of the critical path requires consider-

FIGURE 5-14   Inaccurate I-J diagrams of the bar chart in Figure 5-13.

Total elapsed time 12 days

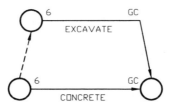

Total elapsed time 6 days

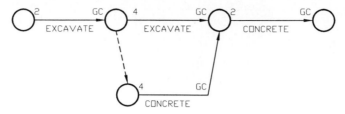

FIGURE 5-15   I-J fragnet showing correct relationships.

ation of the relationship of activity parts indicated on the logic diagram. Only those relationships indicated on the logic diagram are considered in calculating the critical path. Failure to show finish-to-finish relationships among activities that have start-to-start relationships indicated on the logic diagram may change the critical path or result in an inaccurate duration for a chain of activities. An erroneous critical path or inaccurate durations of a chain of activities can also affect the schedule's ability to measure the effect of changes and delay. Usually, activities with a start-to-start relationship will also have a finish-to-finish relationship. To identify the critical path in the network analysis accurately both relationships should be shown. Figure 5-16 shows the combined use of start-to-start and finish-to-finish relationships to depict the logical relationships among parts of an activity.

Activities with both a finish-to-finish and a start-to-start relationship can be either partially or completely concurrent. Figure 5-16 shows two activities which are partially concurrent. Figure 5-17 shows a variety of activities with start-to-start and finish-to-finish relationships. These activities are partially concurrent. The scheduler should be aware that software programs may not be designed to calculate certain combinations of relationships. Figures 5-17a and

FIGURE 5-16   PDM alternative.

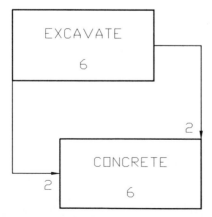

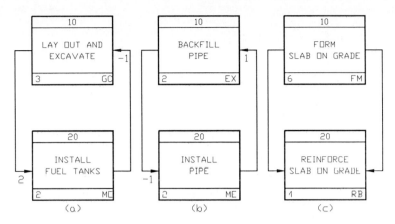

FIGURE 5-17 Combined use of start-to-start and finish-to-finish relationships.

5-17b can be especially troublesome. Many software programs will interpret these two diagrams as loops.

Figure 5-18 indicates two ways of showing complete concurrency between forming a slab on grade and the reinforcing work. The relationships show that reinforcing can start when forming work begins and that the reinforcing can be completed when the forming is completed. The relationships indicate that the two activities can start and finish together.

"Form slab on grade" is not completely concurrent with "reinforce slab on grade," but it is sufficiently concurrent to show the activities as completely concurrent. Completely concurrent activities are defined as two activities, with different durations, where the shorter-duration activity occurs completely within the time frame of the longer-duration activity. In construction scheduling, few activities are actually completely concurrent. Some examples are "test and balance activities" and procurement activities as shown in Figure 5-19a.

FIGURE 5-18 Combined start-to-start and finish-to-finish relationships.

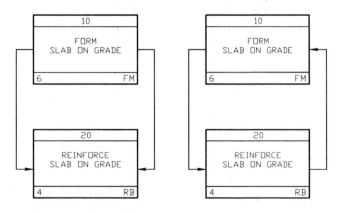

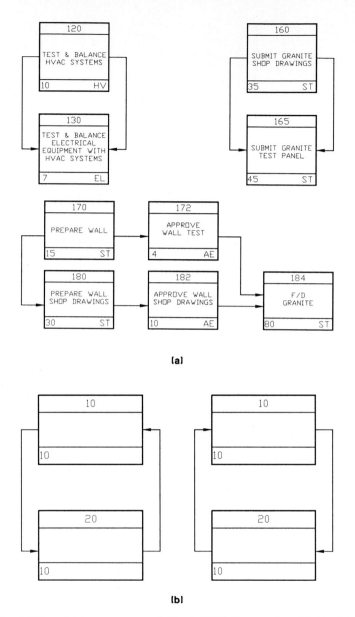

FIGURE 5-19 (a) Completely concurrent activities in PDM diagramming. (b) Absolute complete concurrency.

When a scheduler wishes to place two activities into a PDM schedule with absolute complete concurrency, the best way is to make those two activities into one activity. If for some reason the needs of the scheduler do not allow a single activity (separate crafts), then the two activities must have identical durations with start-to-start and finish-to-finish relationships as shown in Figure

5-19b. Lags or leads cannot be used. As stated earlier many software programs will interpret these activities and relationships as part of a loop, making computer calculation impossible. This is one limitation of the precedence diagramming method.

PDM DIAGRAMS

Characteristics

Just like an I-J diagram, a PDM diagram should be complete and show all logical relationships among activities. The PDM diagram should not include any unnecessary or redundant relationships. A comprehensive diagram will include not only field construction activities, but also mobilization and procurement activities. However, because a PDM diagram can become very cluttered with arrows, only necessary relationships should be depicted on the diagram. Too many arrows will not only make the diagram cluttered but will also unnecessarily complicate calculation of task dates during the forward and backward passes as will be discussed later.[3]

Logic Loops

The scheduler must be careful to avoid accidental logic loops in precedence diagramming. With the complex relationships it is easy to include loops into large PDM diagrams.

In I-J diagramming, the scheduler can minimize the chances of including logic loops by keeping the activity's J node number larger than its I node number. Because of the use of four logical relationships in PDM, however, this procedure cannot always be followed in precedence diagramming. Note that Figure 5-17a shows that "Lay Out and Excavate" is the "I node" and "Install Fuel Tanks" is the "J node" of the start-to-start relationship of the arrow 10-20. However, the illustration also shows that "Install Fuel Tanks" is the "I node" and "Lay Out and Excavate" is the "J node" of the finish-to-finish relationship of the arrow 20-10. Because "J-node" numbers may frequently be lower than "I-node" numbers in PDM due to finish-to-finish relationships, keeping "J-node" numbers larger than "I-node" numbers cannot be used to avoid logic loops. Numbering PDM diagrams following the flow of the work will minimize reverse numbering (or backbranches) but may not eliminate them.

Procedures are available to help avoid impermissible logic loops in the PDM. They will be described in the section that explains how nodes should be numbered. However, the first step in guarding against logic loops in PDM is to avoid using any relationship arrow on the diagram that is not essential.[4]

---

[3] Edward M. Willis, *Scheduling Construction Projects,* New York: Wiley, 1986, p. 168.
[4] Edward M. Willis, *Scheduling Construction Projects,* New York: Wiley, 1986, p. 169.

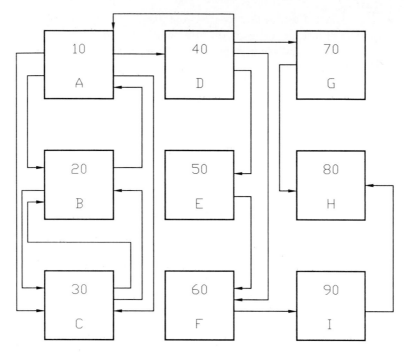

FIGURE 5-20   Precedence diagram with three logic loops.

Figure 5-20 is a precedence diagram that contains several impermissible logic loops. A logic loop exists wherever it is possible to trace a path from a point on the diagram through the diagram back to the starting point. The path of a logic loop can pass through arrows in the direction of the arrows, through nodes from left to right, or through parts of nodes. There are three impermissible logic loops in Figure 5-20:[5] (1) SS 20-30, activity 30, and FS 30-20; (2) FS 10-40, activity 40, and FF 40-10; and (3) FF 30-20, FF 20-10, and FF 10-30. Follow each logic loop on the diagram. Understand that on large project schedules a logic loop could pass through hundreds of activities and relationships.

Dangling Activities

Most schedulers recommend that in a comprehensive schedule all activities except the first and last nodes should have at least one preceding activity and at least one succeeding activity. All nodes except the start node should have at least one arrow which enters at the left, and all nodes except the last node should have at least one arrow that exits at the right. Some precedence scheduling software packages ignore good scheduling practice and permit open ends or dangling activities. In such cases, the software considers dangling activities

[5] Edward M. Willis, *Scheduling Construction Projects,* New York: Wiley, 1986, p. 169.

to be individual completion dates which must be completed on the dangling activity's early finish date. This makes the dangling activities always critical. Numerous dangling nodes can prevent a realistic critical path, inhibit evaluation of the schedule, and prevent the schedule from accurately measuring the impact of changes or delays. When the scheduler uses numerous dangling finish or start nodes, the effects of changes or time extensions on the dangling nodes cannot be measured accurately.

There are many activities that may start at any time after notice to proceed. Similarly, there may be many activities that can finish at any time prior to project completion. In both cases, however, all of the potential dangling start or finish activities will not be completed at the same time. Limitations of manpower, equipment, or materials will usually require these activities to be done one at a time. To reflect these limitations, a sequence should be determined and indicated. All dangling start and finish nodes should be tied into the logic. However, dangling activities should be tied into the logic only where a real construction relationship exists. In cases where no construction relationships can be found, the activity should be tied to the project start and project completion activities.

As will be seen in subsequent sections, the computed early start time will be zero for an activity with a dangling start. Activities A, E, and F in Figure 5-20 are dangling start activities.

An activity without a relationship originating on the right side is known as a dangling finish activity.[6] Activities G and H in Figure 5-19 are dangling finish activities. Recall that the first activity in a schedule is permitted to be a dangling start and the final activity is permitted to be a dangling finish. Schedulers also call dangling activities "open ends."

Figure 5-21a shows a plotted logic diagram with a dangling start activity, and Figure 5-21b shows a plotted logic diagram with a dangling finish activity.

Let us consider some examples. A dangling start node with a 10-day duration that must be completed by the sixth month of the project may have 115 or more days of float. This may not be realistic if the activity cannot reasonably start at notice to proceed. A dangling finish activity with a 10-day duration that starts 6 months prior to project completion will be considered critical by some software packages even though it may have 115 or more days of float. If the activities are affected by changes which prevent the activities' completion or lengthen their duration, the project completion date will not be affected until all the float has been eliminated. This situation contradicts the established understanding that any delay to an activity which is critical will delay the project completion date.

If either dangling activity is dangling because it can theoretically be started or finished at any time but equipment or other restrictions actually require it to be performed in sequence, delays that affect the activity will affect project completion despite a contrary indication on the logic diagram. Failure to tie

---

[6]Edward M. Willis, *Scheduling Construction Projects,* New York: Wiley, 1986, p. 170.

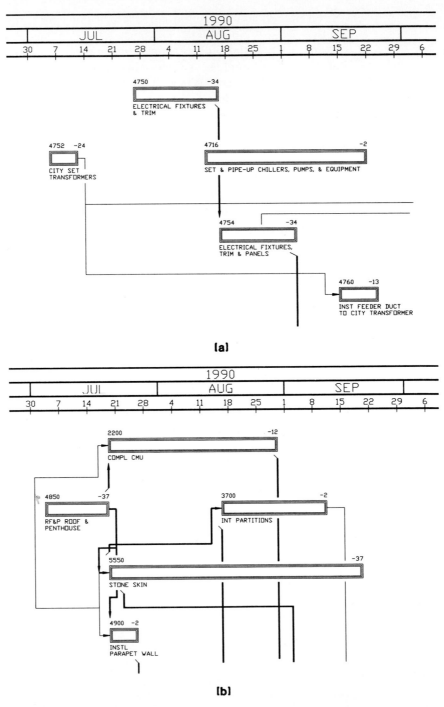

FIGURE 5-21 (a) Plotted diagram with a dangling start activity. (b) Plotted diagram with a dangling finish activity.

the dangling activity into the network prevents the network from accurately depicting the effect of the change on project completion. For example, landscaping the frontage to a building project may theoretically be completed at any time after initial site preparation and before project completion. However, if the same subcontractor who will perform the landscaping is also responsible for grading, paving, sidewalks, and miscellaneous site improvements then a sequence should be shown. Although it may be possible that landscaping the frontage may be completed at any time, showing that activity dangling will not properly represent the contractor's schedule if other activities to be performed by the same subcontractor are also scheduled to be completed at the end of the job. A change that delays the completion of any of the subcontractor's activities may affect not only the landscaping but also project completion. Dangling activities do not communicate the contractor's intent on completing the project; therefore dangling activities are considered poor scheduling practice.

Drawing the Logic

An activity in a precedence diagram is represented by description of the work and a corresponding number enclosed within some kind of symbol. The symbol is usually a rectangle, but it may be circular, hexagonal, or any other convenient shape. Figure 5-1a shows some common symbols for PDM schedule nodes.

A rectangular activity node should be drawn about $1\frac{1}{4} \times 2$ inches (4 cm $\times$ 6 cm). That is big enough to provide space for 24-character activity description, the duration, the activity number, and some additional information such as early start, early finish, late start, and late finish dates within the node. One method for drafting the logic diagram is to use self-adhesive squares for activities, which can easily be arranged, rearranged, or shifted as necessary.[7] Chapter 6 details procedures for developing the logic diagram.

It is suggested that the diagram be structured; that is, the activities should be in equally spaced rows and columns. When drafting PDM networks manually, each relationship should be shown individually if the result will be graphically readable. In some areas, such as project start or procurement listings, individual relationships produce congested diagrams. When this occurs, careful use of staffs will ensure that there will be no chance of relationships being misinterpreted. If staff diagramming is still confusing, modify the graphic technique so that it is clear.[8]

Plotting programs still do not draw I-J or PDM networks as clearly as manually drafted network diagrams. Complex diagrams, especially those with many relationships between activities, become difficult to read when they are plotted. As the scheduler becomes more experienced, the organization and coding of

[7] Edward M. Willis, *Scheduling Construction Projects,* New York: Wiley, 1986, p. 173.
[8] Edward M. Willis, *Scheduling Construction Projects,* New York: Wiley, 1986, p. 173.

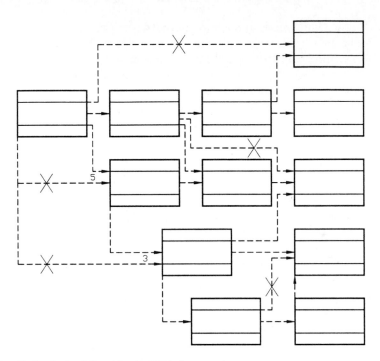

FIGURE 5-22   Redundant relationships in PDM diagramming.

activities can greatly improve the readability of plots. Limiting relationships in the logic to those that are strictly necessary also enhances plot clarity.

Frequently, an activity will be related to two activities which at the same time are related to each other. This situation may result in unnecessary or "redundant" relationship arrows. Redundancies result in needless computations and added diagramming effort. Redundancies should be avoided and, if inadvertently included in the schedule, removed. However, while avoiding redundancies, one must be careful not to oversimplify the schedule; all relationships required to properly reflect the contractor's intent should be included. Figure 5-22 shows some redundant relationships. The redundant ties or relationships are marked with a cross.

Numbering Activities

Activities are numbered for unique identification and in some cases the numbering provides additional "code" information about the activity. Any logical system can be used to number activities as long as each activity has only one number, which is different from that of any other activity in that network. Usually, activity numbers are ordered from lower numbers on the left to higher numbers on the right to match the logic flow or "directionality."

The logic diagram is normally totally completed before the activities are numbered. Activities are usually numbered from left to right and from top to bottom as discussed for I-J diagramming methods. A numbering pattern (numbering by 5s, for example) should be used to permit activities to be added later.

Ideally, one should develop a numbering system which is tied to the work breakdown system for typical activities used. A few of the digits in the number scheme might be reserved for an activity coding scheme (e.g., CSI numbering scheme). Think through your numbering system so that it is an aid to reading the plot and printout. Trying to "code" too much information into an activity number results in confusing printouts and plots.

ANALYZING THE PDM NETWORK

The Forward Pass

Calculation of activity times in precedence diagramming is similar to calculation of activity times in I-J diagramming, except that times for relationship arrows with lags and leads are calculated in addition to times for activities. The forward pass determines early start, early finish times, and project duration, while the backward pass determines late start and late finish times.

The following rules govern the calculation of early start and early finish times during the forward pass. These rules have been adapted from Edward M. Willis, *Scheduling Construction Projects* (John Wiley, New York, 1986), where they first appeared:

1 The smallest possible value for an early start time is zero. Therefore, the initial activity is assigned an early time of zero.

2 The activity's early finish time is the activity's early start time plus its duration.

3 The value for early start trial values of the start of a finish-to-start or finish-to-finish arrow is the activity's early finish time. The arrow's trial-value early finish time is calculated by adding the arrow's lag to the arrow's trial-value early start time.

4 The trial value for early start of the start of a start-to-start or start-to-finish arrow is the activity's early start time. The arrow's trial-value early finish time is calculated by adding the arrow's lag to the arrow's early start time. Note that if an arrow has negative lag, the lag value is added algebraically (subtracted).

5 After the early times for all arrows terminating at the left side of an activity have been calculated, the activity's early start time may be determined. The activity's early start time is the latest early finish of any relationship arrow entering the left side of the activity.

6 After the trial-value early times for all arrows terminating at the right side of an activity have been calculated, the activity's early finish time may be determined. The activity's early finish time is the latest trial-value early finish of any relationship arrow entering the right side of the activity or the activity's early start time plus its duration, whichever is later.

Figure 5-23 shows a precedence diagram before calculation of early times.

The largest value for an early finish time computed during the forward pass is the project duration. For the diagram shown in Figure 5-23, the project duration is 17 days, the early finish time for activity 100. Figure 5-24 shows the forward pass completed.

The Backward Pass

The backward pass is completed by working from right to left in the reverse order of arrows. While performing the backward pass, the late start and late finish times are determined. Several trial values may be calculated for each activity's late times. The smallest trial value is selected as the activity's late finish value.

The following rules, also adapted from Edward M. Willis' *Scheduling Construction Projects* (John Wiley, New York, 1986), govern the calculation of late start and late finish times during the backward pass:

1 The largest possible value for any late start or late finish value is the project duration. Therefore, the last activity is assigned a late finish time equal to the project duration.

2 The value of an activity's late start time is the activity late finish time minus its duration.

3 The value of an arrow's trial-value late finish for the finish of a finish-to-start or start-to-start arrow is the late start for the succeeding activity. The arrow's trial-value late start time is calculated by subtracting the arrow's lag.

4 The value for late finish for the finish of a finish-to-finish or start-to-finish arrow is the late finish of the succeeding activity. The late start for the arrow is calculated by subtracting the arrow's lag. Note that if the arrow's lag value is negative, the late start time must be calculated algebraically by adding its absolute value.

5 After the trial-value late times have been calculated for all arrows starting at the right side of an activity, the activity's late finish time can be determined. The activity's late finish is the earliest trial-value late start time of all relationships starting at the right side of the activity.

6 After the trial-value late times have been calculated for all arrows starting at the left side of an activity, the activity's late start time may be determined. The activity's late start time is the earliest late start of all the relationships starting at the left side of the activity or the activity's late finish minus its duration, whichever is earlier.

Figure 5-24 shows the precedence diagram of Figure 5-23 after calculation of the forward pass and prior to the backward pass. Figure 5-25 shows the forward and backward passes completed for Figure 5-23. The critical path has been darkened.

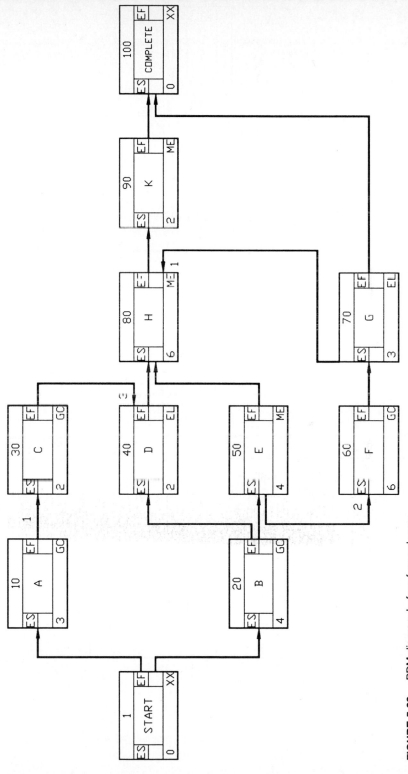

FIGURE 5-23 PDM diagram before forward pass.

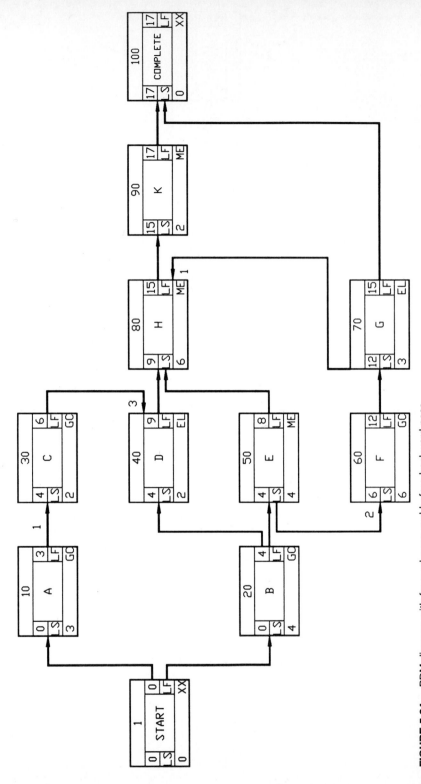

FIGURE 5-24 PDM diagram with forward pass and before backward pass.

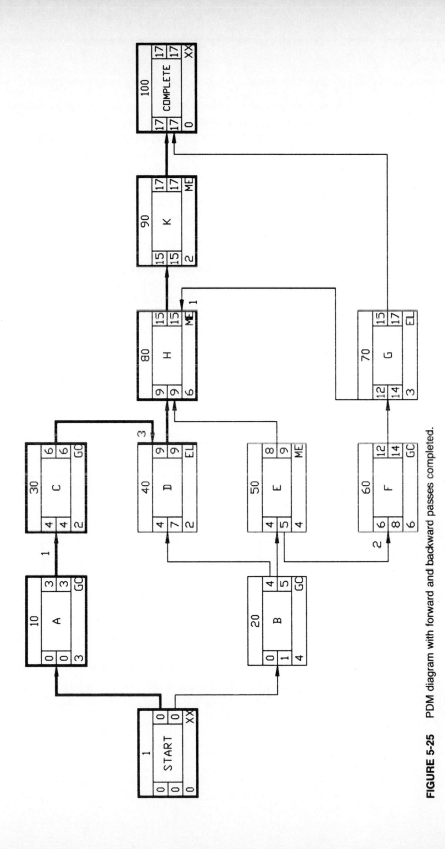

FIGURE 5-25 PDM diagram with forward and backward passes completed.

FLOAT

"Float" is a measure of how much an activity can be delayed without delaying the project completion date. An activity, the start of an activity, or the finish of an activity may be critical. Thus precedence diagramming uses three kinds of activity float: total float (TF), start float (SF), and finish float (FNF).

An activity's total float is calculated by subtracting the early start time and duration from the activity's late finish time. An activity that has zero total float is a critical activity. In order for the project to be completed within its scheduled duration, a critical activity must start on its early start time and complete by its late finish time.

An activity's start may be critical even though the activity itself may not be critical. If the start of an activity is precedent to the start of another critical activity, the start of the first activity may be critical. The start of an activity is critical if its early start time equals its late start time. Any activity with no total float will also have no start float. Start float may be calculated by subtracting an activity's early start time from its late start time.[9]

The completion of an activity may be critical even though the activity itself is not critical. If the finish of an activity is precedent to the completion of another activity (finish-to-finish relationship) that is critical, the completion of the first activity may also be critical. An activity with no total float will also have no finish float. Finish float is calculated by subtracting an activity's early finish time from its late finish time.[10]

As the reader has learned, the proper calculation of float in PDM diagrams is not straightforward. All software programs do not agree on how precedence calculations should be performed. It is possible to obtain different float values and different critical paths with identical PDM diagrams that have been run through different software programs. Minimizing start-to-start, finish-to-finish, and start-to-finish relationships, and minimizing the use of lags and leads, will greatly reduce the variations in float and critical path calculations among different software packages.[11]

THE CRITICAL PATH

As in I-J diagramming, the critical path in PDM is continuous from the beginning to the end of the diagram. The critical path may go through arrows and activities or parts of activities. If an activity's early and late start times are the same, the start of the activity is on the critical path. If an activity's early and late finish times are on the critical path, the completion of the activity is critical. However, just because both the start and finish of an activity are on the critical

[9]Edward M. Willis, *Scheduling Construction Projects,* New York: Wiley, 1986, pp. 164–165.
[10]Edward M. Willis, *Scheduling Construction Projects,* New York: Wiley, 1986, p. 165.
[11]Thomas E. Glavinich, 1990. "Development of a model microcomputer-based project management system." Presented to the Univ. of Kansas at Lawrence, KS, in partial fullfillment of the requirements for the degree of Doctor of Engineering.

path, this does not necessarily mean that the activity is on the critical path. Both the start and finish of an activity must be critical and its total float must equal zero for an activity to be considered critical. Again, different software packages determine activity criticality differently. Some software programs assume that if any part of the activity is critical, the entire activity must be critical.[12]

In precedence diagramming, arrows must also be considered when determining the critical path. An arrow is critical if its relationship float is zero. Following the critical path through a PDM diagram is not as easy as following the critical path through an I-J diagram. Care and patience are required to determine and follow a PDM critical path.

PDM PRINTOUTS

A precedence schedule printout looks different from an I-J printout. In an I-J printout, activities are identified by their I and J nodes. The I and J nodes provide the reader with the ability to determine relationships among activities by observing common I or J nodes on the printout without reference to the logic diagram. Because PDM activities have only one number the reader cannot determine activity relationships just from the activity numbers on the printout. To provide the ability to read a printout and determine relationships without reference to the logic diagram, most precedence software programs permit a sort which includes each activity's predecessors and successors. Figures 5-26a and 5-26b are PDM printouts which show sorts both with and without predecessors and successors. Notice how the length of the sample printout is increased when predecessors and successors are included.

USE OF PDM

Precedence diagramming offers a simpler way to indicate the complex logical relationships among construction activities, particularly when partial concurrences exist between activities. PDM diagrams also tend to be smaller than I-J equivalents. The primary attraction which PDM offers schedulers is its speed of preparation. Precedence diagrams eliminate the need for dummy activities and additional detail to show overlap. The scheduler need not spend as much time preparing a PDM schedule. PDM is very useful when representing repetitive construction activities such as those required in multistory buildings or highway construction. Precedence is able to model the overlapping relationships among activities without having to split activities. The additional relationships available in PDM diagramming, however, may lead the scheduler to assume that the resulting schedule is complete and accurate. Failure to consider all relationships carefully can result in a PDM schedule which is no more accurate than a bar chart.

[12] Edward M. Willis, *Scheduling Construction Projects,* New York: Wiley, 1986, p. 187.

ORIGINAL AS PLANNED -

REPORT DATE 18APR90 RUN NO. 15   START DATE 8MAY86   FIN DATE 28MAR90
10:54   DATA DATE 8MAY86   PAGE NO. 7

NUMERIC SORT

ACTIVITY ID	ORIG DUR	REM DUR	CAL	%	CODE	ACTIVITY DESCRIPTION	EARLY START	EARLY FINISH	LATE START	LATE FINISH	TOTAL FLOAT
2021	9	9	1	0		B7 LINE-.700M DIA-MHB1-20 TO MH2 (PS) 110.7M	1DEC88	11DEC88	1DEC88	11DEC88	0
2023	15	15	1	0		B7 LINE-.600M DIA-MH2 TO MH7 (END B7)(PS) 228.2M	12DEC88	28DEC88	12DEC88	28DEC88	0
2025	25	25	1	0		B8 LINE-.600M DIA-MHB1-19 TO MH15 (TB) 482.8M	20DEC89	17JAN90	20DEC89	17JAN90	0
2027	6	6	1	0		B8 LINE-.300M DIA-MH15 TO MH21(END B8)(TB)153.7M	18JAN90	24JAN90	18JAN90	24JAN90	0
2029	18	18	1	0		B9 LINE-.300M DIA-MHB1-15 TO MH7(END)(TB) 173.4M	29JAN90	18FEB90	29JAN90	18FEB90	0
2031	30	30	1	0		B10 LINE-.375M DIA-MHB7-2 TO MH12(END)(TB)292.5M	22FEB90	28MAR90	22FEB90	28MAR90	0
2033	48	48	1	0		B20 LINE-.600M DIA-MHC1-18 TO MH10(PS) 707.4M	6MAR88	30APR88	6MAR88	30APR88	0
2035	16	16	1	0		B20 LINE-.375M DIA-MH10 TO MH19 (END)(TB) 231.8M	24MAY89	11JUN89	25DEC89	11JAN90	184
2037	19	19	1	0		B21 LINE-.300M DIA-MHB20-4 TO MH6(END)(TB)191.2M	25SEP88	16OCT88	27APR89	18MAY89	184
2039	28	28	1	0		B22 LINE-.300M DIA-MHB20-6 TO MH5(END)(TB)222.3M	20OCT88	21NOV88	23MAY89	24JUN89	184
2041	33	33	1	0		B23-.300M DIA-MHB20-10 TO JCC1-17 (END)(TB) 341.9M	26NOV88	2JAN89	28JUN89	5AUG89	184
2043	49	49	1	0		B30 LINE-1.200M DIA-JCC1-17 TO MH9 (DS) 626.4M	20OCT87	15DEC87	27OCT87	22DEC87	6
2045	15	15	1	0		B30 LINE-1.000M DIA-MH9 TO MH11 (DS) 203.0M	16DEC87	23JAN88	23DEC87	9JAN88	6
2047	27	27	1	0		B30 LINE-.700M DIA-MH11 TO MH17 (END)(PS) 493.6M	4MAY88	4JUN88	4MAY88	4JUN88	0
2051	15	15	1	0		B31 LINE-.450M DIA-MHB30-9 TO MH5 (TB) 269.1M	23JAN89	1FEB89	9AUG89	26AUG89	184
2053	8	8	1	0		B31 LINE-.300M DIA-MH5 TO MH9(END B31)(TB)172.7M	24JAN89	1FEB89	27AUG89	4SEP89	184
2055	12	12	1	0		B32 LINE-.700M DIA-MHB30-11 TO MH4(PS) 142.1M	8JUN88	21JUN88	8JUN88	21JUN88	0
2057	36	36	1	0		B32 LINE-.600M DIA-MH4 TO MH13(END B32)(PS) 448M	22JUN88	2AUG88	22JUN88	2AUG88	0
2061	12	12	1	0		B33 LINE-.450M DIA-MHB32-4 TO MH4 (TB) 208.4M	21JUN89	4JUL89	21JUN89	4JUL89	0
2063	18	18	1	0		B33 LINE-.300M DIA-MH14 TO MH11 (END)(TB)268.3M	5JUL89	25JUL89	5JUL89	25JUL89	0
2067	16	16	1	0		B34 LINE-.225M DIA-MHB33-4 TO MH4(END)(TB)1118.2M	30JUL89	5AUG89	30JUL89	5AUG89	0
2069	16	16	1	0		B40 LINE-.375M DIA-MHC1-21 TO MH2(END)(TB)160.6M	1MAR89	19MAR89	20OCT89	11NOV89	184
2071	8	8	1	0		B41 LINE-.300M DIA-MHB40-2 TO MH1WEST(TB) 163.4M	23MAR89	10APR89	24OCT89	1NOV89	184
2073	21	21	1	0		B41 LINE-.300M DIA-MHB40-2 TO MH8EAST (TB) 74.9M	26JUN89	19JUL89	13NOV89	21NOV89	184
2075	31	31	1	0		B50 LINE-.375M DIA-MHC1-34 TO MH8(END)(TB)285.3M	26JUN89	19JUL89	27JAN90	19FEB90	184
2077	29	29	1	0		B51 LINE-.600M DIA-MHC1-32 TO MH8(END)(PS)345.3M	24JUL88	11SEP88	7AUG88	28MAR90	184
3001	202	202	1	0		B52 LINE-.450M DIA-MHC1-30 TO MH10(END)(TB)1418.8	17FEB87	10OCT87	17FEB87	10OCT87	0
3005	104	104	1	0		C1 LINE-2.750M DIA-START TO JC17 (SP) 2135.8M	11OCT87	8FEB88	11OCT87	8FEB88	0
4001	144	144	1	0		C1 LINE-2.500M DIA-JC24 (DS) 1133.1M	9FEB88	25JUL88	28DEC88	25JUL88	184
4003	100	100	1	0		C1 LINE-2.500M DIA-JC24 TO JC35(END)(DS) 1603.4M	28MAY88	20SEP88	28DEC88	28APR89	0
4005	25	25	1	0		D1 LINE-1.000M DIA-JCC1-22 TO MH17 (SF) 1045.8M	7SEP89	5OCT89	10SEP89	8OCT89	6
4007	32	32	1	0		D1 LINE-.700M DIA-MH17 TO MH22(END D1)(DS)1178.0M	6JAN88	11FEB88	13JAN88	18FEB88	6
4009	42	42	1	0		D10 LINE-2.00M DIA-JCC1-16 TO JC4 (DS) 310.7M	13FEB88	31MAR88	20FEB88	7APR88	0
4010	41	41	1	0		D10 LINE-1.800M DIA-JC7 (DS) 421.2M	18APR88	18MAY88	9APR88	25MAY88	0
4011	37	37	1	0		D10 LINE-1.800M DIA-JC7 TO JC4 (DS) 485.1M	11APR89	23MAY89	11APR89	23MAY89	0
4013	4	4	1	0		D10 LINE-.800M DIA-MH12 TO MH24 (PS) 360.5M	24MAY89	29MAY89	24MAY89	27MAY89	0
4015	35	35	1	0		D11 LINE-.450M DIA-MH24 TO MH25 (END)(PS) 28.9M	15APR89	24MAY89	17APR89	27MAY89	2
4017	42	42	1	0		D11 LINE-.800M DIA-JCD10-4 TO MH-9(END)(DS)1406.2	29MAY89	16JUL89	31MAY89	18JUL89	2
4019	5	5	1	0		D12 LINE-1.800M DIA-JCD10-7 TO JC4 (DS) 462.8M	22JUN89	17JUL89	13JUL87	18JUL87	4
4021	15	15	1	0		D12 LINE-1.800M DIA-JC4 TO JC5(END)(DS) 50.1M	27JUN87	13JUL87	19JUL87	6OCT87	4
4023	69	69	1	0		D20 LINE-1.500M DIA-JCC1-1 TO JC3 (DS) 152.7M	14JUL87	10CT87	19JUL87	6OCT87	4
4024	4	4	1	0		D20 LINE-1.200M DIA-JC3 TO MH12 (DS) 831.2M	26JUL89	30JUL89	29JUL89	1AUG89	2
						D20 LINE-.800M DIA-MH13 (DS) 79.9M	31JUL89	3AUG89	2AUG89	6AUG89	2
						D20 LINE-.800M DIA-MH13 TO MH15 (END)(DS) 32.8M					

FIGURE 5-26a   Primavera Project Planner without predecessors and successors.

PRIMAVERA PROJECT PLANNER

ORIGINAL AS PLANNED

START DATE 8MAY86   FIN DATE 28MAR90
DATA DATE 8MAY86   PAGE NO.   25

ACTIVITY ID	ORIG DUR	REM DUR	CAL	%	CODE	ACTIVITY DESCRIPTION	EARLY START	EARLY FINISH	LATE START	LATE FINISH	TOTAL FLOAT
20146	12	12	1	0		START DEWATERING FOR C1 LINE	25JAN87	7FEB87	3FEB87	16FEB87	8
20145*	42	42	1	0	P FS	INSTALL DEWATERING SYSTEM FOR C1 LINE	7DEC86	24JAN87	16DEC86	2FEB87	8
3001	202	202	1	0	S FS	C1 LINE-2.50M DIA-START TO JC17 (SP) 2135.8M	7FEB87	10OCT87	17FEB87	10OCT87	0
20190	48	48	1	0		DESIGN & SUBMIT SHORING SYSTEMS - GEN SUBMITTAL	29JUL86	22SEP86	29JUL86	22SEP86	0
19150*	10	10	1	0	P FS	SELEC- SOILS/DEWATERING CONSULTANT	17JUL86	28JUL86	17JUL86	28JUL86	0
20200*	27	27	1	0	S FS	AMBRIC REVIEW & APPROVAL SHORING SUBMITTAL	23SEP86	23OCT86	23SEP86	23OCT86	0
20200	27	27	1	0		AMBRIC REVIEW & APPROVAL SHORING SUBMITTAL	23SEP86	23OCT86	23SEP86	23OCT86	0
20190*	48	48	1	0	P FS	DESIGN & SUBMIT SHORING SYSTEMS - GEN SUBMITTAL	29JUL86	22SEP86	29JUL86	22SEP86	0
20210*	18	18	1	0	S FS	SELEC- SHORING SYSTEMS	25OCT86	13NOV86	25OCT86	13NOV86	0
20210	18	18	1	0		SELECT SHORING SYSTEMS	25OCT86	13NOV86	25OCT86	13NOV86	0
20200*	27	27	1	0	P FS	AMBRIC REVIEW & APPROVAL SHORING SUBMITTAL	23SEP86	23OCT86	23SEP86	23OCT86	0
20220*	81	81	1	0	S FS	BUY, SHIP & ASSEMBLE INITIAL SHORING SYSTEMS	5NOV86	16FEB87	15NOV86	16FEB87	0
20220	81	81	1	0		BUY, SHIP & ASSEMBLE INITIAL SHORING SYSTEMS	5NOV86	16FEB87	15NOV86	16FEB87	0
20210*	18	18	1	0	P FS	SELECT SHORING SYSTEMS	25OCT86	13NOV86	25OCT86	13NOV86	0
3001*	202	202	1	0	S FS	C1 LINE-2.50M DIA-START TO JC17 (SP) 2135.8M	7FEB87	10OCT87	17FEB87	10OCT87	0
7069*	50	50	1	31	S SS	CHECKED XNG #9(C1 LINE)- MOB,JACK,DEMOB - 70M	1DEC86	16FEB87	21DEC86	16FEB87	0
20230	12	12	1	0		ZOMAR CANAL BRIDGE CROSSING-SITE SURVEY/INVEST.	7JUL86	30JUL86	11OCT86	23OCT86	73
19010*	60	60	1	0	P FS	SELECT & SET UP PROJECT OFFICES	8MAY86	16JUL86	8MAY86	16JUL86	0
20240*	18	18	1	0	S FS	ZOMAR CANAL - DESIGN BRIDGE CROSSING @ BOULAC PS	13JUL86	20AUG86	25OCT86	13NOV86	73
20240	18	18	1	0		ZOMAR CANAL - DESIGN BRIDGE CROSSING @ BOULAC PS	31JUL86	20AUG86	25OCT86	13NOV86	73
20230*	12	12	1	0	P FS	ZOMAR CANAL BRIDGE CROSSING-SITE SURVEY/INVEST.	7JUL86	30JUL86	11OCT86	23OCT86	73
20250*	21	21	1	0	S FS	ZOMAR CANAL CROSSING - R&A BY IRRIG. DEPT.	21AUG86	14SEP86	22NOV86	15DEC86	79
20255*	27	27	1	0	S FS	ZOMAR CANAL CROSSING - R&A BY AMBRIC	21AUG86	21SEP86	15NOV86	15DEC86	73
20250	21	21	1	0		ZOMAR CANAL CROSSING - R&A BY IRRIG. DEPT.	21AUG86	14SEP86	22NOV86	15DEC86	79
20240*	18	18	1	0	P FS	ZOMAR CANAL - DESIGN BRIDGE CROSSING @ BOULAC PS	31JUL86	20AUG86	25OCT86	13NOV86	73
20260	36	36	1	0	S FS	ZOMAR CANAL BRIDGE - OBTAIN PERMITS & ROW	22SEP86	2NOV86	16DEC86	26JAN87	73
20255	27	27	1	0		ZOMAR CANAL CROSSING - R&A BY AMBRIC	21AUG86	21SEP86	15NOV86	15DEC86	73
20240*	18	18	1	0	P FS	ZOMAR CANAL - DESIGN BRIDGE CROSSING @ BOULAC PS	31JUL86	20AUG86	25OCT86	13NOV86	73
20260*	36	36	1	0	S FS	ZOMAR CANAL BRIDGE - OBTAIN PERMITS & ROW	22SEP86	2NOV86	16DEC86	26JAN87	73

FIGURE 5-26b Primavera Project Planner with predecessors and successors.

Not having to keep track of dummy activities on large schedules may make the schedule easier to read and understand, but precedence diagramming is not as structured as I-J and as a result differences in diagrams, mathematical calculations, and project completion dates abound.

Precedence diagrams which use lag add an additional element of uncertainty. Instead of two assumptions to determine sequence and duration, three are necessary. It can be as difficult to estimate lag as it is to estimate activity durations and sequence. I-J diagrams require only two assumptions.

The variety of logical relationships which PDM permits makes the network analysis more difficult than I-J diagramming. Not only are manual calculations more difficult in precedence diagramming, but also computer calculations are slower because of the variety of calculations necessary.[13] The variety of logical relationships possible has also created differences in PDM software. Not all PDM software follows the same methodology in applying the four logical relationships to network calculations. The way combination start-to-start and finish-to-finish relationships are treated creates some of the variation among precedence scheduling software.[14]

The variety of logical relationships, although able to depict partial concurrences more easily, also makes use of the schedule to measure delay more difficult. PDM permits partial concurrences to be shown at the start of each partially concurrent activity, but one cannot show a continuing concurrency. For example, reconsider Figure 5-16. It was included to demonstrate hc PDM schedule can show the relationships between excavation and concrete work. Start-to-start and finish-to-finish relationships were used to simplify a four-activity I-J fragnet into a two-activity PDM fragnet. In simplifying, however, the PDM schedule also lost some ability to show a continuous relationship between excavation and concrete: Excavation must be performed continuously to permit concrete to be poured continuously. Figure 5-16 shows only a 2-day start dependency and a 2-day finish dependency. No relationship is shown for the 4 days remaining of the 6-day activity. If excavation stops at the beginning of the third day, the schedule does not indicate that concrete must also stop, although the reality in the field is that, without necessary preceding excavation, concrete activities must stop. The Figure 5-16 fragnet shows that it is not until the beginning of the sixth day, when the 2-day finish-to-finish relationship applies, that concrete cannot continue until the final excavation work is completed. According to the schedule, concrete may continue on the third, fourth, fifth, and sixth days regardless of any missing excavation; only the final completion of concrete is delayed. An I-J fragnet, although more time-consuming and difficult to prepare, better illustrates the continuous relationship between the two partially concurrent activities and can therefore better measure delay or impact.

---

[13] Edward M. Willis, *Scheduling Construction Projects*, New York: Wiley, 1986, p. 193.
[14] Thomas E. Glavinich, 1990. "Development of a model microcomputer-based project management system." Presented to the Univ. of Kansas at Lawrence, KS, in partial fullfillment of the requirements for the degree of Doctor of Engineering.

In essence, the lags used in precedence diagramming result in activities being related by time, when the reality of the construction process is that the activities are related by work. I-J scheduling uses only work relationships. For example, a scheduler wishes to indicate that painting of drywall partitions will start after 25 percent of the drywall is erected, taped, and finished. The drywall activity has a 20-day duration. In I-J scheduling the drywall activity would have to be subdivided to show the relationship. In PDM scheduling a start-to-start lag of 5 days would indicate this concept. If, however, the drywall subcontractor got a slow start and hung only 10 percent of the drywall and did not tape or finish any in the first 5 days, could the painter start? In I-J the schedule would correctly indicate that the painter could not start. The PDM diagram would indicate that he should. When the project finished late and the owner accessed liquidated damages, the liability for delay would be incorrectly indicated in the PDM diagram. As the painter, which schedule format would you prefer?

Because of the complexities of the network analysis created by the variety of logical relationships, schedulers are traditionally advised to use only finish-to-start relationships and to avoid overlap and lag in order to keep the diagram easy to understand and analyze. To recognize that in most cases 10 percent of an activity may start before the preceding activity ends may substitute for the use of lag or overlap.

Initial precedence schedules may be prepared with only finish-to-start relationships. It is easier to analyze the network with simple relationships. Start-to-start, start-to-finish, or finish-to-finish relationships should be used only when necessary to reflect a specific relationship on a project which cannot be indicated otherwise.

A more dramatic example of how I-J and PDM vary in their representation of a project is shown in Figures 5-27 and 5-28. Figure 5-27 is a simple precedence diagram showing both start-to-start and finish-to-finish relationships. In developing precedence schedules, many contractors fail to include the finish-to-finish relationships shown on this diagram. Even given the completeness of this diagram, however, the precedence schedule does not depict the reality of the

FIGURE 5-27 Laying an underground pipeline: PDM format.

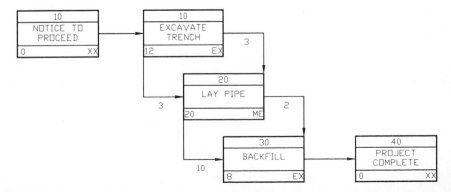

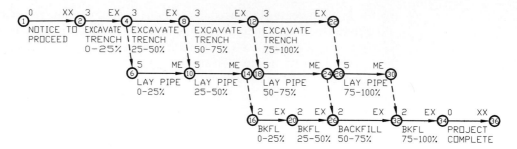

FIGURE 5-28   Laying an underground pipeline: I-J format.

construction process accurately. The schedule indicates that pipe laying can begin 3 days after excavation starts. If, for example, the contractor encountered serious rock problems during the second day of excavation, the PDM schedule as shown would not reflect a project completion delay until 8 days later. The precedence diagram indicates that pipe laying can start 3 days after excavation starts and proceed unimpeded for 20 days. In reality, trench excavation has a continuous relationship with pipe laying. The contractor really wants to indicate that the excavation is to stay 3 days ahead of the pipe-laying process. The precedence schedule does not indicate this.

The second schedule, Figure 5-28, shows the same pipeline project in I-J format. As one can see, a greater number of activities are required to properly show the relationships among excavation, pipe laying, and backfilling. The I-J schedule, however, shows the continuous relationships accurately. Calculation of the associated float for each of the activities in the I-J network shows that as the excavation work progresses, more float is available to absorb the possible removal of rock. This is a very accurate representation of the reality of the construction process. For example, if rock is encountered on the second day of excavation, the I-J schedule shows immediately that the entire process is stopped. If the contractor finds rock on the tenth day of excavation, he will have 6 days of float in which to resolve the problem before it delays pipe-laying work.

If a scheduler using precedence diagramming methods uses only finish-to-start relationships without lags or leads, then the same accuracy can be obtained with the precedence diagramming method. This limitation on precedence diagramming, however, will increase the number of activities in the PDM diagram to that of the I-J diagram.

Various computer software packages use different algorithms to calculate early start, early finish, late start, and late finish dates in the precedence diagramming methods. This can result in different completion dates for the same schedule. I-J or arrow-node type scheduling systems use very similar algorithms in the calculation process, so most such software programs produce identical start and finish dates. The most drastic differences among precedence software programs are found in the use of leads, lags, finish-to-finish relationships, and open starts and finishes. Some software programs allow selection of how these dates are controlled. Different software packages calculate and position float differently. This can obstruct or modify the critical path, resulting in inaccurate delay measurements.[15]

The same difficulties which complex relationships and lag impose on the preparation of the initial PDM schedule also handicap updates and sequence revisions to PDM schedules. The software has many calculation variations when dealing with updates, and it is recommended that the user develop "test networks" to understand and compare computer solutions to manual calculations.

The CPM scheduler should select the format (I-J or PDM) with which he or she is most comfortable. Understanding each format's assets and limitations allows the scheduler to avoid some of the difficulties encountered in the development, updating, and impacting of CPM schedules.

EXERCISES

1 The following table lists some activities, their durations, and their dependencies. Draw a precedence diagram and calculate the ES, EF, LS, LF, and TF for each activity. Identify the critical path.

Activity	Duration	Depends on:
A	14	L
B	7	L, R
C	6	G
D	4	—
E	1	F, M, N, Q
F	3	C, P, R
G	1	D, H
H	4	—
J	2	L
K	8	L, O
L	3	D, G, H
M	2	A, B, L, R
N	5	J, K
O	6	G
P	2	C, R
Q	1	A, J, L
R	4	C

[15] Thomas E. Glavinich, 1990. "Development of a model microcomputer-based project management system." Presented to the Univ. of Kansas at Lawrence, KS, in partial fullfillment of the requirements for the degree of Doctor of Engineering.

2 Draw a precedence diagram for the data in the following table. On the diagram, compute the early start and early finish dates and the late start and finish dates, and indicate the critical path(s). In a table, compute the total float.

No.	Activity	Duration	Depends on:
1	Mobilization	3	—
2	Underground utilities	3	1
3	Foundations	2	1
4	Block walls ext.	6	3
5	Interior framing	2	4, 8
6	Roof framing	4	4, 5
7	Roofing	2	6
8	Floor slab	3	3
9	Carpet & tile	2	14
10	Rough plumb	3	2
11	Finish plumb	2	4, 5, 10
12	Rough elect.	2	4, 5, 7
13	Finish elect.	2	4, 5, 7, 12
14	Wall finishes	5	11, 13
15	Ceiling finishes	3	11, 13

3 Change the I-J diagram you drew for the fuel station shown in Figures 2-3 and 2-4 to PDM. Use start-to-start, finish-to-finish, and start-to-finish relationships if appropriate. Make a forward and a backward pass, calculate activity times and float, and identify the critical path.

4 Redraw the I-J diagram in Figure 5-29 in PDM format.

FIGURE 5-29 I-J diagram for Exercise 4.

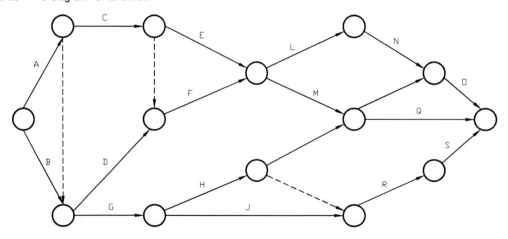

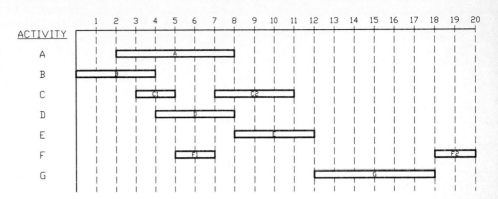

FIGURE 5-30   Bar chart for Exercise 6.

5 Draw the activity-on-node diagram for the following data:

Activity	Succeeded by:	Duration
A	B, E, F,	7
B	K	9
C	H, D	8
D	I, N	11
E	G, M	9
F	L	8
G	C	7
H	I, N	6
I	—	11
J	E, F	10
K	G	13
L	M	17
M	C	10
N	—	13

6 Draw a precedence diagram for the data shown on the bar chart in Figure 5-30.
7 Draw a precedence diagram for the following data with the durations indicated in Figure 5-31.

   Activities A and B begin the project.
   Activity C follows activity A.
   Activity D follows activity B.

FIGURE 5-31   Data for Exercise 7.

ACTIVITY	A	B	C	D	E	F	G	H	J
DURATION	10	12	6	20	14	10	20	8	12

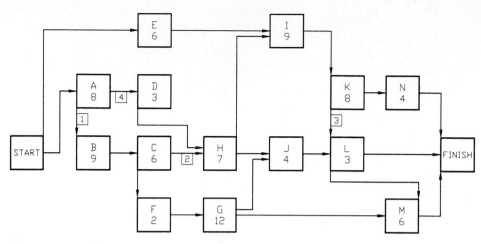

FIGURE 5-32 Diagram for Exercise 9.

Activity F follows activity B.
Activity E can be completed after activity D is completed.
Activity H follows activities C and D.
Activity F can start after activity E begins.
Activity G follows activity F.
Activity J follows activities G and H.

8 How long will the project in Exercise 7 take if activity E is reduced to 8 days and activities A and H are each 12 days long?

9 Perform a forward pass and a backward pass and determine the ES, LS, EF, LF, TF, and FF for each activity on the diagram in Figure 5-32.

10 Perform a forward pass and a backward pass and determine the ES, LS, EF, LF, TF, and FF for each activity on the diagram in Figure 5-33.

11 Discuss possible different methods for determining activity ES, EF, LS, and LF dates. Which calculation methods are most correct?

FIGURE 5-33 Diagram for Exercise 10.

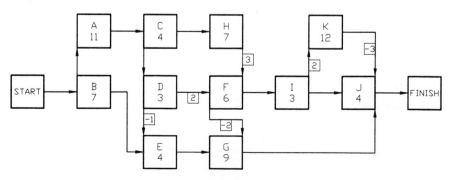

12 Discuss the differences between Figures 5-27 and 5-28. Which format is more accurate? Which format is easier to use? Which format will be more readily accepted by tradespeople in the field?

13 What adjustments would be necessary to the logic diagram in Figure 5-27 if all work were proceeding as scheduled and an update were performed on the seventh day?

14 Review Figure 5-15: Is it a correct I-J diagram for the bar chart shown in Figure 5-13? Why or why not?

CHAPTER
6

DEVELOPING
THE PROJECT
SCHEDULE

INTRODUCTION

Developing a project schedule that will be an accurate and quality tool for use in managing a project's progress is not a simple process. Actual development of the CPM schedule involves a series of construction judgments and coordination among the parties involved in the project. Producing a good schedule requires good people skills, an understanding of design and construction, and available time and inclination for the effort required, in addition to an understanding of network scheduling techniques. There are no shortcuts to producing a quality CPM schedule. While experience in developing CPM schedules will speed the process, a step-by-step method is necessary even for a professional scheduler.

If the CPM fails to portray accurately the contractor's plans for executing the project, whether it be the original schedule or an updated version, not only will the schedule be ignored, but also a valuable management tool will be eliminated. The construction scheduler must make a personal commitment to produce a quality schedule from the outset. A schedule which is inaccurate or does not demonstrate the project plan will not be beneficial and will most likely be a detriment to the project.

Often an activity involves use of a combination of human resources and special equipment. (Photo courtesy of John Deere.)

For the vast majority of us, it is impossible to comprehend all the activities and relationships necessary for an entire schedule on a major project. A detailed schedule for a project such as a multistory office building, a hospital, or a wastewater-treatment plant demands a step-by-step procedure. The more complex and confusing a project is, the more important is a structured approach to developing the original schedule. On a small, simple project such as a gas station or a house, the CPM schedule may be completed with fewer steps.

This chapter explains a structured approach to the development of a CPM schedule. While other approaches may yield equally accurate CPMs, this approach has been tried and tested by many professional schedulers. They have found that this approach anticipates most of the potential problems which can arise during the development of a CPM and that it works equally well for I-J and precedence.

Development of the CPM can be divided into six phases:

Phase 1: Understanding the project
Phase 2: Conceptual approach definition
Phase 3: Physical creation of the schedule
Phase 4: Computerization
Phase 5: Refinement
Phase 6: Production

While these phases can overlap, it is best to attempt to complete each phase before moving on to the next one. Each phase of development becomes the basis for the steps performed in the subsequent phase. While the individual steps within each phase can be performed out of sequence, the phasing is important.

The project used to illustrate schedule preparation is assumed to have been bid competitively and awarded to a single general contractor. The standard architect, owner, general contractor, and subcontractor relationships will also be assumed. Adjustments can be made to the step-by-step procedure when different parties are involved.

The contractor's project manager typically is responsible for developing the project schedule. Because of all the other things that must also be completed at the start of a project, the project manager usually does not have adequate time to prepare a detailed CPM project schedule. If a busy project manager attempts to complete the schedule without help, an inadequate or erroneous schedule may result. Schedule quality is improved if it is completed by an individual who can commit sufficient time and care to its development. Schedule quality is also improved if the scheduler has experience in the preparation of schedules.

The schedule is one of the most important documents that will be created during the project. It is crucial for maintaining progress and productivity, and for documenting the contractor's approach to, and progress on, the project should claims or disputes arise. If it is executed properly, the CPM schedule will become a valuable asset for the contractor and other parties involved in resolving disputes. By facilitating the resolution of problems, the schedule may also further facilitate progress on the project. Sufficient time must be dedicated to the schedule's preparation to ensure a quality schedule.

PHASE 1: UNDERSTANDING THE PROJECT

The first phase of master project schedule development is to understand the project and what is physically required to complete it. This is performed through four steps:

Step 1: Study the scheduling specifications.
Step 2: Study the contract.
Step 3: Study the construction drawings.
Step 4: Study the specifications.

The scheduler must study and understand the project at least as well as those who will be responsible for building the project or the scheduler will lose credibility and, more important, the cooperation of the superintendent, estimator, project manager, architect, and owner. The scheduler cannot expect to give sound scheduling advice unless he or she fully understands the project.

Step 1: Study the Scheduling Specifications

The scheduler needs to understand the scheduling specifications. Do the specifications require I-J, AON, or precedence format? Do the scheduling specifications require a certain level of detail? Are activity durations to be between 3 and 20 work days? Must the durations be expressed in calendar days or work days? Must the logic be drafted, or are plotted schedules sufficient? Do the plots have to be time-scaled? Are summary bar charts or any other documents required to be produced with the project schedule? Is an interim schedule required? How long does the scheduler have before the CPM must be submitted for approval?

Do the scheduling specifications require a specific format for the drafting and/or plotting of logic diagrams? The Veterans Administration, for instance, has very detailed requirements on drafting the logic diagram. They include the positioning of durations, activity descriptions, manpower, dollar loading, area codes, and responsibility codes. The Veterans Administration also requires a very specific method of numbering the logic diagrams. The scheduler must determine what formatting requirements exist, because this will determine how the schedule will be produced.

Finally, the scheduler needs to review the scheduling specifications for any requirements which may affect the software selected and/or the procedure and type of CPM selected. Failure to perform any of these steps carefully can lead to substantial rework, usually when insufficient time is available.

The scheduling specifications will usually define when the schedule must be submitted and whether or not an interim schedule is required. The interim schedule is a schedule of the first portion of the project schedule (usually the first 90 to 120 days) and is required to be submitted shortly after notice to proceed (normally within 30 days). Most scheduling specifications permit the scheduler to develop, refine, and submit the complete project schedule between 30 and 90 days after the interim schedule has been completed.

If an interim schedule is required, the preparation procedure should be followed through the development of the individual subnetworks described later in Phase 3, Step 4. The interim schedule uses summary activities such as "Complete Structure" to indicate work which will not occur during the 120 days of the project. The interim schedule is used to manage the project during the time the complete project schedule is developed. Even though it is intended to be used only for the first 120 days, it is computerized, logic diagrams are prepared, and it is submitted for approval.

Step 2: Study the Contract

The scheduler then needs to review the construction contract. The contract will usually indicate a project duration either in calendar or work days. In some cases, the contract may indicate various intermediate completion dates or other project milestones that are required to satisfy various owner needs. The project contract may also indicate phasing requirements that limit the amount of work which can be done at one time. Indications of crucial completion dates for utilities may be listed within the contract. The contract may list or identify owner-furnished equipment and its scheduled arrival date on the project. The contract may also define various responsibilities and time for architect- and owner-performed functions, such as shop-drawing review and approval, material testing, or on-site inspection. All of this information is important to the project scheduler and needs to be considered in the development of the schedule. If this information is not identified in the contract, the scheduler must either request the missing information or include reasonable assumptions and await revision during the schedule review process.

Step 3: Study the Construction Drawings

The next step in understanding the project is to study the construction drawings. A quick review of each sheet of the construction drawings is helpful to give the scheduler a general concept of what is involved in the project. After the quick review, more detailed review of each drawing is necessary so that the

FIGURE 6-1   (a) Parapet wall construction.

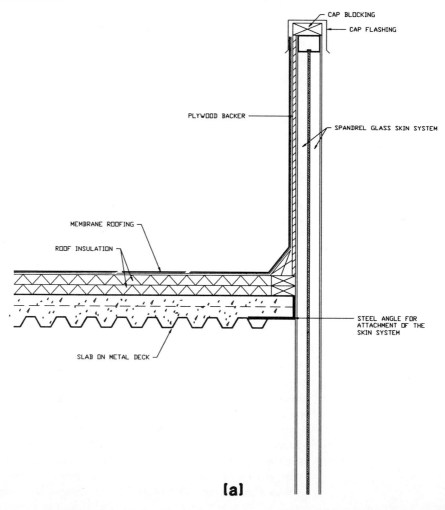

[a]

FIGURE 6-1   Continued (b) Improved parapet wall construction.

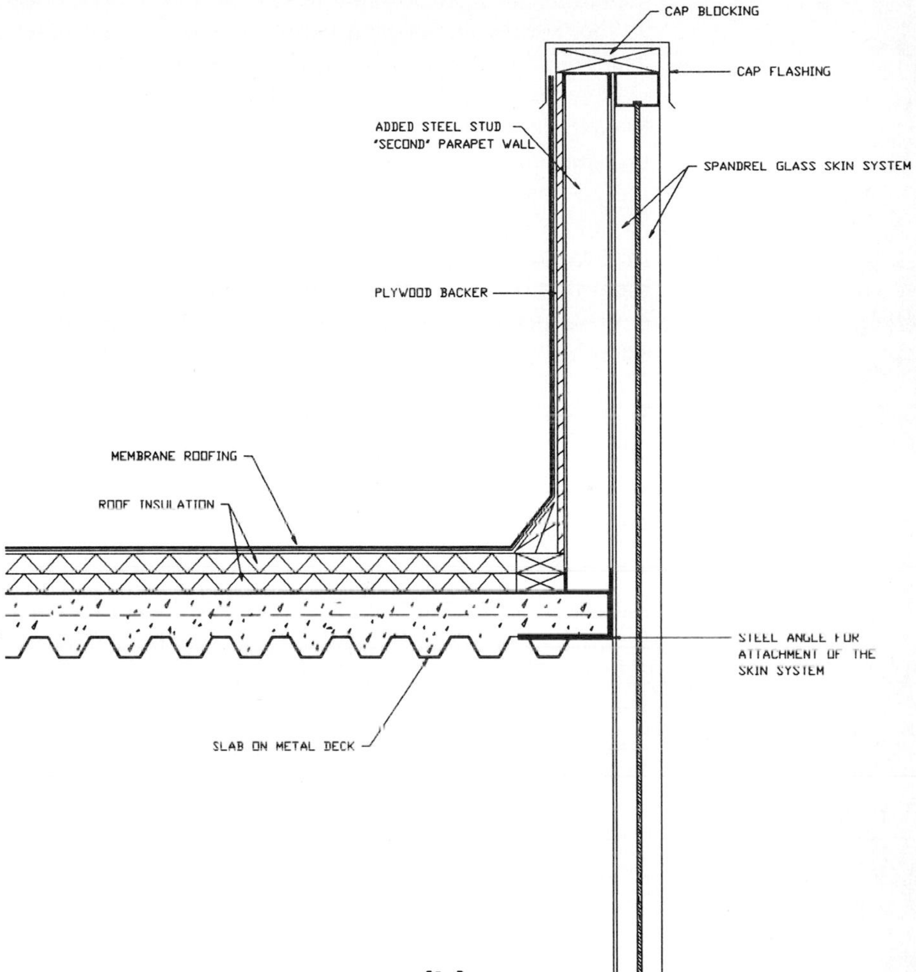

CAP BLOCKING

CAP FLASHING

ADDED STEEL STUD
'SECOND' PARAPET WALL

SPANDREL GLASS SKIN SYSTEM

PLYWOOD BACKER

MEMBRANE ROOFING

ROOF INSULATION

STEEL ANGLE FOR
ATTACHMENT OF THE
SKIN SYSTEM

SLAB ON METAL DECK

(b)

scheduler can gain a thorough knowledge of the various parts of the project. It is during study of the drawings that the scheduler begins to learn about the physical requirements of the various parts of the building so the construction schedule will conform to the physical dependencies shown in the details.

Schedulers should spend sufficient time in drawing study to make suggestions on alternative erection sequences for the various parts of the structure. Plans and wall sections usually provide the scheduler with the most information about the construction sequence. The scheduler should also be alert during study of the drawings to possible adjustments in the details of the wall sections and other parts of the project. Recommendations for revised details could result in more efficient construction and very possibly a better building. For example,

Figures 6-1a and 6-1b show a roof detail as originally presented and how a scheduler's suggestion modified it to improve erection.

Figure 6-1a shows the original detail for the parapet wall on an office building. The roof blocking and plywood parapet backing are attached directly to the skin system framing. This detail requires that the exterior glazing system must be installed before the roof system can be completed. Leaving the roof system in a partially completed state may allow moisture to migrate under the completed portion, causing early deterioration. The uncompleted roof will be subjected to unnecessary abuse while the skin system subcontractor completes that work. Also, the roofing subcontractor may have to remobilize if a continuous flow of work cannot be maintained. When confronted with this detail, contractors will typically postpone any roofing work until the skin system is complete. This can delay the watertight enclosure of the structure and subsequently the completion of the building.

Figure 6-1b shows the addition of a simple "double parapet wall." This allows the roofer to totally complete his portion of the work. Normally tarps are laid over the parapet wall to prevent water infiltration. Once the skin system framing and spandrel glass has been installed, the last blocking and flashing can be completed. This finishing work requires no special equipment and can be performed by one worker. This detail adjustment "breaks" the relationship between roofing and skin system, thus speeding construction.

During review of the construction drawings, the scheduler should continually ask questions related to the sequence of construction:

How much site work needs to be done before the foundation work can start?
How do the utilities interrelate with the foundation work?
How do the underslab utilities affect or interrelate with the slab-on-grade?
How does the structure interface with the foundation?
How do the upper floor slabs interface with the skin system of the building?
Does the skin system have to be completed before or after the roof system?
How do the windows interface with the remainder of the skin system?
Are there any major pieces of equipment which need to be placed before slabs are poured or shell is erected?
Are there difficult interfaces with elevators and structural systems?
How does the masonry interrelate with the concrete work and the floor slabs?
Is there fireproofing, and how will it be applied and how will it be protected?
Are there site limitations which would make certain types of equipment and/or scaffolding difficult or necessary?
Do the attachment procedures for the building skin system require erection from the ground up, from the roof down, or from one side to the other?
Are there on-site restrictions, such as streets or railroad spurs?
Where will on-site storage be located?
What type of crane will be used? Is a tower crane a reasonable option?

The number of questions and potential answers is almost limitless.

When the scheduler is not able to answer each and every one of those questions specifically, he or she should be prepared to give various options to the superintendent and project manager for a final decision. The above list of questions is not intended to be all-inclusive; it is only a small selection of the questions a scheduler needs to consider before proceeding to the next phase. As the scheduler develops experience, relevant questions can be compiled to form a checklist for certain types of projects.

Step 4: Study the Specifications

The scheduler should then review the construction specifications. Specifications may include requirements for phasing various portions of the project, limitations for tie-in of critical utilities, or availability of other portions of the project.

The specifications may also restrict certain construction activities with other types of constraints. A requirement for controlled humidity and temperature prior to drywall installation is an example. This requirement changes the normal sequencing for drywall erection and taping from after the enclosure of the structure to after substantial completion of the HVAC system. Another example is the specified waiting time required prior to stripping forms on various types of concrete. Required constraints must be known by the scheduler before development of the project schedule. Failure to discover these types of constraints on activity sequencing can be troublesome if not disastrous later in the scheduling process. A schedule submitted with these kinds of errors may not be approved or, even worse, the scheduling error may only be discovered in the field during construction. In this situation the schedule would cause a delay rather than preventing one.

Sometimes items that one would expect to find in the contract are listed in the specifications instead. For example, the submittal process is often detailed in the specifications. The specifications will usually indicate what types of submittals, samples, or testing will be required. Of specific importance is the review of mockups and approval of samples by the architect prior to purchase. The scheduler needs to be sure to include necessary activities for these items in the project schedule.

PHASE 2: CONCEPTUAL APPROACH DEFINITION

During Phase 2 of project schedule development the scheduler will learn the planned sequences intended by the various managing parties on the project. During this phase the scheduler's ''people'' skills become a crucial part of the development process. The scheduler will need to exercise humility and salesmanship in order to develop a project schedule that not only represents the intended construction sequence, but also ensures that the intended sequence is both efficient and safe. The scheduler must exercise care when recommending various construction approaches so as not to alienate the managing parties who will control both project completion and the project schedule.

The steps involved in this phase of project schedule development are:

Step 1: Discussion with the project estimator
Step 2: Discussion with the project manager
Step 3: Meeting with the superintendent and subcontractors
Step 4: Meeting with the owner and the architect

During this phase the scheduler collects ideas and concepts from a wide variety of people. This information is crucial to developing the right schedule for the project.

One main principle to be communicated from the scheduler to the other managing parties is that the schedule developed is the schedule of the superintendent, the subcontractors, the project manager, the owner, and the architect. The scheduler must emphasize that the scheduler's involvement is merely to coordinate and integrate all of those interests into one coherent schedule. If the scheduler develops a schedulde without considering the intentions of the superintendent, project manager, and others involved in the project, the schedule may not be used by those key project people. A schedule not used is worse than no schedule at all.

Step 1: Discussion with the Project Estimator

The first individual with whom the scheduler should discuss the project is the project estimator. It is important that the scheduler discuss the project with the estimator before the estimator becomes heavily involved in another esfimate. Early discussions are important so that the estimator will be able to recall assumptions. Some typical items that may be discussed are:

How much winter protection has been allowed?
Is concrete to be poured through the winter?
What types of cranes or other equipment are included in the estimate?
What specific commitments and guarantees have been made by the low-bid subcontractors?
What limitations on construction have been included by the low subcontractor bidders on the project?
Was the project overhead bid for less than the contract duration?
Have any decisions been made about how the project will be built?
Were any additional indications given by the owner or the architect during prebid meetings about how the project should be executed?
Is there any other information which would be valuable to the scheduler in developing an accurate CPM schedule?
Did the contractor make any representations as to its scheduling ability or performance?
Was the method or approach to the project discussed with the owner?

The scheduler should interview the estimator first so that anticipated financial and construction limitations on the project are properly included in discussions with other parties.

The scheduler should obtain a copy of the project estimate if one is available. The contractor's method of bid estimating will determine the detail level of the material quantities. Later in the scheduling process the scheduler can refer to the estimate to determine quantities of work. For example, a scheduler will not normally do an accurate take-off of concrete quantities when scheduling (and determining durations for) a foundation installation. Access to accurate take-off information can increase the accuracy of the durations given the foundation activities. If the estimate also provides anticipated manhours for finite portions of work, the scheduler can use this information in determining manpower loading information and anticipated productivity.

The scheduler should, however, not rely exclusively on the estimate no matter how well it has been put together. The scheduler should still thoroughly review the plans and specifications and determine appropriate project sequencing and durations. For example, a concrete slab and a concrete foundation wall may have similar concrete cubic yard totals and similar manhour totals, but the duration for performance of these activities may be dramatically different. While a detailed estimate will improve and speed the CPM schedule process, it does not replace the need for informed construction sequencing and duration decisions.

Even the best planning may still result in congested work areas during some stages of construction.

Step 2: Discussion with the Project Manager

The scheduler should then discuss the project with the project manager. This discussion must identify any obstacles which the project manager may foresee in completing the project. Questions might include:

Is there difficulty in securing a subcontract with any subcontractor? If so, this may indicate that the scheduler should allow additional time for the procurement of that subcontractor's materials and equipment.

The project manager may have decided to use a single tower crane instead of two tower cranes as estimated. This could change the erection sequence and slow the process slightly, but be more cost-efficient.

Does the project manager have any new insights into the best method for constructing this project?

Is there any special information about how the contractor will allocate equipment and materials to this project?

Does the contractor have sufficient concrete form materials to form the entire project?

Are there other projects that will require form work materials at the same time as this one?

The scheduler should also determine if there is a desire to finish the project in less than the contract time.

Has notice to proceed been received, and if so, when? A specific date is required for entry into the schedule. If a notice to proceed has not been received, an anticipated date should be used until an actual date can be provided.

At this time some preliminary discussion with the project manager concerning work sequencing is in order. These discussions should be very preliminary and should only be considered as options. The final determinations on construction sequencing will normally be made by the superintendent.

Step 3: Meeting with the Superintendent and Subcontractors

The scheduler must exercise great care when meeting with the superintendent. The scheduler must not appear to be dictating how the project should be built. While the superintendent should be open to advice and possible alternatives from the scheduler, the scheduler must make the superintendent believe that the superintendent has prepared the schedule. The scheduler should never insist on one approach to constructing the project. The scheduler may think that the superintendent's approach is not as good, or even impossible, but he or she still should not insist on another approach. After the meeting, the scheduler should meet again with the project manager and discuss the meeting with the superintendent. (The project manager may wish to be involved in the superintendent and subcontractor meetings, thus eliminating this second project manager meeting.) At no point should the scheduler risk alienating the superintendent. The superintendent must have complete confidence that the scheduler will produce the superintendent's schedule for the project. Without the confidence the superintendent will refuse to look at or use the schedule, making it worthless as a field management tool.

Traditionally, construction project superintendents have advanced through the trades to control on-site project work. These people have extensive experience in the execution of construction contracts. They may appear stubborn but actually possess a necessary leadership quality: conviction. Their conviction comes from successful experience on past projects in the face of uncertainty and adversity. It is only natural that such people distrust the "educated scheduler-type" trying to dictate how the project should be built without benefit of the same experience in the workplace. Good people skills are important when meeting with the superintendent.

It is best to meet with the project superintendent before meeting with the other subcontractors. This permits the scheduler to review in detail the process of construction as intended by the superintendent without being scrutinized by the various subcontractors.

The scheduler must be prepared to ask a tremendous number of questions to understand fully how the superintendent intends to construct the project. These questions must include every portion of the project from the footings all the way through final finishes. Examples of questions to ask include:

Which way will the form work move?
Will the foundations move from north to south or from east to west?
Will the footings be poured in short or long sections?
How many days are anticipated to do each of these jobs?
Will the foundation wall work follow directly after the footings, or will the footings be totally complete before the foundation walls are started?
At what point will the slab-on-grade be installed? Will it be done before the roof is put on, or immediately after underground utilities are completed?
In what sequence will the exterior skin be installed?
How will the interior finishes relate to that exterior sequence?
Will the project proceed without weathertight enclosure, or will total weathertight enclosure be accomplished prior to drywall installation?
How many square feet of slab-on-deck can be poured per week?
Will the fireproofing be installed before or after enclosure?
Will the subcontractors be expected to work concurrently or consecutively?
In what time frame and sequence will subcontractor work be performed?

Regardless of how many questions the scheduler asks, there will always be several important questions missed. During development of the subnetworks it will be necesary for the scheduler to contact the superintendent again, usually by phone, to work out additional minor details. This is normal to the process and helps the superintendent believe that the scheduler is truly developing the superintendent's schedule.

After the subcontractors join the meeting, the scheduler should review the general project sequence discussed earlier with the superintendent. The superintendent and scheduler should work together to indicate to the subcontractors what they anticipate the subcontractors' roles to be in the schedule development. The scheduler also needs the cooperation of the subcontractors. This cooperation is developed much the same as it is with the superintendent: The subcontractors are asked how they intend to complete their work. Asking the subcontractors how they intend to construct the project is necessary to maintain subcontractor commitment to work within the completed schedule's sequence and duration. As long as the subcontractors' approach conforms to the rest of the project schedule, the subcontractors' approach should be used. Only if the subcontractors' sequence or duration does not meet the requirements of the project schedule should adjustments be discussed.

At the subcontractor meeting it may be difficult to determine whether or not a specific subcontractor's approach will fit within the project schedule. For example, the plumbing subcontractor may demand 30 days to install 15 fixtures. Even though the scheduler believes that this is not a reasonable duration, the scheduler cannot determine at the subcontractor meeting whether or not the plumber's 30 days will unreasonably affect the rest of the project schedule. It is important that the scheduler use the subcontractor's durations if at all possible. If it is later determined that the subcontractor's intentions cannot all be incorporated, the scheduler should contact the subcontractor and attempt to work out a mutually agreeable alternative. Normally the project manager and/or the superintendent should also be present at any discussions to resolve subcontractor scheduling conflicts.

Once all the necessary information has been collected, the scheduler should spend a short period of time reviewing sample printouts and logic diagrams from another project with the subcontractors. Time spent showing the subcontractors and superintendent how to read the printouts will make them more comfortable with the project schedule. Extensive instruction in the CPM technique is not necessary. At this point, the scheduler merely wants to develop familiarity with what the schedule will look like and how it can be read.

Schedulers should also spend time explaining the benefits of CPM scheduling to the superintendent and subcontractors. The scheduler should explain to the subcontractors how it will assist them in resource leveling and produce a more efficient construction project. This should result in more profit for all contractors. This type of discussion early in the project helps foster a team approach to executing the contract.

The meeting with the subcontractors can be difficult. Subcontractors tend not to want to commit to any sequence or duration for their work and, if forced, normally ask for substantially longer than they really need. The scheduler needs people skills to persuade the subcontractors to agree to reasonable durations and sequences. On most projects the scheduler will find one or two subcontractors who will merely say, "I will keep up with you." This is a commitment that they will meet whatever schedule is produced, because their work is minor or they feel they can match any speed of the general contractor. Usually, however, there will be one or two subcontractors who feel that commitments to a sequence and duration in a CPM schedule have been used to their disadvantage in the past. Too many schedulers indiscriminately adjust subcontractor sequences or durations as the project proceeds, compressing the subcontractors' work into shorter and shorter time frames, without the subcontractors' approval. Subcontractors who are hesitant to commit to a sequence or duration may have encountered such a scheduler. It is important for these subcontractors to understand that the schedule will reflect their durations, and that no adjustments will be made during the project unless they agree to them. On occasion a scheduler will run into a subcontractor who resists providing any information for the CPM. Producing an accurate original CPM schedule despite uncooperative people will be discussed later.

Subcontractors and the superintendent must feel confident that the schedule will be developed in everyone's best interest (not compromising their bidding position) and that the schedule will be used and implemented fairly throughout the project. It is the responsibility of the scheduler to establish this attitude. On occasion, a scheduler will encounter people who just refuse to work with a CPM. This special challenge requires unique communication and persuasive skills beyond the scope of this text.

Step 4: Meeting with the Owner and the Architect

If at all possible, the scheduler should meet or discuss the schedule with the architect and the owner prior to completing the schedule. The purpose of this contact is both to determine what the owner and the architect want from the CPM schedule and to gather additional schedule data. Involving the owner and the architect in the scheduling process helps them understand how the contractor will proceed on the project and helps them adjust their activities to support that procedure.

If the specifications and contract do not specify a return time for shop drawings, this time frame should be requested from the architect. This is crucial to the procurement process. It is a good idea to show the architect a sample printout, identify the procurement portion, and explain what activities in the submittal/approval process are critical. This will assist the architect in determining review priority among shop drawings as they are submitted. Prompt processing of shop drawings is always an asset to timely project completion. Using the CPM to encourage the architect to help the contractor with timely information and approvals is good scheduling practice and good business sense.

It is wise to discuss the project schedule with the owner to identify any additional sequencing requests, phasing, or restrictions on construction. This develops an important level of cooperation beyond what is normally experienced between the contractor and the owner. The schedule should be modified if there is any reasonable way to incorporate these additional owner desires. These modifications will go far in providing cooperative relationships on the project.

Review with the owner any required delivery dates for owner-furnished equipment or material and any other owner responsibilities which will affect the contractor's construction progress. A short time spent showing the owner how to read the CPM schedule will help the owner interpret the data. Slight modifications in the format or layout of the CPM may make the schedule more compatible with the owner's and the architect's needs. Techniques in presenting the CPM data to assist the owner in reading the schedule are discussed later in this chapter.

At this point the scheduler should have a fairly complete idea of how the project will be organized and sequenced. While every possible sequencing question may not have been asked or every duration determined, the vast majority of major relationships should have been defined. The scheduler is almost ready to put pencil to paper and start developing the actual logic.

PHASE 3: PHYSICAL CREATION OF THE SCHEDULE

Physical creation of the original project schedule is divided into eight steps:

Step 1: Select software.
Step 2: Divide the project into subnetworks.
Step 3: Develop responsibility codes.
Step 4: Develop information codes.
Step 5: Develop specific subnetworks.
Step 6: Draft or plot the logic diagram.
Step 7: Number the activities.
Step 8: Tie the subnetworks together.

These steps constitute the majority of work required to develop the original project schedule. Successful completion of these steps requires an individual who understands both network scheduling and construction methods and procedures. These steps also require concentration. They are best accomplished in as few uninterrupted periods as possible. Piecemeal development may result in a fragmented, incoherent schedule.

Step 1: Select Software

At this point the scheduler must select the computer software. The scheduler should select the most appropriate software to meet the specification requirements, the needs of the contractor's personnel, and the desires of the architect and owner. If the scheduler does not have the flexibility to select among several appropriate programs, the scheduler needs to be sure that the available software will meet the requirements of the scheduling specifications. If available software does not meet the specifications, approval is necessary for the variation from the owner or architect, or the appropriate software must be acquired.

Most package scheduling software will produce a quality schedule if the scheduler understands its limitations and abilities. The scheduling style should be adjusted to fit the software. If the scheduler is unfamiliar with the intended software, some time should be spent learning its abilities and peculiarities. Running a small "test" schedule will inform the scheduler about most of the software's abilities.

Step 2: Divide Project into Subnetworks

For the vast majority of projects, it is difficult, if not impossible, to schedule the entire project, from beginning to end, in one continuous thought process. Instead, schedulers subdivide the project into units called "subnetworks." Each subnetwork covers a group of work activities related to a specific function or part of the project. For example, the enclosure subnetwork includes brickwork, glass, roofing, parapet, flashing, exterior doors, and exterior trim. Dividing the project into subnetworks allows the scheduler to concentrate on the important relationships among activities and avoids confusion with other parts of the project. Division of the project into subnetworks also provides natural breaking points for the plotting of large schedules in a usable format.

The following is a list of typical subnetworks for a small, two-story hospital addition:

Project start
Project completion
Procurement
Site work and utilities
Excavation and foundation
Superstructure
Roofing and enclosure
Rough-in and finish first floor
Rough-in and finish second floor
Mechanical room equipment
Rooftop HVAC and accessories
Hydraulic elevators
Medical room finish and equipment
Tie-in and renovation at existing hospital first floor
Tie-in and renovation at existing hospital second floor

While subnetworks may vary among schedulers, the essence of the subnetworks will be in everyone's schedule. If the project will be complex and difficult to sequence, it is best to divide the project into more subnetworks. This will facilitate modification and adjustment of the interrelationships should this be necessary. Possible further divisions of the typical hospital subnetworks could include separating Excavation and Foundation, and Roofing and Enclosure. In a very complex or difficult schedule, one may also separate the rough-in from the finish on the individual floors, or divide the floors into north or south or east and west according to how the interior finishes will be executed.

At this point, the scheduler should number the subnetworks sequentially in the order they will most likely be performed in the field. Project Start will be number 1, and Project Completion will be last. Procurement could be number 2, and Site Work number 3, and so on through the logical progression of construction. Site work and utilities work may in fact start very early in the project but not be totally complete until very near the end of the project. Sequencing the site work subnetwork early in the schedule is correct and a reasonable scheduling practice. If the schedule is planned to be plotted in time-scaled format, it is advisable to break long-duration subnetworks, such as site work, into two or more subnetworks. This will improve the readability of the plots.

Subnetworks will vary from project to project. A typical listing of subnetworks for the construction of a new wastewater-treatment plant facility might be as follows:

Project start
Procurement
Grit building
West primary tank
Nitrification facilities
Aeration control building
Nitrification tanks
Blower building
Final tanks
Effluent building
Garage
Storage building
Maintenance shop demolition and renovation
Service building modifications
Blower building modifications
Miscellaneous improvements to distribution structure
Junction chambers
Existing effluent storage tank modifications
Remodeling of existing office facilities
New employee parking area
Miscellaneous site work
New effluent meter vault
New propane storage tank
New site piping
Demolish existing sludge beds
Demolish existing administration building
Demolish existing final tanks
Demolish existing trickle filter
Demolish existing grit tanks
Demolish existing primary tanks
Project completion

While subnetworks may vary among wastewater-treatment projects according to the equipment and modifications required, each structure is treated as an independent subnetwork.

The scheduler may wish to further divide each of the wastewater project's structures into two subnetworks, one for the structural work—such as foundation, concrete, steel, and roof enclosure—and one for the mechanical and electrical equipment and systems. On many wastewater-treatment plant projects the structure and enclosure work is performed by the general contractor and the mechanical and electrical work by subcontractors. Dividing each structure into two subnetworks permits a continuous flow of general contractor work from building to building while also permitting the same flow of mechanical and electrical work from building to building.

Before sequencing the project, the scheduler should discuss the operating requirements of the wastewater-treatment plant with the owner/operator. Parts of the plant may have to remain operational while other parts are demolished and replaced. Normally the specifications will require that the contractor limit interruptions to the daily operations of the wastewater-treatment facility. Although a good specification will provide the phasing required to construct the new plant with minimal disruption to the existing operation, the owner/operator may provide more detail.

The main purpose of dividing any project into subnetworks is to reduce the amount of work to be interrelated by the scheduler at any one time. Subnetworks provide three advantages. First, breakdown of the project into subnetworks facilitates accurate, detailed scheduling of a group of related activities. Second, it simplifies the interrelation of all the various work activities into the entire project schedule. Relating 10 subnetworks of 200 activities each is easier than relating 2000 separate activities simultaneously. Third, while the sequence of events may be firmly defined at the outset of the project and the relationships among the subnetworks may be positively known, changes on the project site, unpredictable weather conditions, or other unknowns may require modification of the project schedule. Developing the project in subnetworks greatly eases the process of rescheduling whenever it may occur.

The development of a subnetwork normally requires between 30 minutes and 4 hours. Once a subnetwork is fully developed, the scheduler can temporarily set aside that work and concentrate on another subnetwork. If the entire project cannot be scheduled in a continuous effort, subnetworks allow coherent points at which breaks to complete other work may be taken.

Step 3: Develop Responsibility Codes

Before the scheduler actually places a pencil to paper, it is best to develop a list of abbreviated responsibility codes. This is no more than a list of codes used to identify who is responsible for individual work activities. The responsibility list in Table 6-1 shows both two-digit and four-digit abbreviations for various parties on the construction site. Typically, off-site suppliers and fab-

TABLE 6-1   TYPICAL RESPONSIBILITY CODES

Two-digit code	Description	Four-digit code
GC	General contractor	GCON
EL	Electrical contractor	ELEC
EV	Elevator contractor	ELEV
ME	Mechanical contractor	MECH
HV	HVAC contractor	HVAC
LS	Landscape contractor	LAND
UT	Utilities contractor	UTIL
EX	Excavation contractor	EXCV
FD	Foundation contractor	FOUN
FL	Flatwork contractor	FLAT
FM	Concrete formwork	FORM
CR	Carpentry contractor	CARP
ML	Millwork contractor	MILL
WD	Wood doors and frames contractor	WDDR
MD	Metal doors and frames contractor	MTDR
SF	Storefront contractor	STFT
MM	Miscellaneous metals contractor	MISC
FR	Flooring contractor	FLOR
CP	Carpeting contractor	CRPT
DW	Drywall contractor	DRYW
ST	Studding contractor	STUD
CK	Caulk and sealants contractor	SEAL
FP	Fireproofing contractor	FPRF
FS	Fire sprinkler contractor	FSPR
PL	Plumbing contractor	PLUM
AC	Acoustical contractor	ACOU
MS	Masonry and brick contractor	MASN
PT	Painting contractor	PAIN
TB	Test and balance contractor	TEST
GZ	Glazing and skin system contractor	GLAZ
HC	HVAC controls contractor	HVCT
SS	Structural steel contractor	STRU
RB	Reinforcing steel contractor	REBR
SP	Swimming pool contractor	SWIM
PV	Paving contractor	PAVG
SP	Lawn sprinkler contractor	LSPR
VW	Vinyl wallcovering contractor	VWCV
CF	Computer floor contractor	COMP
OW	Owner	OWNR
AE	Architect	ARCH
XX	Milestones	XXXX

Mobilization and site work mark the beginning of on-site construction activities. (Photo courtesy of John Deere.)

ricators are not coded but are listed under the responsible party who will be on-site. Whether one uses two- or four-digit responsibility codes is up to the scheduler and the capabilities of the software selected. Table 6-1 includes most of the potential contractors on a project site. While all contractors listed will not be encountered on all projects, the table is a ready reference of commonly used abbreviations in network diagrams.

Table 6-1 also includes abbreviations for project milestone activities. The use of double X's or four X's as the responsibility code for the milestones facilitates analysis of original and updated CPM schedule printouts. When sorting the printouts by responsibility code, the milestones will normally appear on the last page. This permits anyone to turn to the last page of the printout and quickly find out how well the project is doing. Updating a network schedule is discussed in detail in Chapter 8.

Step 4: Develop Information Codes

It may also be necessary to add various information codes to the activities in the CPM network. Information codes can facilitate analysis of the schedule and increase the information which can be derived from the schedule. Coding activities can be as simple as the numbering of the subnetworks. One may also use alphabetical codes for the various subnetworks.

The most common information codes are building location identifiers. Another common code is to identify the value of labor and material for each activity. Veterans Administration and Corps of Engineers projects typically require "cost or dollar loading." Cost loading is the assignment of dollar values to every work activity. This code can be used for cash flow projections and for monthly draw requests. An activity may be required to have only one value or be required to list labor and material cost information separately for each activity. This subject is discussed further in Chapter 11. The value of each activity can be shown on both the logic diagram and the printout.

Another common information code is craft requirements. If an activity is "man-loaded," each activity is assigned the number and type of craftspeople it will take to perform that activity. As with other information codes, manpower can be shown on both the printout and the logic diagram.

Other possible codes include accounting codes for categories of cost, equipment codes, and codes for the contractor's or subcontractors' productivity analysis.

By defining the various codes before drafting the logic, the scheduler can avoid combining activities which should be separated in the code sorts. For example, a scheduler may use one activity for curbs and sidewalks on a site-work subnetwork. However, if the bid separates curbs and sidewalks into separate bid items and the specifications require separate recording of their costs, separate activities are necessary. Anticipation of the required codes prevents the scheduler from having to rework the schedule later.

The number of possible codes for an activity is almost limitless. However, some scheduling software packages limit the number of information codes that can be put on each activity. Some permit only two, while others permit substantially more codes per activity. The scheduler needs to be aware of software capabilities before using any codes.

Step 5: Develop Specific Subnetworks

Preparation   Developing specific subnetworks is the next step in the process of preparing the master project schedule. Before actually drawing the logic diagram, the scheduler should again review the project drawings which relate to that specific subnetwork. A quick check of the specifications relating to the work in that part of the project is also helpful in developing an accurate subnetwork. Normally the subnetworks are developed in the rough chronological order that they will be performed on the job site. Developing subnets out of order is possible, but does not yield the coherent logic which more easily flows from an orderly approach.

The scheduler is now ready to start developing the master project schedule. Some schedulers prefer to use large sheets of paper, others prefer to use smaller 18-inch × 24-inch paper. The paper should be transparent, so that copies of the logic subnets can be made. The rough subnetwork diagrams the scheduler prepares are commonly called "scratchies."

Once the subnetworks have been selected, the individual activities to be included within the subnetworks are identified. As discussed in Chapter 3, the scheduler first prepares a list of the work activities to be included. Once the activities have been identified, they are arranged in the order the activities will be performed. Experienced schedulers generate the activities, assign durations and responsibility codes, and draw the subnetworks in one continuous process. This ability is gained with experience.

The following sections outline what should be included in typical subnetworks for a three-story office building schedule.

Project Start A project start subnetwork should be included in every CPM schedule. This should include an activity for notice to proceed, contractor mobilization, and other activities necessary to get a construction project started, such as mobilization of subcontractors, hookup of utilities, project survey and layout, bond and insurance, and possibly an activity for the architect's supervision on the job site. These activities normally initiate a project and provide a point from which to tie in the succeeding subnetworks. An activity in the project start subnetwork for the general contractor's project supervision with a duration to match the project duration can be used immediately to tell the scheduler and other users the number of days remaining in the project and the percentage of time consumed to date.

If the project schedule is cost-loaded, a project start subnetwork allows the contractor to place a value on mobilization and supervision.

Normally, the project start subnetwork is numbered with single- and two-digit numbers. This permits the project start subnetwork to appear first in a computer-generated numeric sort printout. While this numbering system does not have to be followed, it has been found to be an effective communication method. Figure 6-2a is a sample project start subnetwork in I-J format; Figure 6-2b is a sample project start subnetwork in PDM format.

On some projects it is advisable to list owner release dates in the project start subnetwork. For example, if the clarifier in a wastewater-treatment plant will not be available for demolition until spring of the following year, the scheduler may use an imposed start date for the start of the clarifier activity, indicating the owner release date. Imposed starts and finishes are a function available in most software programs. Essentially, the imposed date for any activity may be placed into the computer's database. The software then uses the imposed date in the calculation process. Most software programs indicate imposed dates on the printout in some manner, sometimes with an asterisk or an "I."

The scheduler, however, has a better way to indicate an imposed date. The scheduler should use an activity to indicate, and explain, the purpose of the imposed date. The activity's duration is the appropriate time to the start date of the following activity. For our example, an activity with the description "clarifier available for demolition 3/15/92" with the appropriate duration to 3/15/92 and a responsibility code of "OW" would fulfill the scheduler's needs. This specifically identifies what restrains the start of the clarifier work. If the scheduler elects to use an imposed start date instead of an activity, the diagram or the printout will not indicate the reason for the imposed start date. The

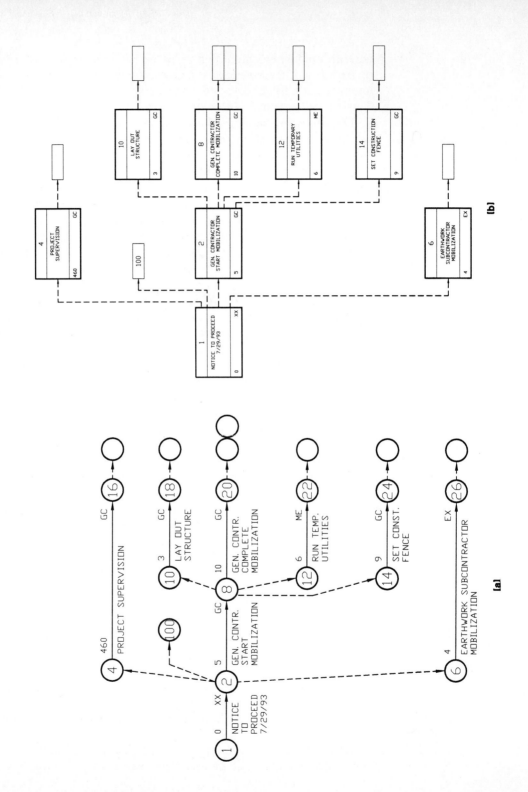

[a]

[b]

146

essence of scheduling is communication. If the scheduler wishes to communicate what is actually occurring, an activity is a much better way to show an imposed start date.

Project Completion   The next subnetwork drawn is the project completion subnet. Even though this subnet is chronologically the last subnet in the sequence, it is usually drawn second to define the end point of the project. The project completion subnetwork is normally numbered with 9000- or 9900-series numbers. It includes activities such as final project punchlist and cleanup, contractor demobilization, a milestone activity called project complete, and possibly owner occupancy and other similar activities. This subnetwork has the only open end recommended on the diagram, the last activity in this subnetwork. Normally, the last activity has a flag placed on it indicating the project completion date. Figures 6-3a and 6-3b are typical project completion subnetworks in I-J and PDM formats.

The final activity normally also has an imposed finish date. This date is placed into the computer software and matches the contract completion date. Contract completion and any other interim contact completion dates are the only imposed finish dates that should be used in the CPM schedule. Use of imposed finish dates which are not contractual can create multiple critical paths that are not related to project completion. Multiple critical paths unrelated to the contract completion dates can cause confusion should the contractor seek time extensions for delays. The critical paths caused by imposed dates may be more critical than the contract completion critical path. Without detailed explanations, a designer or owner may reject requests for time extensions. Alternate critical paths from imposed finish dates within the project may also distort the true critical path from which delays should be measured.

Should the contractor wish to schedule the project for a duration shorter than the project contract duration, a supplemental activity should be added to the project completion subnetwork. This supplemental activity should indicate "contractor's time to complete." Its duration should be the difference between the contractor's intended completion date and the contractual completion date. The "contractor's time to complete" activity should indicate the contractor's intended project completion date with a flag, but no imposed finish date. The activity for "contractor's time to complete" is placed after all work items such as demobilization and final project cleanup and before the contractual completion date activity. The contract completion date activity should have an imposed finish date which matches the contract completion date. See Figures 6-4a and 6-4b.

If the contractor fails to make progress as anticipated, the duration on the activity "contractor's time to complete" can be reduced to absorb contractor delays. The purpose of "contractor's time to complete" is to announce the

---

FIGURE 6-2   Project start subnetworks in (a) I-J format and (b) PDM format.

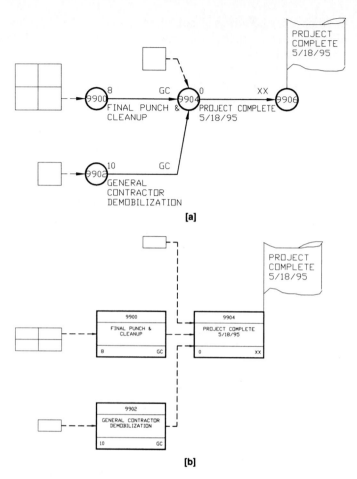

FIGURE 6-3 Project completion subnetwork in (a) I-J format and (b) PDM format.

contractor's early completion but maintain flexibility if the anticipated early completion should turn out to be incorrect.

It is important that the contractor inform the owner very early in the project if the project may be completed early. It is also important that the contractor tell the owner not to plan to use the facility until the contract completion date. Otherwise, the owner may reasonably interpret the contractor's intended early completion date as the time when the facility will be completed. Any failure to meet the early completion date may be subject to liquidated or actual damages. The authors recommend that contractors discuss early completion alternatives with legal counsel prior to issuing the original project schedule.

Procurement The procurement subnetwork is normally drawn as a "ladder," in staff format as shown in Figures 6-5a and 6-5b. The procurement subnetwork is usually easy to complete and is normally prepared early because

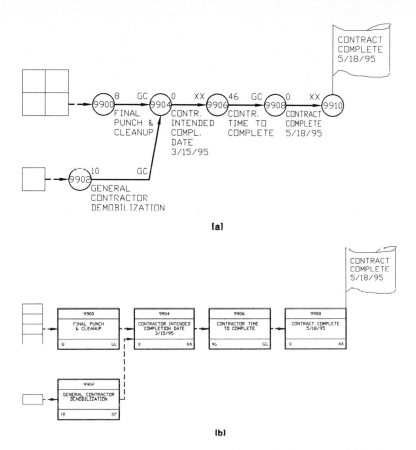

FIGURE 6-4 Project completion subnetwork for projects scheduled to be early in (a) I-J format and (b) PDM format.

of its ease. The procurement subnetwork is normally started with a restraint (or relationship in PDM) from notice to proceed into the first node (or activity) on the procurement ladder. In the first column of the ladder are listed the various items to be submitted to the designer by the contractor. The second column provides time for approval by the designer. The third column is for the fabrication and/or delivery of that item. Note that it is not necessary to rewrite the description under each heading on the scratchies. When the descriptions are put into the software database they must be typed in for each activity.

Do not combine submittal approval and delivery into a single activity. Each of these activities is the responsibility of a different party. Submittal, approval, and delivery must be separated to portray the process correctly and alert the individuals responsible to the time in which their responsibilities must be completed.

A restraint from the fabrication and delivery activity ties to the appropriate place in the logic diagram where that item is to be installed.

There are two ways to make the connection from the fabrication delivery

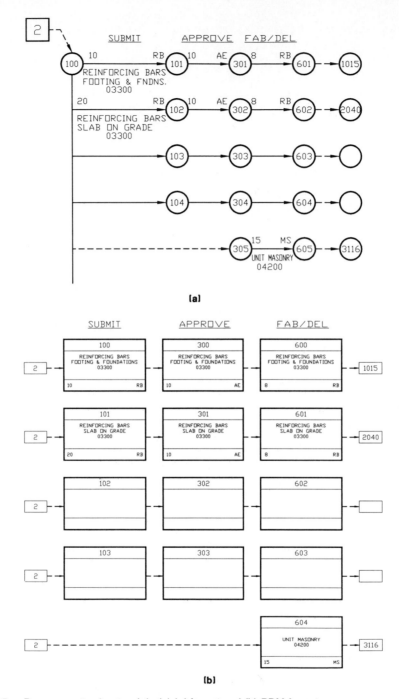

FIGURE 6-5 Procurement subnetwork in (a) I-J format and (b) PDM format.

activity to the main logic diagram if the schedule is prepared in I-J format. One is to make the J node of the fabrication/delivery item the I node of the installation activity. This procedure works well and minimizes the number of restraints when the logic diagram is to be hand-drafted. However, if the logic diagram is to be plotted by computer, it is better to have a separate J node for the fabrication and delivery item and a restraint to the I node of the installation activity. This procedure produces a cleaner, more readable plot and makes subdivision of the plot (if necessary) easier. Since a scheduler cannot anticipate whether the schedule will ever be plotted, it is advisable to stick to the latter drawing technique. A restraint to the I node of the installation activity is shown in Figure 6-5a. If the scheduler is using the precedence diagramming method, the relationship format automatically eliminates any choice.

To identify the activities that belong in the procurement ladder, the scheduler should review each section of the specifications. Most specifications will indicate at the head of each section which items need to be submitted for approval by either shop drawing, certification, or sample. Each individual submittal should be a separate activity on the procurement ladder. Although the scheduler cannot always know in advance how the submittals will be made, he or she should make an educated approximation so that the CPM schedule updates will follow the actual submittal process.

The scheduler also needs to subdivide certain submittals. An example is reinforcing bars. The specifications may require "all" reinforcing steel to be submitted. However, on a concrete building with concrete footings, foundation walls, slabs, and columns, a single delivery of rebar is unlikely. The shop drawings for rebar will most likely be submitted in several packages. The scheduler should anticipate in what packages the rebar shop drawings will be submitted. Typical packages include footings and foundation walls; columns and slab, first floor; columns and slabs, second floor, columns and roof; and miscellaneous rebar. A separate activity for each of these submittals should be listed in the procurement subnetwork. A separate approval and fabrication delivery activity is also necessary. By separating the submittals, the relative priority of each of the various rebar submittals can be determined.

Another example is the fabrication and delivery of wood doors for a large office building. Assume a 38-story office building that requires wood doors on each floor. The scheduler may not wish to divide the submittal, approval, and fabrication and delivery of wood doors into 38 separate groups of activities. However, dividing the doors into groups of five floors can both separate the submittal to determine priority and reduce the number of delivery activities. The scheduler may also only tie in the delivery to the earliest floor in which doors are needed and let the door supplier understand that a continuous flow of doors is necessary on the project. The single activity provides the door supplier with a date for when the first doors are needed. While this approach is simple and direct, it does have disadvantages. If the project schedule is changed, rearranging floor-completion priority, the delivery time for the first group of doors will be inaccurate. If the project is accelerated, the door supplier will not know that the shipments are needed more quickly.

Another option for the door example would be to show a single submittal and a single approval activity, but multiple fabrication/delivery activities tied into each group of floors. The scheduler should select the best method for the project and item in question.

It is advisable to include the specification section in the description under each submittal activity. This assists the contractor and the designer in determining for what specification section the submittal is made. It also assists in determining who is responsible for that submittal. While responsibility codes can be included on the procurement activities, this is not always done. Some submittals are required from a variety of sources. It is not always easy for the scheduler to determine exactly who is responsible for these items. If the scheduler does know who is responsible for each submittal, then responsibility codes should be placed on the submittal activities. The designer is responsible for all approval activities.

The duration agreed upon for return of shop drawings should be placed on all approval activities. Reasonable fabrication and delivery times for each specific delivery activity should be included. These activity durations can be estimated by the scheduler unless more accurate information is available from the project manager. The scheduler should not place similar durations on all the submittals for the project. Typically, the landscape submittals will not be made during the first 2 months of the project, while the rebar submittals are required promptly. At the same time, the scheduler should not expect the rebar fabricator to submit all shop drawings on the same day. The scheduler should use durations to establish priority among the rebar shop drawings. For example, the foundation rebar may be required 20 days after notice to proceed, the columns and first floor rebar 30 days after, and so on. This allows 2 weeks between each submittal for the preparation of the next set of shop drawings.

Sometimes it is necessary for related submittal, approval, and fabrication and delivery activities to tie into multiple points in the project schedule. For example, electrical switchboards will all be submitted together, approved together, and fabricated together, but their installation may be performed at different times at different locations in the building. Therefore, a separate restraint from fabrication and delivery should tie to each corresponding installation activity.

Sometimes the project specifications do not require submittals for all materials on the project. For example, drywall or masonry block submittals are often not required. The scheduler can either not include those items for which submittals are not required or list them only for fabrication and delivery. As a general rule, if the material is readily available in the quantities needed for the project, including them in the procurement subnetwork is not necessary. However, if a certain delivery is necessary even though a submittal is not required, it is advisable to include the needed material on the procurement subnetwork.

The procurement subnetwork is normally the last page when the logic diagram is drafted by hand. This makes it easy for anyone reviewing the logic diagram to turn directly to the last page to review this procurement listing. Chronologically, the procurement subnet occurs very early in the project. When plotted by computer, the procurement subnetwork will usually show up among the first few pages of the plot.

A three-digit number is normally used for the procurement subnetwork. Typically, the first node is 100. The remaining nodes are numbered from top to bottom for submittals first, approvals second, and all the fabrication and delivery activities third. This process produces an easily readable procurement list on the printout.

An independent I node is not necessary for each submittal, and using only even-digit numbers is not recommended on this type of subnetwork. Typically, procurement subnetworks can include as many as 200 or more separate items. Saving numbers for additional procurement activities is good scheduling practice.

Site Work and Utilities   The site work and utilities subnetwork is somewhat different from other subnetworks typically encountered in a construction schedule. The site work subnetwork may include work that will occur very early in the project, work that will occur during the main portion of the project, and work that must be completed after a substantial part of the structure and enclosure work is complete. Thus the subnetwork for site work, unlike other subnetworks, includes work to be performed throughout the project. The scheduler may elect to divide site work into two or three separate subnetworks. This can help in scheduling a project with complex site work. Subdivision is also advisable if the logical diagrams are to be plotted in time scale. It is difficult to compress a subnetwork which covers several years onto a single sheet.

On a typical office building project, there will be site-work activities which must be completed before foundations can be started. These typically include relocation of utilities, running new utilities under the building, and possibly clear and grub, removal of trees, and demolition of miscellaneous structures. This type of work clearly must be done before the structure is started.

Site work such as fine grading away from the structure, miscellaneous concrete, utility work not adjacent to the structure, or remote small buildings can usually be done concurrently with the erection of the building. When scheduling this site work, the scheduler must consider work access, material storage, trailer locations, and the potential for damage if the work is done too early in the project.

Typically, the last group of site-work activities occurs after the building is enclosed. Waiting until the building is enclosed to complete the remainder of site work permits heavy equipment to be removed which was used to erect the exterior skin and distribute materials into the structure. This equipment usually will damage sidewalks, curbs, and paving.

Site-work activities included in the last group may include placement of

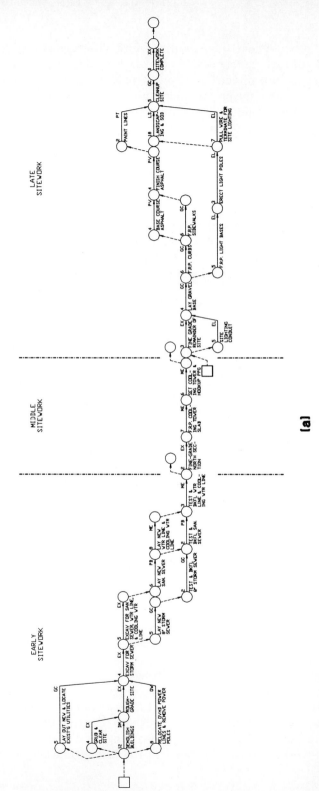

FIGURE 6-6a Site-work subnetwork in (a) I-J format.

154

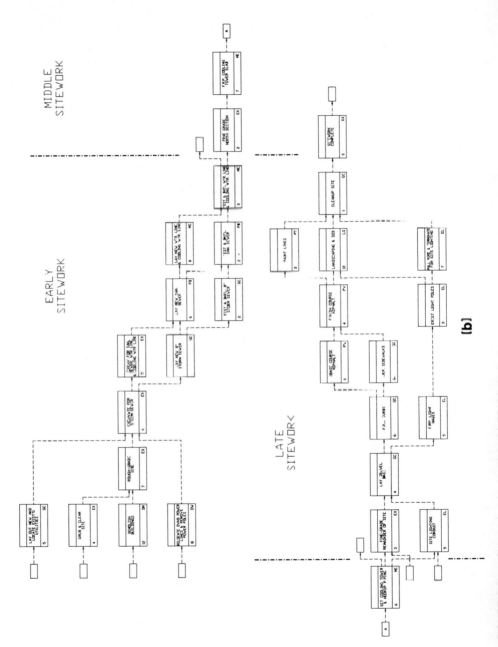

FIGURE 6-6b Site-work subnetwork in (b) PDM format.

Placing concrete is a labor-intensive operation.

transformers, cooling towers, grade-mounted HVAC equipment, and lawn-sprinkling systems; underground conduit and site lighting; curbs, signs, side-walks, concrete paving, pads, asphalt paving, and gravel surfacing; and land-scaping. Figures 6-6a and 6-6b are typical site-work subnetworks. Figures 6-6a and 6-6b show early, middle, and late site work in a single subnetwork.

It is good scheduling practice and improves the readability of both the logic diagram and the printout to include activities known as milestones within most subnetworks. For most subnetworks, a milestone at the end of the subnetwork indicating completion of that subnetwork is adequate. For site work, for ex-ample, an activity with a description ''site work complete'' is an appropriate addition to the scheduling logic. This activity has a responsibility code of XX and a duration of 0. It is the last activity in the subnet. By placing this activity into the subnet, it will show up in the printout as a key point in the timing of the entire project. It becomes an important reference point for ensuring timely progress of the project work.

On hand-drafted logic it is customary, but not absolutely necessary, that site work appear on the first sheet of the logic diagrams. This is because site work is the first major work undertaken on the project.

Foundations The foundations subnetwork includes activities such as mass excavation; foot excavation; and forming, reinforcing, and pouring footings, walls, columns, and other structural concrete supports. Except for the smallest of structures, it is best to divide the footings and foundation walls into segments, and identify them either by column line, building side, or some other recog-nizable delineation.

Activities for underslab plumbing, electrical, granular base, and slab-on-

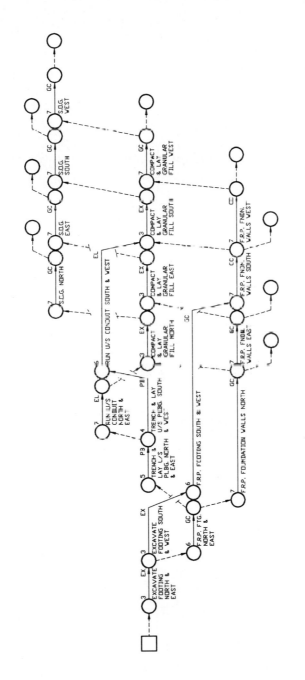

FIGURE 6-7a Foundation and slab-on-grade (S.O.G.) subnetwork in (a) I-J format.

[a]

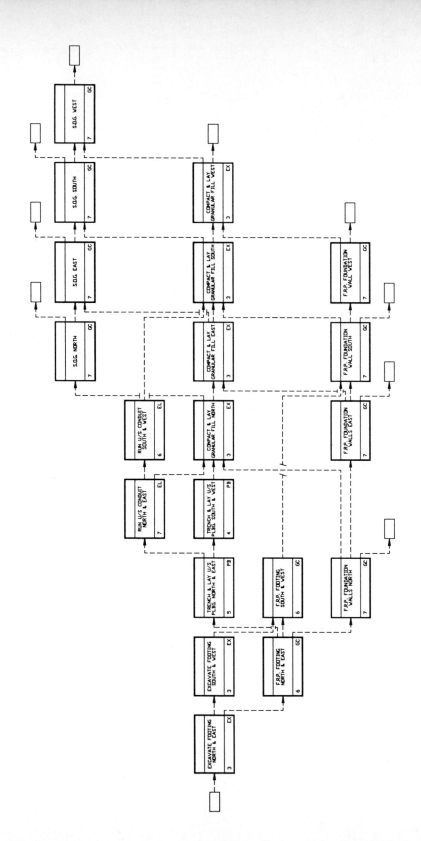

FIGURE 6-7b Foundation and slab-on-grade (S.O.G.) subnetwork in (b) PDM format.

[b]

grade may or may not be included in this subnetwork. The decision as to whether to include this work is partially related to the size of the structure and when the contractor intends to place the slab-on-grade. If there is a minimal number of underslab items and the weather permits slab-on-grade installation, the contractor may elect to complete the slab-on-grade during or shortly after foundation construction. In this case, the work should be included in the foundation subnetwork. However, if there are a large number of underslab items and/or difficult slab-on-grade configurations, the contractor may elect to wait until upper floors are installed to provide both time and protection to pour the slab-on-grade at a later date. If so, the slab-on-grade work may be included in a separate subnetwork. This procedure allows the structure to continue to be erected vertically without waiting on the slab-on-grade. Figures 6-7a and 6-7b are typical foundation subnetworks with slab-on-grade (S.O.G.) work included.

While the concept of subnetworks is to concentrate on one portion of the construction process at a time, the scheduler must also understand how the building is to be constructed. For example, if the contractor intends to finish 25 percent of the foundations (or the northwest corner) prior to erecting steel, then the scheduler cannot lump all the foundation work into a single activity. The foundation work should be divided into at least four activities so that either a restraint, or a finish-to-start relationship in PDM, can be used to indicate when the structural steel may proceed.

When developing the "scratchies" for this subnet, a small note indicating where relationships go can be included in the diagram. (See Figures 6-7a and 6-7b.) For example, site work could tie "in" to the foundation work, indicating which site work activities have to be completed before the foundation work can start. Similarly, an "out" from each quadrant of the foundation work would tie into the structural steel erection. At this point the scheduler has not developed the structural steel subnetwork, but an "out" will indicate a relationship that needs to be tied in when the structural steel subnetwork is completed. Connection of subnetworks will be discussed in detail later in this chapter.

As discussed earlier, it is advisable to include milestone activities at the end of the subnetwork. For the foundation subnetwork, an activity "Foundation Work Complete" is desirable. On major project schedules, it is also helpful to include a "kickoff" milestone at the beginning of each subnetwork. This milestone activity may carry a description such as "Start Foundation Construction." A kickoff activity typically has an XX responsibility code and a 0-day duration. Sometimes, instead of a kickoff milestone, the scheduler uses an activity such as "Lay out" with a short duration indicating the start of specific work. This activity can also be used for finding the starting point of that subnetwork.

Structure   The structure subnetwork (sometimes called superstructure), is usually an independent network and almost always a single subnetwork unless the building is extremely large or the project constitutes more than one building. From discussions with the superintendent, project manager, estimator, and

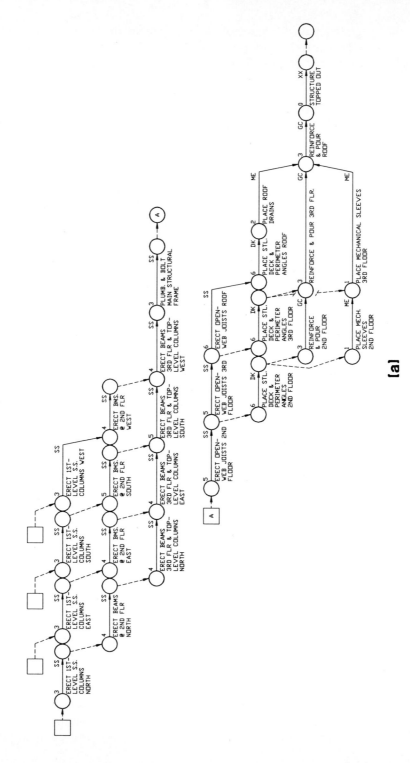

FIGURE 6-8a Typical structure subnetwork in (a) I-J format.

160

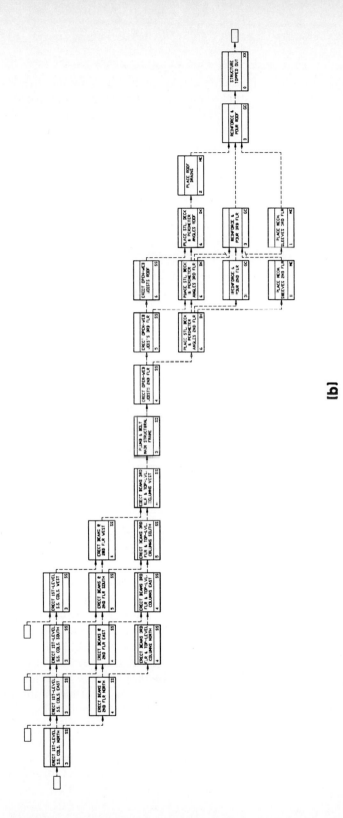

FIGURE 6-8b   Typical structure subnetwork in (b) PDM format.

(b)

161

subcontractors, the scheduler understands how the structure will be erected. If the structure is steel, typical activities might include erection of columns and beams; miscellaneous steel; plumb and bolt; open-web joists; steel deck; angles and framing for slab-on-deck; and reinforcing and placement of slab-on-deck. Sometimes the subnetwork will also include items for structural masonry or elevator-shaft masonry.

If the structure is concrete, typical activities might include forming, reinforcing, pouring, and curing of columns; and structural walls, beams, and slabs. Figures 6-8a and 6-8b are typical structure subnetworks.

At the end of the structural subnetwork, a milestone activity indicating "Structure Topped Out" is typically used to indicate the completion of the structure subnetwork. This is a key point in the construction process, and an important milestone to be included.

If the structure is concrete, it is also important that the scheduler specifically allow time for the curing of structural concrete in the structure subnetwork. Scheduling a project with little or no time for cure can be a serious hazard when a contractor or a subcontractor attempts to stay with that schedule in the field. Specific activities for cure time also remind the contractor not to remove shores and reshores too early. Many schedulers using PDM use lags to allow for cure time. Without specific cure activities, however, a scheduler may adjust or reduce lag times later if the project falls behind schedule. Because the lag is not identified as a cure activity, it may be mistaken as a time frame that the scheduler can adjust.

The importance of concrete cure activities is another example of why it is important for the scheduler to represent all conditions in the logic. The use of lags and imposed starts or finishes instead of activities is generally poor scheduling practice and should be avoided whenever possible.

Enclosure   The enclosure subnetwork typically is prepared by studying the building wall sections and elevations. Typical activities included in an enclosure subnetwork are the installation of exterior masonry and facebrick; the installation of framing and glazing materials; parapet wall construction; roof topping slab; insulation; built-up or membrane roofing; flashing and sealing; caulking; installation of exterior doors; canopies; and miscellaneous metal or concrete panel installation. Sometimes, on a smaller project, rooftop HVAC equipment is included in this subnetwork. Whenever possible, however, it is advisable to have a separate subnetwork for rooftop or penthouse HVAC and mechanical equipment. The enclosure subnetwork usually includes the roofing work, although a separate roofing subnetwork is somtimes appropriate. Figures 6-9a, 6-9b, and 6-9c illustrate typical enclosure subnetworks. Figure 6-9c uses leads, lags, start-to-start, and finish-to-finish relationships in the PDM format.

Miscellaneous items, including roof drains, roof walks, and gravel ballast, should also be included on the enclosure subnetwork if there is not a separate subnetwork for roofing. The scheduler should not forget the interior roof drain

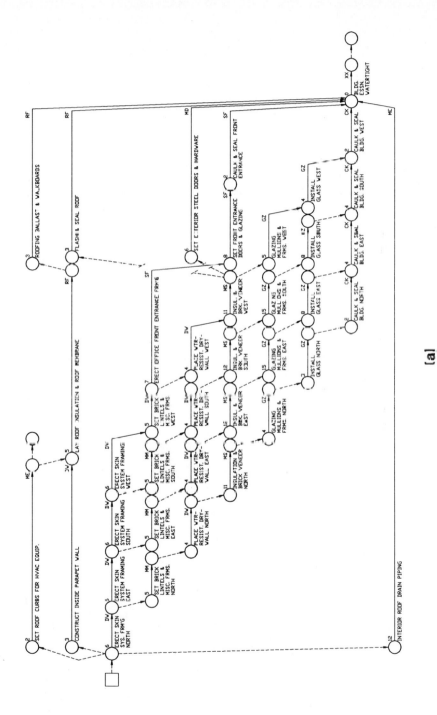

FIGURE 6-9a Enclosure subnetwork in I-J format.

[a]

163

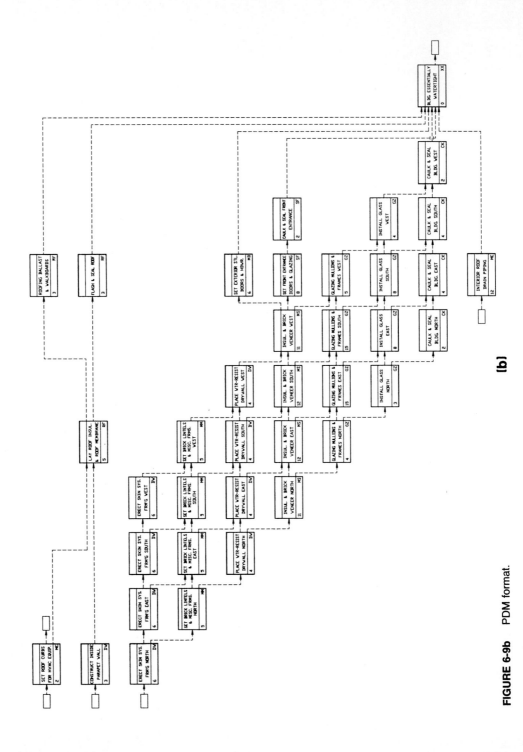

FIGURE 6-9b PDM format.

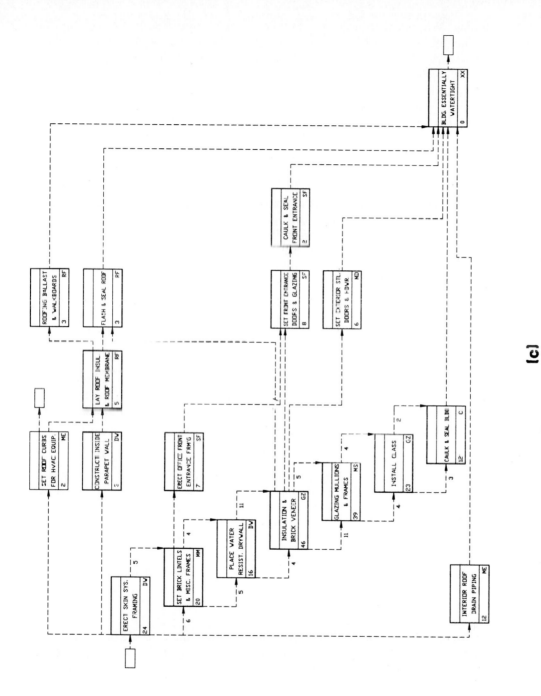

FIGURE 6-9c PDM format using leads and lags.

The boxes in the diagram contain the following:

- ROOFING BALLAST & WALKBOARDS — RF — 3
- FLASH & SEAL ROOF — RF — 3
- CAULK & SEAL FRONT ENTRANCE — SF — 2
- SET ROOF CURBS FOR HVAC EQUIP. — ME — 2
- CONSTRUCT INSIDE PARAPET WALL — DV — 3
- LAY ROOF INSUL & ROOF MEMBRANE — RF — 5
- SET FRONT ENTRANCE DOORS & GLAZING — SF — 8
- SET EXTERIOR STL. DOORS & HDWR — ND — 6
- BLDG. ESSENTIALLY WATERTIGHT — 0 — XX
- ERECT OFFICE FRONT ENTRANCE FRM'G — SF — 7
- INSULATION & BRICK VENEER — GZ — 46
- GLAZING MULLIONS & FRAMES — MS — 39
- INSTALL GLASS — GZ — 23
- CAULK & SEAL BLDG — C — 12
- ERECT SKIN SYS. FRAMING — DV — 24
- SET BRICK LINTELS & MISC. FRAMES — MM — 20
- PLACE WATER RESIST. DRYWALL — DV — 16
- INTERIOR ROOF DRAIN PIPING — ME — 12

[c]

165

piping. Without that piping the building is not watertight. It is very important to include a milestone activity indicating "Building Essentially Watertight." This is probably the most important individual milestone in the project schedule.

The enclosure subnetwork includes the exterior skin system and the roofing system required to make the building "essentially watertight." The term "essentially watertight" is the point at which one can reasonably install sheetrock and other materials that could be damaged by rain or other moisture. A building is substantially watertight when the majority of glazing is in place, the roof system is in place and most of the flashing installed, and other systems are sufficiently complete to preclude any major infiltration of rainwater. It is sometimes necessary or desirable to leave sections of the skin system open while waiting for delivery of interior materials or major equipment. If the structure can be made reasonably watertight through the use of plastic sheeting or other material, then the structure can still be termed "essentially watertight."

Many interior materials not only require that the structure be essentially watertight but also require "conditioned air." Conditioned air limits temperature and humidity extremes. High humidity may not only damage drywall but also affect ungalvanized ductwork, metal ceiling grid systems, and other interior materials. The scheduler needs to identify the construction materials that may be affected and understand any specification requirements for their installation prior to sequencing them in the project schedule.

The scheduler must study the parapet wall detail carefully when scheduling the enclosure subnet. Double-wall parapets allow either the roofing or the exterior skin system to be installed first. (See Figures 6-1a and 6-1b.) The separate walls allows completion of either item with only the draping of a tarp or plastic sheeting material over the parapet wall until the second trade completes. The final cap flashing connects the two parapet walls and seals the entire system.

When developing the subnetwork for enclosure, the scheduler should know whether the skin system will be erected vertically, with an entire face of the structure done at one time; floor by floor from the ground up; or floor by floor from the top down. The scheduler needs to understand what structural framing must be complete before miscellaneous framing and support for the skin system can be installed. The scheduler should consider what site work will occur around the structure and the necessity to complete that enclosure work so that site work can proceed. Prevailing winds and time of year when the enclosure work will be performed should also be considered.

Rough-in and Finish   The rough-in and finish subnetwork can be divided into two separate subnetworks, one for the rough-in and one for the finishes. Typically, these two are included as one subnetwork because the rough-in and the finish work in any particular area are closely related.

Normally, rough-in work is considered to be any work that can be done before the building is substantially watertight. While this work will vary slightly from project to project, it typically includes layout; interior masonry and door-

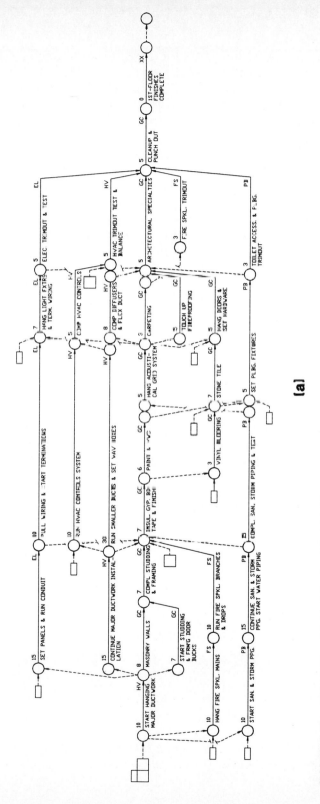

FIGURE 6-10a Rough-in and finish subnetwork in (a) I-J format.

167

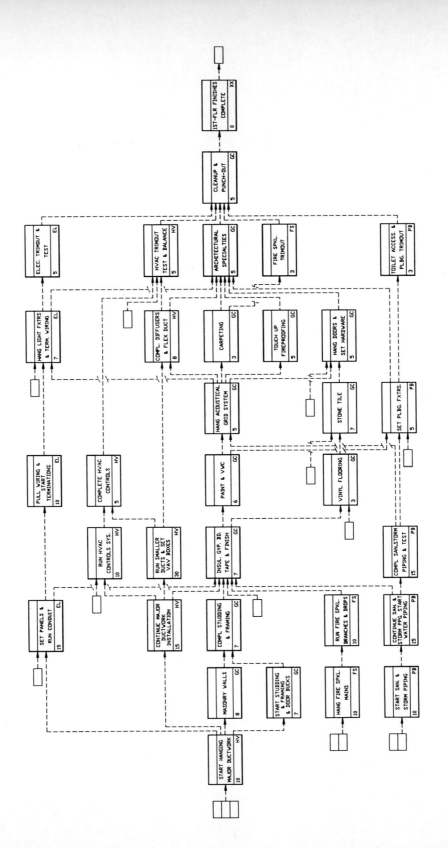

FIGURE 6-10b Rough-in and finish subnetwork in (b) PDM format.

bucks; studding and framing; rough-in of plumbing, fire sprinklers, power conduit, and some wire pulling; hanging of HVAC ductwork; fire-sprinkler drops and heads; some equipment installation; and miscellaneous other conduit work.

Items which typically must wait for watertight enclosure include drywall, completion of wire pulling and termination, light fixtures, ceiling grid, ceiling tiles, floor tiles, carpeting, ceramic tile, fixture installation, HVAC trimout, most equipment, architectural specialty items, painting, vinyl wallcovering, doors and hardware, and, of course, punch and cleanup.

Rough-in and finish subnetworks will usually be subdivided into identifiable areas. A three-story office building will be divided into three rough-in and finish subnetworks, one for each floor. A single-floor rough-in and finish subnet usually covers approximately 25,000 square feet. Larger floor areas should be subdivided in some appropriate fashion. Trying to cover too large an area with a single rough-in and finish subnetwork leads to difficulty in the phasing or sequencing of the work. If the studding and framing activities will take more than 20 workdays, a subdivision of the subnets is usually advisable. While scheduling a structure that has only miscellaneous finish work, such as an open-shell office building, the square-footage rule would not apply. On the other hand, the squre footage covered by the rough-in and finish subnetwork might be reduced for a hospital with a wide variation in finish requirements. Figures 6-10a and 6-10b are typical rough-in and finish subnetworks.

The scheduler should be careful to avoid the tendency to use standardized subnetworks for rough-in and finish. Many times a standardized subnetwork will not only fail to include needed activities but will also include unneeded activities and unreasonable relationships. For an office building with identical floors, a rough-in and finish subnetwork developed for the first floor can be reproduced and used for each identical floor. However, it is important to develop a separate "scratchie" or make a copy for each and every subnetwork. This will allow the proper numbering and crew ties (restraints) on the subnetworks. This is discussed in detail later in this chapter.

Mechanical Rooms Mechanical rooms should typically be a separate, small subnetwork of activities not included within the rough-in and finish for the particular floor on which they are located. By isolating this subnetwork, it is easier to show how mechanical equipment is installed and its proper relationship to other activities, without confusing that work with the other mechanical work and finishes on the same floor. These subnetworks typically include items such as forming, reinforcing, and pouring housekeeping pads; setting HVAC fans, equipment, coils, and air-handling units; and hanging ductwork. Other activities may include heating and cooling piping and conduit; and HVAC controls, wiring, panels, and unit heaters. While these subnets are usually small, they are appreciated by the tradespeople in the field, who normally complete the mechanical equipment room work with separate crews from the main mechanical installations. Figures 6-11a and 6-11b are typical mechanical room subnetworks.

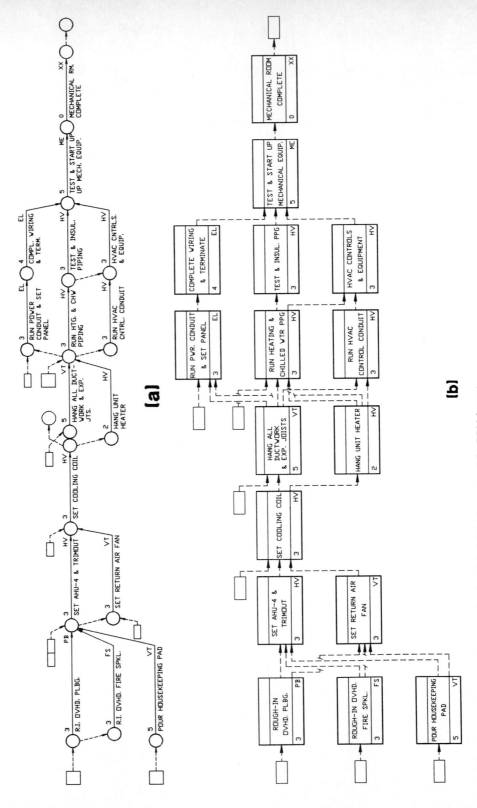

[a]

[b]

FIGURE 6-11 Mechanical room subnetwork in (a) I-J format and (b) PDM format.

Step 6: Draft or Plot the Logic Diagram

Most schedule specifications require a logical diagram to accompany all schedule submissions. The logic diagram may be either hand-drafted or plotted. The scheduler needs to determine whether time scaling or readability is important in determining how the logic diagram will be produced. Many times manual drafting of the logic network is desirable. Plotted schedules on larger projects are not always readable. Schedule software for producing plots has improved dramatically in the past few years. With careful planning, appropriate subnetting, and some trial and error, the scheduler should be able to produce readable plots for most project schedules. Manual drafting allows a much cleaner and more understandable network for reading both in the field and in the office. One disadvantage to manual drafting is that it is slow and time-consuming—and exceptionally slow if it must be time-scaled. If the updated logic diagram must be time-scaled, it must usually be completely redrafted each time it is updated. Thus, time-scaled diagrams are typically not prepared manually.

It has been the author's experience that CPM schedules that are in excess of 300 activities tend to produce plots that are difficult to read. While readable plots of 500 or more activities can be developed, additional time is spent adjusting plots to improve their readability. As more time is invested in making the plotted diagram readable, the option of hand-drafted logic becomes more attractive. In some cases, however, both a hand-drafted and a plotted CPM diagram are produced.

At this point, the scheduler has a small stack of "scratchies" ready for the next step in the scheduling process. If he or she decides to have the CPM hand-drafted, the subnetworks should be numbered in the order they should be drawn. The logic diagram is drafted from the accumulated "scratchie" diagram. If the scheduler has developed the CPM subnetworks chronologically, drafting can proceed concurrently with the development of the remaining subnets. Once completed, hand-drafted logic allows careful concentration on how the project interrelates and careful checking of the various relationships. While the quality of a plotted schedule can be just as good, it requires a higher level of skill to read and check for all important relationships.

Once the "scratchies" have been developed, the scheduler may distribute copies of these subnetworks to the project superintendent and other involved parties for an intermediate check of the scheduling assumptions and relationships. This intermediate check is not always necessary, but it is a good place to resolve problems which may exist within the subnetworks. Sometimes, between the time the scheduler discusses the project with the superintendent and the time the subnetwork is developed, the superintendent may change the intended approach. It is easier to adjust the sequences involved within a subnetwork at this point than later in the schedule development process. This also allows the scheduler to provide an indication that the development of the schedule is proceeding. The scheduler should continue with the manual drafting (if that is the option that was selected) while the superintendent and other parties are carrying out their review.

Step 7: Number the Activities

The scheduler should then number the activities in the individual subnetworks. As indicated earlier, activities in the project start subnetwork are numbered with one- and two-digit numbers, and the procurement subnetwork is normally numbered with three-digit numbers. Activities in subnetworks other than project start and procurement should be numbered with four-digit activity numbers. If the scheduler has elected to use hand-drafted logic it is advantageous to co-ordinate the numbering with the sheet numbers for the logic. On the first logic sheet, all four-digit activity numbers should start with the number 1. All four-digit activity numbers on the second sheet should start with the number 2, and so on so that the sheets follow with 1000, 2000, 3000, etc. Typically, on hand-drafted logic more than one subnet will be drawn on a page. When this is done, the best way to number subnets is to start with easily discernable numbers. For example, if five subnets are drafted manually on page 2, the numbering for each subnet would be as follows:

First subnetwork	2000, 2002, 2004, 2006, . . .
Second subnetwork	2200, 2202, 2204, 2206, . . .
Third subnetwork	2400, 2402, 2404, 2406, . . .
Fourth subnetwork	2600, 2602, 2604, 2606, . . .
Fifth subnetwork	2800, 2802, 2804, 2806, . . .

Skipping 2100, 2300, . . . leaves room to add a subnetwork at a later date.

If, however, the total number of hand-drafted logic sheets exceeds nine, the scheduler must decide whether to go to a five-digit system or to adjust the

Although winter concrete work is quite common today, winter weather may delay, or, at a minimum, significantly reduce productivity.

numbering and not match the first digit to the sheet number of the logic diagram. If the logic diagram is to be plotted, other numbering schemes can be used.

It is best to number the individual nodes within the subnets with gaps in the numbers, to allow for the addition of activities and/or nodes at a later date. The activity nodes should be numbered by 2's in an up-and-down, left-to-right format, making sure that all activity numbers have I nodes smaller than the J nodes. This should be achievable without tying the subnetworks together. If it is not, the scheduler risks a loop. A scheduler should proceed to number all of the nodes throughout all of the "scratchie" sheets or drafted diagrams as the case may be. Once all the nodes have been numbered properly, the scheduler can move on to the next step of tying the project schedule together.

Step 8: Tie the Subnets Together

The scheduler is now ready to tie the subnetworks together and produce a coherent project schedule. The first step is to mentally consider what relationships exist among the various subnetworks. It is best to work in chronological order. At this point, the scheduler is using "ins" and "outs" (see Figures 6-12a, 6-12b, 6-13a, and 6-13b) to connect the various subnetworks.

The use of "ins" and "outs" is merely a method of drawing a restraint or relationship from one subnetwork to another. Instead of drawing lines all the way across a large sheet to indicate a restraint or relationship, an "in" and corresponding "out" are used. The same procedure is used for restraints or relationships between sheets. The numbering system for nodes automatically tells the reader what logic sheet the other end of the restraint or relationship

FIGURE 6-12   "Outs" in (a) I-J format and (b) PDM format.

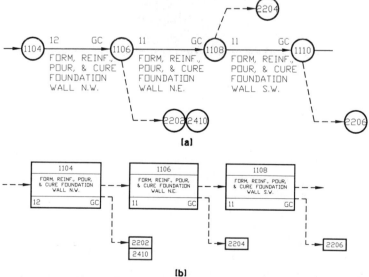

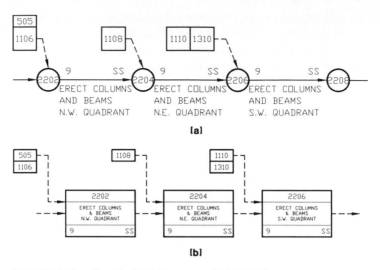

FIGURE 6-13 "Ins" in (a) I-J format and (b) PDM format.

is on. "Ins" and "outs" are also used by some software programs when plotting portions of schedules. Plots can be organized to plot by subnetwork just like hand-drafted logic.

The scheduler starts the connecting process by considering what work will immediately follow the mobilization indicated in project start subnetwork. Typically, some site work will start shortly after mobilization. Usually some site work is required before foundation work can proceed. However, some of the foundation work may be able to proceed concurrent with the site-preparation work. An "out" from mobilization is provided with a corresponding "in" at the beginning of Site Grading, Grub and Clear. Another restraint or relationship may be required from mobilization at the beginning of certain excavation for foundation work. Most of the time, a restraint or relationship is included from Notice to Proceed directly to the procurement subnetwork. Once all the obvious successors have been considered for the project start subnetwork, the scheduler moves on to the next subnetwork.

From the site-work subnetwork there will be restraints or relationships to the activities for excavation and foundation work in the concrete foundation subnetwork. A single restraint or relationship may take care of all the physical ties, or several restraints may be necessary. The scheduler must consider each of the work items within the site-work subnet, and how they interrelate or do not interrelate with the foundation subnet. Sometimes the scheduler must make minor adjustments to the "scratchie" or drafted subnet diagram during this phase. This is reasonable and normal in the development process.

The scheduler then moves to the foundation subnetwork, and considers how it relates to the erection of structural steel. As discussed previously, the foundation subnet will be divided into activities which roughly correlate with the way the structure will proceed. Typically, 25 percent of the foundation is required before the start of structural steel erection. Restraints or relationships must show the *continuous* relationship between the foundation and the structural steel. One restraint is not usually adequate. If the steel is to follow in 25 percent increments, it will take four restraints or relationships to tie the foundation work adequately to the structural steel. Depending on how the scheduler has divided the foundation work, including interior column pads, etc., it may take substantially more ties to show adequately the physical relationship between the foundation and the structural steel erection.

Do not make the mistake (which is common in precedence scheduling) of including only one relationship tie from a foundation subnet to a structural subnet. There is a continuous working relationship between these two subnetworks. A single tie may properly indicate the starting point for work in the following subnet, but it does not indicate the real working relationship of construction in the field. The foundation work must continue until completion so that following structural work can continue. While this seems obvious, it is not always shown properly in precedence diagrams.

The scheduler then moves on to the structural subnetwork and considers its ties back to the foundation subnetwork. All the restraints or relationships may not have been picked up, requiring additional ties. The scheduler then considers any additional ties from the site-work subnetwork to the structural steel erection. For example, structural steel erection may require a level place for a tractor-type crane. This leveling is not required for the foundation work, but it is required for the structural steel. These additional ties must be considered.

The scheduler then looks from the structural steel subnetwork to the enclosure and roofing subnetwork. How do these different subnetworks relate? At what point can the enclosure work start? How does the skin system relate to the completion of the structural steel? All of the physical restraints or relationships from structural steel to the enclosure network should be drawn. The scheduler continues placing "ins" and "outs" to the roofing subnetwork (if it is a separate subnetwork) until he or she is satisfied that the structural work is correctly related with those subnetworks.

The scheduler next considers the enclosure subnetwork, looking backward to the structural subnet. Are all the activities in the enclosure subnetwork properly related? Is there any site work that is required prior to enclosure, possibly some additional grading, or backfill of the foundation, to allow the exterior skin system to proceed properly?

The scheduler then looks forward and considers the relationship of the enclosure subnetwork with the interior finishes. Tying the substantially watertight milestone activity to each floor at the start of dry finishes is an important tie that is usually required on the schedule.

Remembering that a certain amount of site work was to follow the enclosure subnetwork, the scheduler then places restraints or relationships that are appropriate from the enclosure subnetwork back into the site work subnetwork. This tie from enclosure back to site work may create a backbranch. Careful consideration of the ties is necessary to prevent loops.

The scheduler next moves to the rough-in and finish subnetworks. The scheduler must consider what work is required that will allow layout and rough-in ductwork, conduit, and plumbing to start. Typically, the structure must be erected and slabs-on-deck must be in place and at least partially cured before the rough-in work can proceed. These ties need to be shown. Each rough-in and finish subnetwork should be tied to the structural work and to the enclosure subnetwork. Completion of the rough-in and finish subnetworks can then be tied to the project completion subnetwork. The scheduler should look back to the site-work subnet and tie its completing activities to project completion.

Schedulers should next spend some time reviewing all the subnetworks, looking for open ends or incomplete ties from subnet to subnet. These activities need to be tied into either their own subnetworks at a logical point or other subnetworks with the use of "ins" and "outs." Any work that does not reasonably need to be completed before project completion should be tied into the project completion subnet. At this point the "scratchies" (or drafted diagrams) should have no open ends or dangling activities. The only open ends permitted are "Notice to Proceed" and the final "Project Complete" activity.

The scheduler has now completed all the physical restraints that belong in the execution process of a construction project. The scheduler has not, however, included restraints or relationships for all the other reasonable construction procedures and limitations followed in efficiently executing a construction contract. Restraints or relationships in a CPM schedule can indicate much more than just a physical tie. Restraints or relationships can indicate crew ties, equipment flow, preferred construction sequencing, contract requirements, weather restraints, etc. The use of restraints or relationships to represent these needs is not detrimental but is in fact an asset to the project schedule.

The scheduler is now ready to add these other types of ties to the CPM schedule. Usually, equipment ties are added next. The limitations of equipment may require that the structure be totally complete before the skin system is erected, as when only a single tower crane is available on a project. Limitations of formwork and other equipment items are usually considered in the original sequencing within the subnetworks. Limitations on equipment necessary for rough-in and finish, however, are not necessarily included within a single subnet. Our example project has three rough-in and finish subnetworks. The scheduler needs to relate these to any equipment limitations (such as a manlift) that may restrict the ability of the subcontractors or general contractor to proceed on more than one floor at a time.

Crew ties are possibly the most important restraints or relationships, other than physical ties, in a project schedule. For the example project, the most obvious place that crew ties will be necessary is in the sequencing of labor from floor to floor during the rough-in and finish phases of the project. Unless crew ties are added, it is conceivable that all three floors could be scheduled for rough-in at the same time. Most contractors would consider that an inefficient and unnecessary construction sequence. Crew ties are added from one floor to the next in the sequence anticipated. In our sample schedule, the first floor is roughed in before the start of the second floor rough-in, and so on. Typically, crew ties for major subcontractors are tied from floor to floor. HVAC ductwork, electrical conduit, plumbing, fire sprinklers, studding, and framing are tied by crew from floor to floor. Tying these work activities from floor to floor allows an orderly sequence of work and an automatic ''leveling'' of manpower on the project. Typically the ties are placed from completion of an activity on the first floor to the start of the same activity on the second floor, and so on. During this process the assets of the independent I node rule—if I J scheduling format was selected—will become obvious to the scheduler.

Crew ties within the finishing portion of these subnetworks are also important. These also are tied from floor to floor through the building. Drywall, painting, ductwork, electrical fixtures, wiring, ceramic tile, and plumbing fixtures are typical finish activities that can be tied by crew from floor to floor. The crew tying of the finish activities satisfies the same need as the crew tying for the rough-in activities, to indicate the manpower limitations for that portion of the work.

Should circumstances later in the project require the contractor to finish the building from the top floor down, instead of from the bottom floor up, it is a relatively simple matter to adjust the crew ties to show work from the top down. The ability to change sequences easily is one of the basic reasons for the subnetwork development procedure.

While the simple project used for this example does not lend itself to the use of other types of ties, many more exist. One example is phasing restraints. Contract requirements may require that only one or two structures in a wastewater-treatment plant be demolished and rebuilt at one time. A tie between the completion of the structures and the demolition of following structures may be necessary. Another example of a phasing restraint would be a limitation of a hospital renovation project to a certain number of rooms or areas that can be taken out of service at one time. Other contractual limitations may require similar phasing restraints.

Another type of limitation is weather. Restraints or relationships can be used to indicate weather limitations on a project. For example, in reviewing the project, the scheduler may determine that some of the concrete work will take place in the winter. The scheduler may desire to limit or totally stop all concrete work during the winter season. Limitations may be provided by ties between various concrete activities. Total shutdown of all concrete work could be done with an individual activity, such as a milestone activity, but with a duration indicating winter work concrete shutdown. The scheduler may wish to wait until computerization of the schedule or the manual forward pass to locate and provide a duration for this additional activity.

Now that the scheduler has completed the installation of all the physical, crew, phasing, contract, and other types of restraints or relationships, the entire project schedule should be reviewed, either in its drafted format or in its "scratchies" format. The scheduler should check again for open ends. All activities must have a predecessor and a successor, except the very first activity in the project and the very last activity in the project.

Should there be more than one contractual completion date, successors are not absolutely required from the intermediate milestone flags. However, it is desirable from a scheduling format basis to add a restraint or relationship from the completion of intermediate milestones to the final project completion flag if no other logical restraint exists.

The next step is for the scheduler to perform a manual forward pass. The purpose of the forward pass is to help the CPM scheduler conceptualize and understand how the various parts of the project will progress. It also allows the scheduler to study how the various seasons and weather conditions will affect the project. The forward pass also provides a manual check of the logic and correctness of the project sequence. During the manual forward pass, approximate completion dates for each subnetwork can be noted on the "scratchies" or drafted logic. This allows a first check of whether the scheduled project duration will fit within the contractually required completion date.

Even if the scheduler is computerizing the network, a manual forward pass should be performed. While the scheduler can skip this step and allow the computer to calculate the forward pass, this eliminates the logic check and jeopardizes a quality CPM schedule. A manual forward pass, even on large project schedules, should not take more than a couple of hours.

The manual forward pass also allows the scheduler to make adjustments early. Once the forward pass is completed and the scheduler has made minor adjustments to make the project schedule approximately fit within the contract time frame, schedule preparation moves to the next phase.

PHASE 4: COMPUTERIZATION

The next phase is entering the CPM logic information into the computer for processing. The actual input is a manual process which takes great care to produce an accurate database. Thousands of numbers are entered into the computer. A single error in the input can cause loops or inaccurate schedule information. While a data check will be performed at the completion of this step, it is far better to enter accurate information in the first place and not be confronted with a serious number of data input errors.

The first step in computerizing the schedule is to enter the project name and various identifying information about the project schedule into the computer. Many software programs will ask you to identify the work week. Monday through Friday is the common work week in Western countries, while Sunday through Thursday is the standard work week in others. The software package typically also asks the scheduler to identify nonworking holidays. Schedulers should enter these days into the software program and be aware that the CPM calculations will skip over the various weekend and holiday days. Be sure to indicate whether holidays that occur on Saturdays or Sundays are to be shifted to Friday or Monday. If the scheduler has elected to schedule the project using calendar-day durations, then he or she needs to enter whether all calendar days will be considered work days.

In both PDM and I-J format, the typical procedure for data input is to follow the instructions in the software package for entering activity information. Each activity is entered into the database by its description, its I and J nodes or precedence number, its duration, its responsibility code, and any other codes that may have been assigned to it. Once all the activities for the CPM have been entered into the database, the data processor and/or scheduler returns to the logic diagram and adds all the restraints. (With PDM schedules the predecessors and successors are put in at the same time as the activities.) This process takes approximately 1 hour per 50 activities, depending on the speed of the data processing individual and the quantity of data for each activity.

When the entire database is complete, the scheduler merely needs to add the imposed finish date on the Project Complete activity and the project start date on the Notice to Proceed activity. Once this is done, the schedule dates and durations can be calculated. Follow the steps provided in the software program to calculate the project schedule. If you have done a quality job of scheduling and have made no data processing errors, you should not get any loops. Loops, however, are not uncommon. The vast majority of loops are created by a data processing error. Considering the tens of thousands of numbers entered in the data processing stage, it is not surprising that a reversal or misplaced number can cause a loop. Have the software program print out the loop so you can sit and check the loop through your logic diagram. If there is a data processing error, correct that error. If there is a relationship error that is actually your scratchie or hand-drafted diagram, correct the error and then correct the corresponding data in the computer base. Recalculate the CPM schedule. It is not really unusual for more than one loop to exist because of more than one data entry error. Correct the loops until no further loops are found. Recalculate the schedule and print out a numeric listing of your CPM schedule. This printout is your network analysis.

After the CPM has been produced in I-J or activity-number format (also known as a numeric listing), the scheduler needs to proceed with what is commonly known as a "data check." It is imperative that the scheduler base adjustments and modifications on an accurate printout. If there are data errors hidden within the CPM database, an erroneous schedule can be produced. A line-by-line, manual check of the CPM data (which has ben entered into the computer) should be made to verify that the computer schedule matches what is on the "scratchie" sheets. While minor error in the descriptions and spellings will not cause schedule problems, numerical transformations or duration errors can cause serious problems. Responsibility code or other code errors can also create serious problems.

PHASE 5: REFINEMENT

Refinement of the CPM schedule is carried out in two parts. The first step is the independent refinement done by the CPM scheduler to modify and adjust the CPM to be an accurate representation of the plan of approach. Included in the scheduler's modifications are adjustments for seasonal weather and efficiency adjustments to make the project flow more smoothly. The second step of the refinement process involves meetings with the superintendent and subcontractors to make sure that the scheduler has interpreted their directives accurately. Minor adjustments are also made to the schedule for minor plan changes by the superintendent or subcontractors that have occurred since the original scheduling meeting.

Scheduler Refinement

Compare the printout with the logic diagrams to determine if the project schedule has overrun the contract time. If the schedule is long, a negative number appears in the slack or float column, indicating that there are more days in the CPM schedule than are allotted in the contract time. If the CPM schedule shows the project just critical, check the Project Complete activity to determine if the early start and late start are exactly the same date. If the early start is before the late start, then you have fewer workdays in your project schedule than the contract completion time. Normally, if you have done a reasonable job of scheduling the project, keeping in mind the time frame required for the contract and being reasonable with durations, the project schedule will be within 5 percent of the contract duration. Many times the CPM will be within 10 to 20 days of the contract time.

If the schedule has too many days in it, look through the CPM printout, especially the critical path, and see where durations might be slightly long. Reducing durations by one or two days on a variety of different activities can usually reduce the schedule to conform to the contract completion date. If the project is slightly short, follow the same technique but add days to a few specific activities that may appear slightly short. Do not arbitrarily cut or add days if activity durations are correct. This adjustment is not to be taken lightly. This effort will usually bring the project to the project-completion date. If the project is just slightly short and the scheduler believes the durations are all reasonable and are not in need of adjustment, the duration of the final punch and cleanup activity can be increased to bring the project just to critical status.

If the project is substantially long and minor adjustments to project-activity durations will not bring the project within the contract time, then more serious scheduling revisions must be considered. Project sequencing must be studied. The relationships among the various subnetworks must be reconsidered, especially along the critical path. Look at the critical path to see if some work could be started slightly earlier, or if more work could be accomplished concurrently. By making adjustments to the start points, end points, and the sequential relationships, the schedule can usually be brought into alignment with the contract completion time. Should this not be possible through minor adjustments in durations, sequencing, and relationships, then the project schedule must be "crashed" or expedited. Refer to Chapter 10 for the proper techniques for expediting a schedule.

The Critical Path

Once the project schedule is within the contract completion date, and it is just critical, the CPM scheduler needs to review the critical path carefully. Winter work must be considered. Possibly other durations should be shortened to permit interior work to be scheduled during the winter, when better-quality manpower may be available. It is also important for the CPM scheduler to spend some time reviewing the critical path, and make sure it follows a reasonable and logical path through the project. Just because the critical path does not flow where you thought it should when you developed the schedule does not mean it is erroneous. If the critical path follows a path which you did not expect, you should check each and every relationship and activity's durations. Possibly you misconstrued some relationship, and caused a false critical path. If an item such as erection of a flagpole or toilet accessories shows up on the critical path, it is an indication that some reconsideration of the project sequences is required.

Float Analysis

The next step for the scheduler, once he or she is assured that the critical path is reasonable, is to check the float or slack in the project. Excessive float on various activities normally indicates a failure to provide sufficient ties within the work sequence. Work that is not tightly scheduled indicates a lax approach to the project schedule. Sometimes excessive float indicates a failure to include crew ties or equipment limitations in the project schedule. Although some activities, such as planting trees around the perimeter or painting the sign in front of the project, may show high float numbers, these types of activities do not cause alarm. However, if interior rough-in or finish activities show excessive float, this may indicate an error in the project schedule. Typically, average float should be 15 to 30 workdays on a project similar to the one used as an example in this chapter. While some projects may experience critical paths on almost every path through the project, this is an unusual circumstance. An excessively high number of critical paths could be an indication of a schedule manipulated for future delay claims. This is not to say that a project which is mostly critical has necessarily been manipulated.

Float analysis can be performed mathematically by taking the float number from the printout for each and every *work* activity in the project schedule and totaling the numbers. Then divide that number by the number of work activities. The result is average float. While this number by itself is not meaningful, its comparison to other typical project schedules will give the scheduler a feel for how "flexible" the project schedule is. The higher the average float number, the more flexible the project is. The lower the average float number, the less flexible the project is. For typical office building construction with 1- to $1\frac{1}{2}$-year project durations, typical average float should range between 15 and 30 workdays. Excessively higher numbers strongly indicate a poorly constructed schedule. Significantly lower numbers indicate a very tightly scheduled project. A happy medium with control of manpower flow and equipment will normally produce a project schedule within the parameters indicated above.

Different types of projects will have different types of average float. A structured project with severe phasing and limitations on construction work will normally have average slack in the 5- to 15-workday range. Major projects with large amounts of unrelated work, such as wastewater-treatment plants, large power plant projects, and other projects, can sometimes reveal large average float numbers. Average float provides the scheduler with a simple check on the quality of the completed original schedule. Average float is not a conclusive indication of a schedule's quality, but it does provide a benchmark for a quality review.

Refinement with Superintendent and Subcontractors

The CPM scheduler should produce a numeric-listing printout and a responsibility-by-early-start printout. The numeric printout is produced by listing the I and J nodes or PDM activity numbers in numerical order. The responsibility-by-early-start listing is a sort which lists the various responsibility codes that have been placed on the activities in chronological order. For example, this printout will show all the electrical activities, grouped together in chronological order. If time permits, a responsibility-by-early-start computer-generated bar chart is a nice asset for this meeting with the superintendent and subcontractors.

If hand-drafted logic has not been elected, the CPM scheduler should run project schedule plots. These plots should be organized and readable to the maximum degree possible. Copies of each of the printouts and the plots should be provided for the superintendent, the owner, the architect, and the various subcontractors at least 2 days but preferably a week before the refinement meeting is held. This will allow these people some time to review the project schedule and develop their final input.

Again, the scheduler should meet with the project superintendent prior to meeting with the various subcontractors. Any final comments and adjustments to the schedule will need to be integrated into the project schedule as smoothly as possible. After meeting with the superintendent, the scheduler should meet with the superintendent and subcontractors to get their final input and adjustments to the CPM schedule. If the scheduler has done a good job of developing the CPM, the number of adjustments will be minimal. However, subcontractors regularly change their minds between the original schedule meeting and the final development of the schedule.

It is a good idea for the scheduler to have the notes from the previous meeting to remind the subcontractors of the commitments they made to the CPM schedule. If possible, the CPM scheduler should make the modifications requested by the subcontractors. Sometimes, through the bar chart development process, the subcontractor can see that the schedule anticipates working in more than one or two areas at a time. The subcontractor may have anticipated only two crews, whereas the schedule may reflect three or more crews working in different areas at the same time. It is important for the scheduler to point out the various floats that may be available to allow the subcontractor to level manpower forces within the project schedule. If possible, the scheduler should add crew ties to reflect this desire. However, if the CPM schedule cannot fit within the contract time with these additional ties, then discussions should be held between the subcontractor, the superintendent, and possibly the project manager in order to work out difficulties.

Advanced Fine-Tuning Techniques

There are a wide variety of fine-tuning techniques available to the CPM scheduler. Depending on the amount of information the scheduler has elected to include in the CPM, there are several different techniques for resolving and fine-tuning scheduling problems. If the CPM scheduler has included manpower as a resource within the schedule, then manpower may be leveled or constrained through the use of various software functions. Other resources which were included on various activities can also be leveled or constrained.

"Resource leveling" is a software function that uses available float and adjusts the time frame for performing those activities to minimize peaks and valleys in resource usage. For example, assume that the scheduler selects the electrical trade contractors' work to be leveled. The scheduler selects this option in the software program and has the CPM schedule recalculated to show the most optimum time frame for the various electrical activities. This process does not change the end date of the project schedule. All calculations are limited to available float time within the original project schedule. Some programs allow leveling a wide variety of resources and/or trades simultaneously. With this program option, the total manpower can be leveled to the optimum degree for all trades. Resource leveling is explained in detail in Chapter 11.

"Constraining" is very similar to leveling of a resource; however, constraining specifically limits the number of craft personnel or other resources available to the project. Using a similar example to the above reference, the scheduler may enter a limit of 10 electricians available on a construction project. If the CPM schedule software finds that certain activities must be delayed because of a limitation on manpower, the schedule will automatically be extended. This extension of the project can change the end date. This procedure can be helpful in analyzing a project when a limited number of resources or craftspeople are available for a specific part of a project.

The use of leveling and constraining is helpful in fine-tuning the CPM schedule. However, it is not advisable to use the resulting printouts with the leveling or constraining procedures for distribution as the original CPM schedule. The scheduler should study the printouts and add restraints or make adjustments in the planning to produce the same effects as the resource-constraining or leveling procedure. The problem with using the software packages to perform these leveling procedures and not adding ties or making adjustments is that the scheduler (or anyone else reading the CPM) cannot tell where these "invisible ties" are being placed in the logic. Each time the CPM is updated, the software may select a different location to provide these "invisible ties." It will be difficult, if not impossible, to prove to an owner why these "invisible ties" rationalize the need for a time extension (should one be requested). Considering that the basis for CPM scheduling is communication, an "invisible tie" placed by the software package is not useful.

As scheduling software packages become more and more sophisticated, other options will undoubtedly become available to the scheduler to assist in refining CPM schedules. However, schedulers will always have to remember that true schedule quality comes from careful conceptualization of how the project will be constructed and accurate interpretation of that information into a CPM diagram. Regardless of the sophistication of the software, the software does not produce quality schedules. It can only process information in a format which is easily understood and facilitates quick and easy data entry. Computers and software do not generate quality schedules—only schedulers can.

After meetings with the superintendent and subcontractors, some input via phone or a meeting with the owner and the architect is always desirable. After this discussion, the scheduler can return to the office and make final adjustments to the CPM. Making the adjustments requested by the superintendent and subcontractor may have resulted in a project schedule which again needs minor adjustments in activity durations or relationships. These minor adjustments should be performed with permission of the responsible contractor, and then a final CPM schedule can be produced.

PHASE 6: PRODUCTION

The scheduler should now produce final production printouts, plots, and/or hand-drafted logic blueprints, as the case may be, for distribution to the various contract parties. The scheduler should produce a numeric-sort-with-restraints (relationships for precedence) printout, a responsibility-by-early-start printout, and any other printouts required by the specifications. A computer-generated responsibility-by-early-start bar chart is always a nice production printout which is easily readable by a wide variety of different people on a construction project. Other computer-sort printouts may also be desired. This final production should be distributed with ample copies to the project manager, the superintendent, the subcontractors, the owner, and the architect.

At this point the scheduler is normally finished with the development of the total original project schedule. The scheduler's next responsibility occurs during the first update, which is normally scheduled for approximately 2 months after the notice to proceed date. Many times the original schedule development will have passed this point in time and an update will have to be performed immediately after production of the original schedule. If approval is required on the original project schedule, the scheduler should wait until after it has been approved by the owner and the architect if at all possible before updating the CPM schedule. Chapter 8 discusses updating the CPM schedule.

EXERCISES

1 You have been awarded the contract to add a concourse to the city's international airport. The concourse is to consist of two levels. The lower level will be for baggage handling, support services, offices, food preparation, and pilot lounges. The upper level will be a typical concourse waiting area with moving walkways, loading ramps, and large waiting areas for departing flights. A utility tunnel will run underneath the lower level of the concourse. The contract also includes paving the aprons out 100 feet from the concourse structure. There will be a parking lot and departing-flight drop-off area for automobiles, and an attached structure for rental cars and a baggage pickup area. The new construction is to be attached to an existing concourse, which will require minor renovation at the tie-in point. The new structure will be approximately 100 feet wide, two stories high, and approximately $\frac{1}{2}$ mile long. Break the project into subnetworks as if you were preparing to schedule this project.

2 You have been retained by a municipal wastewater-treatment plant operator to provide the scheduling on a major new wastewater-treatment plant. The project has been let to multiple prime contractors.

The general prime contractor has subcontractors for excavation, vertical concrete work, concrete flatwork, metal building erection, masonry, structural steel, roofing, drywall/painting, flooring, toilet partitions, mechanical/HVAC, and electrical.

The mechanical prime contractor has the majority of the sophisticated mechanical systems in the wastewater-treatment plant. However, he has a subcontractor for concrete inserts, underground piping systems, and setting of major equipment. The mechanical contractor also has a subcontractor who will take care of miscellaneous concrete and structural steel work.

The third prime contractor is an electrical contractor, and he has only two subcontractors: one for the underground duct bank and one to set major electrical transformers.

As the scheduler, you need to develop a four-digit responsibility-code listing that will segregate the printout results, first by prime contractor and then by each prime's subcontractors, so that all the prime contractors' work can be listed in a single group and each of the subs under those primes can be listed in a single group. Develop a four-digit responsibility-code listing that satisfies these requirements.

3 Activity codes can include responsibility, manpower, divisions of work, equipment, labor, materials, and bid item codes. List three additional codes that could be placed on activities in a CPM schedule.

4 Select a local building for which you have complete access. Examine one floor on that building and determine finish materials, construction, etc., in sufficient detail to develop a complete rough-in and finish subnetwork for that floor of the building. Number that subnetwork starting with the digit 3200.

5 Select a building of moderate size for which you can obtain complete contract documents. Review these documents and develop a complete CPM schedule using the arrow diagramming method, including subnetworks, numbering ties, and relationships, and perform a final forward pass. Select a time frame for the building and meet that time frame with your schedule.

6 Repeat Exercise 5 using the precedence diagramming method.

7 Take the schedule developed in Exercise 5 or 6 and enter it into your available computer software to produce complete and verified numeric-listing printouts, responsibility-by-early-start printouts, and bar chart printouts.

8 Use the data developed in Exercise 7 and plot your diagram in the most readable format possible, using any available plotter.

9 Dollar-load the schedule produced in Exercise 5 or 6 and run a plot of cash flow based on early finish and another cash flow curve based on late finish. Use a monthly basis for your curve points. If your software will not plot the curves, draft the curves manually from the calculated data.

ADVANCED EXERCISES

10 Manpower-load the schedule produced in Exercise 5 or 6 and produce a manpower curve with your available software.

11 With the manpower-loaded schedule produced in Exercise 10, run the CPM schedule through both the leveling and constraining operations of your software. Try at least three different resource limits with the constraining option of your program.

DISCUSSION TOPICS

1 Perform a manual forward pass on Figures 6-9a, 6-9b, and 6-9c. Discuss the similarities and differences. Which method produces the most accurate schedule?

2 Discuss alternative methods or additional steps that could be used in developing the original project schedule.

CHAPTER 7

OTHER SCHEDULING METHODS

PROGRAM EVALUATION AND REVIEW TECHNIQUE (PERT)

Introduction

The project completion date for some projects, such as the space shuttle and various military projects, is extremely critical to the success of the overall mission. For these projects it is essential to have, during the early planning stages of the project, a tool for evaluating the likelihood of achieving project completion on time. The "Program Evaluation and Review Technique" (PERT) was developed for this purpose. Lack of project planning and control on early U.S. Navy programs to develop the Polaris missile resulted in actual costs and project durations that exceeded estimated times and costs by as much as 50 percent. In an attempt to get a better estimate of cost and project duration, the Navy assembled a joint research team consisting of Lockheed Aircraft Corporation (the prime contractor on the Polaris Missile Systems Program), the U.S. Navy's special project office, and the consulting firm of Boos, Allen and Hamilton to develop a better program planning and control system.

PERT was intended to assist in planning when no historical cost or time data are available with which to estimate overall project duration and cost. After development and application to the Polaris missile system in 1958, PERT was estimated to have saved nearly 2 years in the completion of the weapons system. PERT is used in planning and controlling new product development and research projects where the uncertainty associated with cost and schedule

Some projects will involve many repetitions of the same basic operation. (Photo courtesy of Northwest Engineering Co.)

must be evaluated. PERT, rarely used on today's construction projects, has the potential to be used on any construction project where quantities or productivity is uncertain, such as asbestos removal. PERT is a statistical treatment of uncertain activity performance time. It estimates the probability of meeting specified completion dates. PERT assumes a variability in activity durations based on a variability in production rates. The critical path method assumes that activity durations do not vary or vary so little that they can be assumed to be deterministic. PERT was not originated to plan a project comprised of known events, but rather to evaluate likely completion times of programs consisting of unknown events.

PERT Defined

PERT, like CPM scheduling, uses logic diagrams to analyze performance times. PERT charts are drawn as activity-on-arrow diagrams. PERT focuses on the event.

PERT enables the scheduler to estimate the most probable project duration and the probability that the project or any portion of the project will be completed at any particular time. PERT is therefore considered a probabilistic method because it introduces probability into the computation procedures.[1]

A probabilistic system would ideally use a historical activity time-duration frequency distribution. In construction, no such historical frequency distribution is usually available. Since the distribution does not exist, a method has been developed for creating a model distribution that can be used for activities. The distribution requires only three estimates of duration: an optimistic, a most likely, and a pessimistic duration. A formula reduces the three estimates to one time estimate and a measure of dispersion for each activity. This approach is applicable to estimating the durations of activities for which there is no historical experience. The scheduler can use probable durations combined in a way that proportions the likelihood that any of the three estimates will be the actual activity duration.[2]

PERT focuses on events (the nodes) that must occur prior to successful completion of a project. The probability of achieving performance of all activities which define an event is the outcome of PERT calculations. Because of PERT's emphasis on events, PERT is considered event-oriented. CPM, in contrast, focuses on completion of the individual activities that comprise the project. CPM is thus considered activity-oriented.

PERT uses three time estimates for each activity. An optimistic duration is estimated which presumes that everything involved in the performance of the work goes well (high productivity). A pessimistic duration is estimated assuming that almost everything involved in performance of the work goes poorly (low productivity). Pessimistic durations are not based on cataclysmic disasters such as a once-in-500-years flood, but rather on events that a scheduler might reasonably anticipate could go wrong—such as late material delivery, incomplete design, equipment breakdowns, or bad weather. The third estimate that PERT uses is the most likely duration, which anticipates the amount of time the activity will require most of the time if the activity is repeated many times. The most likely duration is not the average of the optimistic and pessimistic estimates. Most schedulers can anticipate more pessimistic occurrences than optimistic occurrences when estimating durations. However, most schedulers also understand that the occurrence of all the possible pessimistic occurrences is not very likely.

A PERT network is shown in Figure 7-1. Several different formats for drawing the nodes are possible other than the one shown here. However, the drawing technique is not as important as the method of calculating the effective duration. A standard I-J format can be used to draw a PERT schedule.

---

[1] Edward M. Willis, *Scheduling Construction Projects,* New York: Wiley, 1986, p. 200.

[2] R. B. Harris, *Precedence and Arrow Networking Techniques for Construction,* New York: Wiley, 1978, p. 327.

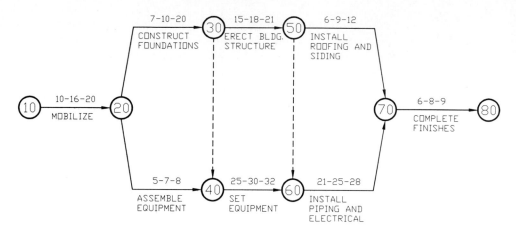

FIGURE 7-1 PERT chart.

Computing PERT Durations

Before showing the PERT calculations, it is necessary to review a few basic principles associated with probability that will apply. The first of these principles relates to the frequency distribution.

Table 7-1 and Figure 7-2 show the distribution of the amount of time it takes to place concrete columns for a classroom building. The number of minutes for each column was measured and the times were recorded. Table 7-1 lists the results, showing that the most frequent time that it took to place a concrete column was 50 minutes. This occurred in 16 of the total of 64 observations.

Plotting the frequency of occurrence in Figure 7-2 results in a distribution of the times for placing a concrete column. From this frequency distribution, it takes between 46 and 52 minutes to place a concrete column. Of the 64 columns, 47 columns were placed in this amount of time. In this case the "mean" would be equal to the sum of all times divided by 64, or in this case 48.3 minutes. The "mean" is simply the arithmetic average of all of the observations. One can also determine the "median" time, the value for which there are half as many observations higher and half as many observations lower in value. The median for this activity is 48 minutes. The "mode," or the value for which there is the greatest frequency, is 50 minutes. The mode is also the most likely value.

The mean, median, and mode are all measures of the central tendency of the distribution. Observing more columns being poured and developing a larger

TABLE 7-1 CONCRETE PLACEMENT—COLUMNS—MOLECULAR BIOLOGY BUILDING

Frequency	0	1	1	4	6	9	12	16	10	4	1	0	0	0
Minutes to place	36	38	40	42	44	46	48	50	52	54	56	58	60	62

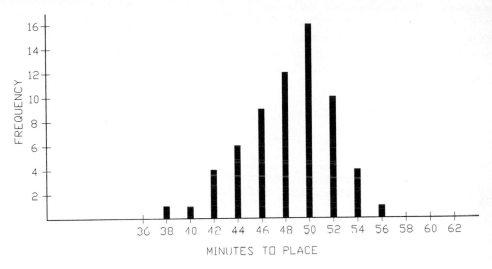

FIGURE 7-2   Distribution of concrete placement time—columns—Molecular Biology building.

number of observations (at the same time measuring more precisely the time that it takes to place a column to the nearest second) would result in a continuous distribution similar to that shown in Figure 7-3. This continuous distribution, although it has the same overall shape, is much smoother in form than our histogram of 64 observations.

This frequency distribution can be converted into a probability distribution by dividing by the total number of observations. Setting the area underneath this curve equal to 1 makes it possible to determine the cumulative probability

FIGURE 7-3   Continuous distribution.

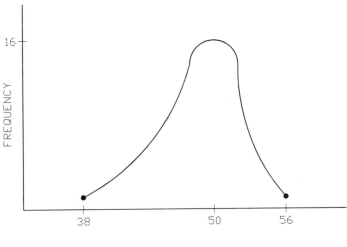

at any point along the axis for a particular time estimate. The distribution can be defined by using two probabilistic terms, the "mean value" and the "standard deviation." The calculation of the mean was shown earlier. The standard deviation is a measure of disperson calculated as:

$$\sigma = \frac{\sqrt{\sum\limits_{1}^{n}(x_i - \bar{x})^2}}{n - 1}$$

where

n = number of observations
x_i = observation value
$\bar{x}$ = mean value
σ = standard deviation
σ^2 = variance

The "variance" (also a measure of disperson) is equal to the standard deviation squared. For the example shown in Table 7-1, the mean is 48.3, the standard deviation is 3.68, and the variance is 13.54.

Another measure of disperson is the range, which is represented by the difference between highest and lowest observations. The range is a poor measure of dispersion because it does not capture the fullness or thinness of the distribution (the relative number of observations at various levels).

FIGURE 7-4 Normal probability distributions.

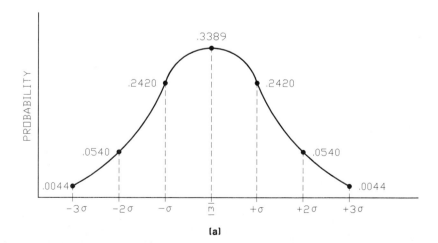

[a]

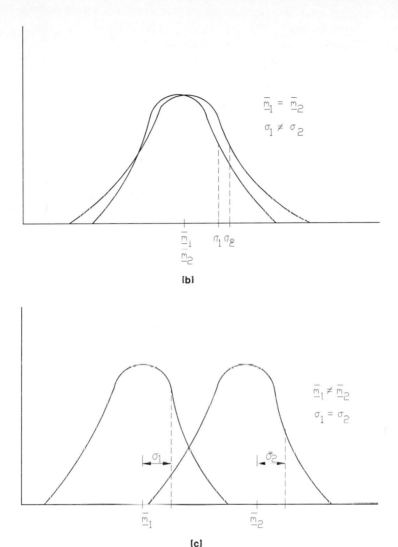

[b]

[c]

Figure 7-4a shows a normal probability distribution and illustrates typical values for a normal distribution for the standard deviation. The area under the total curve is equal to 1 in this distribution, and approximately 68 percent of the area is between $+1$ standard deviation and -1 standard deviation. Figures 7-4b and 7-4c illustrate other normal distributions. In Figure 7-4b, the distributions have the same mean value but different standard deviations. In Figure 7-4c the distributions have different mean values but the same standard deviation.

A final concept that is used in PERT computations is known as the "central limit theorem." This theorem provides a means to combine activity duration distributions. The theorem has three parts that apply to the adding of independent probability distributions:

1 The mean of the sum is the sum of the means.

2 The variance of the sum is the sum of the variances.

3 The distribution of the sum is a normal distribution regardless of the shape of individual distributions.

By reducing all of the distributions to a common normal distribution, one can quickly identify the probabilities associated with these distributions from standard tables such as the one shown in Table 7-2. In Figure 7-5 a cumulative normal probability distribution for project duration illustrates how this information can be used to determine the likelihood of completing a project before a particular date. In this figure it is shown that for a particular duration related to $-.67$ standard deviation from the mean project duration, there is a 25 percent chance that the project can be completed by this time. Later sections of this chapter will illustrate how these cumulative normal probability distributions for project duration can be developed.

TABLE 7-2 CUMULATIVE PROBABILITIES OF THE STANDARD NORMAL DISTRIBUTION
Entry is area $1 - \alpha$ under the standard normal curve from $-\infty$ to $z(1 - \alpha)$

z	.00	.01	.02	.03	.04	.05	.06	.07	.08	.09
.0	.500	.504	.508	.512	.516	.519	.523	.527	.531	.553
.1	.539	.543	.547	.551	.555	.559	.563	.567	.571	.575
.2	.579	.583	.587	.591	.594	.598	.602	.606	.610	.614
.3	.617	.621	.625	.629	.633	.636	.640	.644	.648	.651
.4	.655	.659	.662	.666	.670	.673	.677	.680	.684	.687
.5	.691	.695	.698	.701	.705	.708	.712	.715	.719	.722
.6	.725	.729	.732	.735	.738	.742	.745	.748	.751	.754
.7	.758	.761	.764	.767	.770	.773	.776	.779	.782	.785
.8	.788	.791	.793	.796	.799	.802	.805	.807	.810	.813
.9	.815	.818	.821	.823	.826	.828	.831	.834	.836	.838
1.0	.844	.843	.846	.848	.850	.853	.855	.857	.859	.862
1.1	.864	.866	.868	.870	.872	.874	.877	.879	.881	.833
1.2	.884	.886	.888	.890	.892	.894	.896	.898	.899	.901
1.3	.903	.904	.906	.908	.909	.911	.913	.914	.916	.917
1.4	.919	.920	.922	.923	.925	.926	.927	.929	.930	.931
1.5	.933	.934	.935	.937	.938	.939	.940	.941	.942	.944
1.6	.945	.946	.947	.948	.949	.950	.951	.952	.953	.954
1.7	.955	.956	.957	.958	.959	.959	.960	.961	.962	.963
1.8	.964	.964	.965	.966	.967	.967	.968	.969	.969	.970
1.9	.971	.971	.972	.973	.973	.974	.975	.975	.976	.976
2.0	.977	.977	.978	.978	.979	.979	.980	.980	.981	.981
2.1	.982	.982	.983	.983	.983	.984	.984	.985	.985	.985
2.2	.986	.986	.986	.987	.987	.987	.988	.988	.988	.989
2.3	.989	.989	.989	.990	.990	.990	.991	.991	.991	.991
2.4	.991	.992	.992	.992	.992	.992	.993	.993	.993	.993
2.5	.933	.994	.994	.994	.994	.994	.994	.994	.995	.995

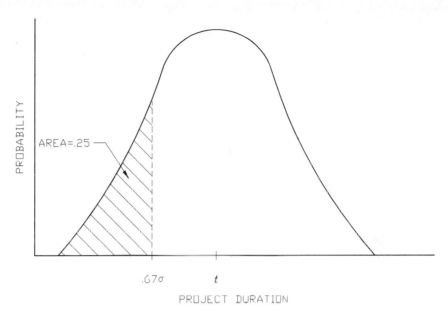

FIGURE 7-5   Cumulative normal probability distribution.

Selected percentiles							
Cumulative probability, $1 - \alpha$	.90	.95	.975	.98	.99	.995	.999
$z(1 - \alpha)$	1.282	1.645	1.960	2.054	2.326	2.576	2.090

Example: $P(z \leq 1.74) = .9616$ so $z(.9616) = 1.74$

The PERT model uses the three estimates of activity duration which are made to compute the statistical limits that describe the activity duration and the distribution. The formulas which describe this distribution have been derived from approximations of statistical distribution and have been simplified for ease of calculation. The parameters that are needed to describe the distribution and to utilize the central limit theorem are the expected duration, the standard deviation, and the variance. The expected duration or the mean is calculated by adding the optimistic duration, the pessimistic duration, and four times the most likely duration together and dividing the sum by 6. The standard deviation is estimated as the difference between the pessimistic duration and the optimistic duration divided by 6. The variance is equal to the standard deviation squared.

$$t_e = \frac{a + 4b + c}{6}$$

$$\sigma = \frac{c - a}{6}$$

$$v = \sigma_2$$

where

t_e = expected duration

a = optimistic duration

b = most likely duration

c = pessimistic duration

σ = standard deviation

v = variance

Calculating PERT Event Times and Probabilities

PERT charts can be constructed similar to activity-on-arrow logic diagrams. The project is broken into individual work elements and the logic is prepared using the arrow techniques. After the logic diagram has been completed, estimates of durations are made. These are listed on the diagram above the arrows as shown in Figure 7-6. Using a tabular format like that in Table 7-3, the expected duration for each activity is computed and then noted on the arrow diagram as shown in Figure 7-6. Table 7-3 can also be used to compute and list the standard deviation for each activity using the formulas developed earlier.

After the expected activity durations have been computed, a forward pass and a backward pass can be performed as was done for other types of CPM diagrams. The critical activities are those activities with matching forward and backward pass values which have no float. Float is again calculated by subtracting the early event time plus the duration from the late event time.

As the PERT network is a probabilistic approach, it is possible to determine the probability of completing either the project or the individual activities at any specific time. It is also possible to determine the time duration for a given

FIGURE 7-6   PERT chart.

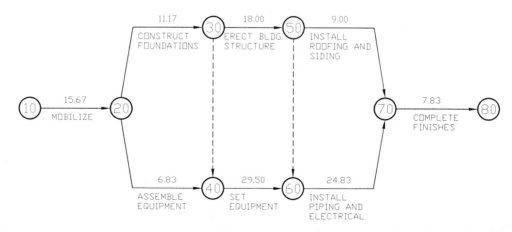

TABLE 7-3   PERT CALCULATION TABLE

Activity	Optimistic	Most likely	Pessimistic	t_e	T	$V(T^2)$
10–20	10	16	20	15.67	1.67	2.79
20–30	7	10	20	11.17	2.17	4.71
20–40	5	7	8	6.38	0.50	0.25
30–50	15	18	21	18.00	1.00	1.00
40–60	25	30	32	29.50	0.83	0.69
50–70	6	9	12	9.00	1.00	1.00
60–70	21	25	28	24.83	1.17	1.37
70–80	6	8	9	7.83	0.50	0.25

probability. This is done using the central limit theorem concepts and the cumulative normal distribution assumption and tables.

For the example shown in Figure 7-6, it is possible to calculate the probability of achieving various durations at each event along the network. To do this the central limit theorem is applied. The expected or mean time for each event is simply the sum of the mean times for each activity leading to that event. The variance for the total time elapsed is equal to the sum of the variances for each of the activities in the chain of activities leading to the event. Therefore the standard deviation for the event time is equal to the square root of the total variance of the event time. It should be noted that it is not possible simply to add the standard deviations of the individual activities to arrive at the standard deviations of the event time. The central limit theorem states that the variance of each activity is added to arrive at the total variance. The standard deviation is the square root of the event variance.[3]

Having found both the event-time standard deviation and the mean event time, the third concept in the central limit theorem can be applied. The distribution of the chain of activities to an event is a cumulative normal distribution, and it is possible to calculate the probability of various event times other than the mean event time. Before the cumulative normal table can be used, it is first necessary to convert the expected time and standard deviation to normal standard deviation values. This is accomplished by taking the event under study and subtracting from it the expected event time and dividing this difference by the event-time standard deviation. This is sometimes termed a z value. The z value is searched for in the standard normal distribution table (Table 7-2). Using the z value, a fraction or a percentage of the area under the normal curve can be obtained. If the z value is positive, then the percent is read directly from the table. If the z value is negative, then this percent is subtracted from 100 and the difference represents the probability of achieving the event in the specified time or less. Example 7-1 will use the values shown in Figure 7-6 and Table 7-3 to illustrate these calculations.

[3] Edward M. Willis, *Scheduling Construction Projects,* New York: Wiley, 1986, p. 203.

One drawback of assessing the probability of meeting the scheduled dates with this method is that only critical activities are considered. In determining probability it is assumed that the critical path is not affected by alternative paths no matter how near critical the other paths may be. If a delay occurred on a near critical path, it could become critical and all probability calculations would no longer be applicable. When using PERT to assess probability, it is necessary to examine near critical paths carefully and to make an assessment of the likelihood of them becoming critical. It may also be useful to employ a simulation technique such as Monte Carlo simulation to determine the probability of other paths becoming critical.

Example 7-1. For the network shown in Figure 7-6, the following probabilistic values are first determined:

Event	T_e	V_e	T_e
10	0	0	0
20	15.67	2.79	1.67
30	26.85	7.50	2.74
40	26.83	7.50	2.74
50	44.83	8.50	2.92
60	56.33	8.19	2.86
70	81.17	9.56	3.09
80	89.00	9.81	3.13

The expected event times (T_e) are computed using the expected durations in the same manner that a forward pass in activity-on-arrow diagramming is done. This requires choosing the longer path total at each merge event. This value corresponds to the early start time for subsequent activities. The critical path is determined by performing a backward pass through the network and identifying the activities with identical and late event times.

The event variance is determined by adding the variances for the activities that comprise the longest path to the event. For example, the variance for event 60 is 2.79 (activity 10-20) + 4.71 (activity 20-30) to (dummy activity 30-40) + 0.69 (activity 40-60) = 8.19. The standard deviation for the event is the square root of the variance. If two or more paths are all equal in value, the longest path to an event with the larger variance value is used. For event 60, the standard deviation is ÷ 8.19 = 2.86.

The standard deviation and the expected event times can be used in a normalized cumulative normal distribution form to determine the probability of meeting a schedule date.

Assume that it is desired to know the probability of reaching event 60 by day 52. The first step would be to determine the z factor which would correspond to day 52:

$$z^{52} = \frac{52 - 56.33}{2.86} = -1.51$$

Table 7-2 shows for $z = -1.51$ that the α value is $1 - .934 = .066$. This means that the probability of reaching event 60 by day 52 would be about 6.6 percent.

Assume now that it is desired to determine the probability that event 60 will be reached in less than 60 days. Then:

$$z = \frac{60 - 56.33}{2.86} = 1.28$$

$$\text{For } z = 1.28 \qquad \alpha \text{ Probability} = .899$$

This means that there is an 89.9 percent chance that event 60 will be reached by day 60.

Finally, it is desired to determine the time duration for reaching event 60 that corresponds to an 80 percent probability. For this probability it is necessary to find the corresponding z value. The probability (d value) .800 is located in Table 7-2. The corresponding z value is .84 (.799 is the closest α value).

The value of T can be found by substituting the known values of T_e and z as follows:

$$.84 = \frac{T^x - 56.33}{2.86}$$

$$T^x = 56.33 + .84(2.86)$$

$$= 58.73 \text{ days}$$

Similar calculations are often made for project completion and other milestone events.

Knowing the probability of completing a particular activity on a given date can be very helpful in making decisions about the sequencing of construction. For example, consider the development of a new residential development. Designers and constructors can use past experience to predict how much time it will take to design and build the development, but neither may be able to accurately estimate how long it might take to sell the completed homes. A PERT approach with probabilistic analysis can be used to predict the amount of time necessary to build and sell the completed homes. This type of information can be extremely useful in making financial decisions about the viability of the project.

Advantages and Disadvantages of PERT

The primary use of PERT is for projects which have not been done before. PERT provides a basis from which time and cost performance can be estimated. PERT also permits more information than other deterministic methods. PERT provides an assessment of the probability of reaching certain milestones by specified dates or of achieving overall project completion within a specified time period.

Highway projects involve several continuous activities which must be carefully sequenced for efficiency. (Photo courtesy of Barber-Greene.)

One of the major drawbacks of PERT in construction applications is that it requires multiple time estimates, which can be time-consuming to develop. A second deficiency of the PERT method is that the computed values of early event times and of the probability that events can be completed at a specified time are based solely on the duration of activities along a single path. If the PERT schedule contains several concurrent critical paths, there is a chance that a near-critical probability and its variability may negate the accuracy of the analysis. Also, the probability that alternative critical paths cannot be completed within their respective durations is not considered in PERT probability analysis. A practical disadvantage of the PERT method is that although it is not difficult, the mathematical calculations are much more complex than other scheduling methods and therefore must be offset by additional benefits. The perception in today's construction industry is that sufficient additional benefits do not exist to justify the added complexity.[4]

LINEAR SCHEDULING METHOD (LSM)

Network schedules assume that construction activities can be divided into a number of relatively small, discrete activities that can then be sequenced in the order of their performance. Although many construction projects can be divided into such sequential discrete activities, other projects cannot. In some cases, the same construction activities performed by the same crew progress continuously for the duration of the project. Transportation projects exhibit

---

[4] Edward M. Willis, *Scheduling Construction Projects,* New York: Wiley, 1986, p. 219.

this characteristic because of their linear nature (one operation or crew follows another sequentially). Highway construction involves activities for clearing, grubbing, grading, subbase, base coarse, and paving. Each of these activities must be repeated by the same crew from one end of the project to the other. Often the only distinguishing feature for these linear-type activities is their rate of progress. Network scheduling techniques can be used to schedule repetitious activities, but the resulting schedules are either very small (if the durations of the activities are large) or boringly repetitious (if the durations of the activities are subdivided by physical placement or location). Often bar charts are used to schedule linear projects, but they have their own disadvantages. Bar charts only relate the listed activities to the time scale, do not indicate activity inter-dependence, and—specifically for linear-type projects—cannot indicate variations in rate of progress.

An alternative method available for projects with long-duration activities is the "linear scheduling method" (LSM). The origin of linear scheduling is not clear. Linear scheduling has some relationship to line-of-balance schedules used in industrial manufacturing and production to evaluate a production-line flow rate. The name attached to linear-type scheduling varies throughout the literature that describes it. This may contribute to the LSM schedule's unclear origin. Linear scheduling can also be called "vertical production method" (VPM) due to its applicability to high-rise building construction. In addition to transportation projects, repetitive work in the typical floors of a high-rise building, airport runways, pipelines, mass transit, precasting or fabrication, and tunnels are projects well suited for a LSM approach to scheduling.

In addition to scheduling an entire project, the linear scheduling method can be used to evaluate interrelationships among a few select activities from a larger group of activities included in a network schedule. One of the disadvantages of network schedules is their inability to distinguish rates of progress among activities. A network schedule activity's duration is the total time required to complete the activity. The number of units that can be completed within any period of the duration is not apparent. LSM, in contrast, displays the number of units that will be completed within any period of the activity's duration. Even when the project has been scheduled by network methods, it may be very useful to be able to study the rate of progress among a few interrelated activities. LSM schedules can be used to evaluate activities in order to adjust, slow, or speed progress among the interrelated activities regardless of the project scheduling method employed.

Definitions

Figure 7-7 shows the basic format of a LSM schedule. The LSM diagram is used to plan and record progress on multiple activities performed continually over the duration of the entire project. The horizontal axis plots time, while the vertical axis plots location or distance along the length of the project. Individual activities are plotted separately, resulting in a series of diagonal

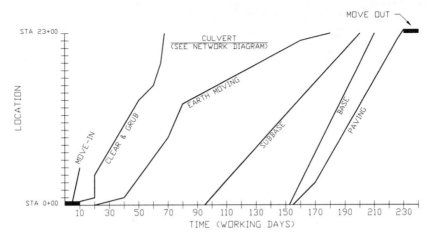

FIGURE 7-7   Linear scheduling method diagram.

lines. Activities may represent trade subcontractors moving from one floor to another in a high-rise building or the clearing, excavation, stringing, pipe laying, welding, and backfill operations on an underground pipeline. The slopes of the diagonal lines representing activities show the planned rate of progress at any location along the length of the project.

Elements of LSM Schedules

The LSM diagram compares time and location or sequence number of repetitive units. The change in location over time is a measure of the project's progress. Location can be measured in many ways. In high-rise building construction, the appropriate location measure may be stories or floors. In housing construction, the appropriate measure might be subdivisions, or apartments. In transportation projects, the convenient location measure is distance, usually expressed in stations (100 feet), kilometers, or miles. Time is most commonly measured in workdays, although hours, weeks, or months may be used.

Determining an activity's progress to form points on the activity lines in LSM is determined as durations for activities in network schedules are. Progress is estimated by the scheduler with necessary assistance from the estimator or manager and considering equipment, labor, or material restrictions. Once activity progress has been determined, it is plotted on the LSM schedule based on location and time as shown in Figure 7-8. The completion time for each activity is a function of the rate of progress and the amount of work to be accomplished.

Initial determination of the rate of progress should be based on the minimum direct unit cost of completing the activity. If it is necessary to compress the schedule, the rate of progress may be increased. Generally, either increasing or decreasing the rate of progress will increase the direct unit cost and therefore the completion cost for the activity. Caution should be shown in situations

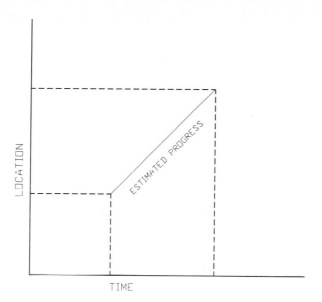

FIGURE 7-8   Plotting activity progress lines.

where indirect project cost may cause a shift in the optimum rate of progress (see Chapter 10).

The rate of progress for an activity may vary due to location or time. When beginning an activity, craft productivity is usually lower, reflecting the well-known learning curve. Individual production on an activity may vary as conditions vary. For example, the production rate for clearing and grubbing will vary with the density of forestation. Known progress variations can be shown on the LSM schedule at the appropriate location. Figure 7-7 shows a variation in progress at various locations for "clear and grub" and "earth moving." Repetitious task lines can be subdivided into activities grouped by common progress rates at given sections of the project. Activities are identified by a combination of numbers (I, J), with the first number representing the line and the second number the activity on the line.

Progress may also be affected by interference from other activities such as equipment maintenance or material restrictions. These and similar interruptions may be indicated on the LSM schedule by restraints. Figure 7-9 shows a restraint caused by equipment restriction, in this case a slip-form paver. Paving must be completed on one street before paving on another can begin. This schedule shows a sequence of Street A followed by Street B. The restraint indicates the time necessary to transfer and assemble the slip-form paver at the new location. The restraint is drawn as a dashed line.

Some spacing among interrelated activities may be required. The spacing serves as a buffer in order to prevent one activity from interfering with another or to accommodate differences in unit rates. For example, excavation for a new highway may take longer to perform than the installation and compaction

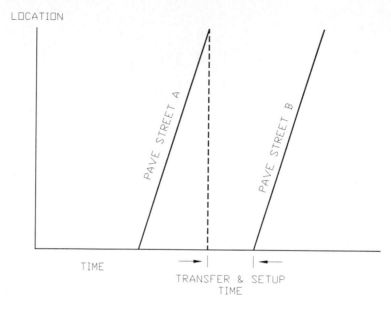

FIGURE 7-9   Use of restraint on LSM diagram.

of subbase material. Since excavation precedes subbase work, to prevent earthwork from stopping subbase installations, subbase activities may be delayed from starting until excavation is sufficiently ahead to permit the subbase work to be performed continuously. Alternatively, the progress rate may be slowed to avoid interference or interruption. Figure 7-10 shows activity interference between excavation and subbase work which can be avoided by use of "buffers."

FIGURE 7-10   Activity interference.

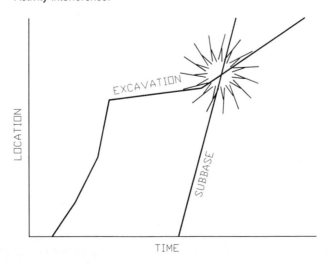

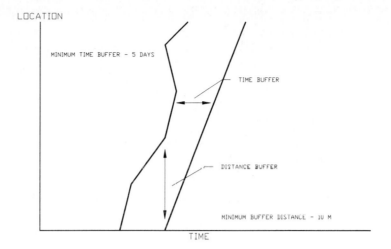

FIGURE 7-11   Use of activity buffers in LSM schedules.

Buffers may indicate required distances or time between activities. Figure 7-11 shows the use of minimum time and distance buffers on a LSM schedule. The buffers are drawn as solid lines.

Buffers are also used in LSM schedules to identify critical activities. A critical activity in a LSM schedule has a minimum buffer at both the start and completion of the activity. The buffers can influence both the start and finish of one activity, the finish of the activity and the start of a succeeding activity, or the start of the activity and the finish of a preceding one. In all cases, however, a critical activity will have a minimum buffer at each end of it. In Figure 7-11, minimum buffers at either end of activities indicate them to be critical.

FIGURE 7-12   Activity intervals for LSM schedules.

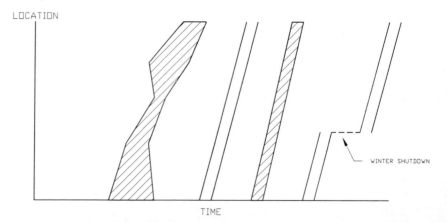

LSM scheduling can also be used to show the start and finish of an activity at any location. "Activity interval" is the term used to refer to the period of time between the start and finish of an activity at a particular location. The activity interval can be indicated in many ways. Figure 7-12 shows several ways that activity intervals can be indicated.

Preparation of LSM Schedule

A LSM schedule may be prepared for any series of sequentially related activities. Consider the activities necessary to excavate; form and pour foundations and slab; or frame a four-unit duplex development as shown in Figure 7-13.

To prepare a LSM schedule, the rate of progress for an activity must be known. The logic shown in Figure 7-13 is sufficiently detailed to indicate a rate of progress. One can see that excavation of each duplex's foundation is planned to require 2 days; forming and paving the concrete for each duplex's foundations and slab is to take 7 days; and framing each duplex, 8 days. Many network schedules, however, do not indicate a rate of progress. The duplex schedule in Figure 7-13 could have been drawn to indicate only one excavation activity for all buildings. If only one activity for excavation had been provided, the scheduler would have been required either to calculate a progress rate or to ask the estimator what rate of progress was anticipated. Of course, if a LSM schedule is the only schedule to be used, a rate of progress for each activity must be determined initially in order to complete the schedule.

Each activity's progress line can be plotted on the LSM schedule from the durations shown on the network schedule's logic. For example, from the network diagram it is apparent that each duplex's foundation excavation is planned to require 2 days. The LSM schedule will show the excavation for one of the

FIGURE 7-13 Four-unit duplex I-J fragnet.

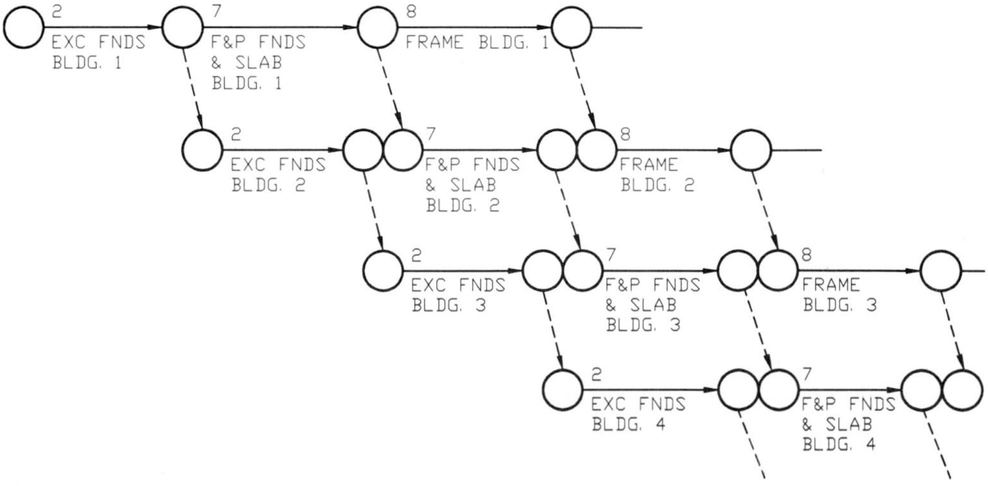

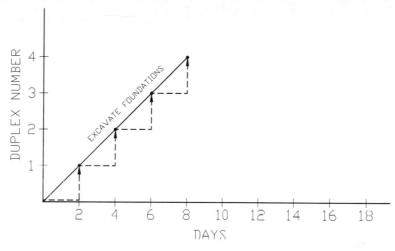

FIGURE 7-14   LSM foundations.

duplexes completed every 2 days. Figure 7-14 shows how foundation exca-
vations' line can be prepared.

The logic diagram also shows that forming and pouring foundations and slab
can begin immediately after completion of the foundation excavation. The
network schedule assumes that no buffer is necessary between the completion
of excavation and beginning to form the foundation for concrete pour. In other
words, forming and pouring the foundations and slab for the first duplex can
begin immediately after the excavation is completed. If the LSM schedule is
to be prepared from scratch, the scheduler must determine what buffer is
required between sequential activities. Forming and pouring each building's
foundation is expected to take 7 days. Figure 7-15 shows the addition of an

FIGURE 7-15   LSM "Form and Pour Foundations" activity line.

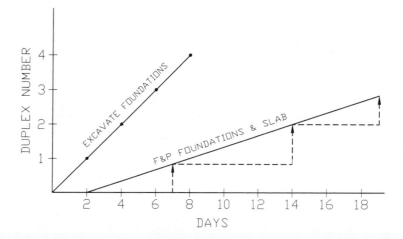

FIGURE 7-16   LSM schedule from logic in Figure 7-13.

activity line for forming and pouring foundations beginning at day 2, when excavation has been completed.

Note that the rate of progress to form and pour the foundations is slower than that for foundation excavation. Progress lines for the remaining activities are prepared similarly. Since the network schedule indicates that no buffer is necessary, the framing line can begin as soon as the foundation and slab have been poured on the first duplex. Figure 7-16 shows the addition of the framing activity.

As indicated earlier, the progress for the activities shown on the network schedule from which the LSM schedule was prepared had already been determined. Suppose, however, that the scheduler is required to add an activity for brickwork to the LSM schedule—an activity for which progress or buffer had not been determined earlier. In order to add the line to the LSM schedule, the scheduler must determine progress and buffer.

Working with the estimator, the scheduler determines that each duplex is estimated to require 160 manhours to complete the required brickwork. The estimator also indicates the crew will be composed of three people, two masons and one assistant to mix mortar and distribute materials. The crew will work a regular 8-hour shift. The scheduler can thus determine that each duplex will be completed in 6.6 days:

$$3 \text{ men} \times 8 \text{ hours/day} = 24 \text{ manhours/day}$$
$$160 \text{ manhours} \div 24 \text{ manhours/day} = 6.66 \text{ days}$$

The scheduler will use 7 days for the completion of each duplex.

A progress line for the brickwork can be added showing completion of each duplex's brickwork every 7 days. Before adding the brickwork line, however, the scheduler must determine whether a buffer is required. The scheduler determines that a 1-day buffer is necessary after the completion of the frame-work for delivery of brick, cement, and scaffolding equipment. Figure 7-17 shows the LSM schedule with the addition of the brickwork progress line and buffer.

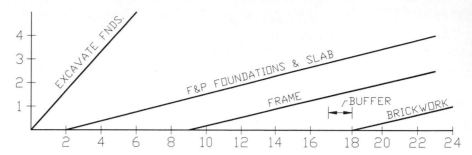

FIGURE 7-17   LSM schedule with brickwork.

Use of LSM Schedules

Each line on a LSM schedule indicates the rate of progress for an activity. In using a LSM schedule, the objective should be to keep all the activity progress lines as close to parallel to each other as possible, considering the economics of slower or faster production. In this way project completion can be achieved as quickly as possible. Similarly, progress lines that show a wide variation among themselves may be an indication of insufficient manpower for activities with low slopes or overmanning on activities with high slopes.

The start and finish dates for each activity can be determined by assigning the appropriate calendar day to the start of the initial activity as was done when dates were determined in network schedules. The dates can be very useful in scheduling subcontractors, material, equipment, and labor. Similarly, the start and finish dates for each trade can be determined, and by combining activities which require similar craftspeople with crew requirements for each activity, a labor histogram can be prepared in order to balance manpower.

Advantages of LSM Schedules

Planners and schedulers recognize a considerable void between a bar chart and network schedules. LSM schedules may fill that void. The LSM schedule is easier to prepare and use than a network schedule but presents more information than a bar chart. On certain projects, the LSM schedule may be preferable to a network schedule. Further, LSM schedules show rate of progress, something neither bar charts nor network schedules are capable of indicating. In the repetitive portion of projects, LSM schedules convey the nature of the problem more quickly, aid in its solution, and present the intent of the schedule even if the project is scheduled using another method. On projects involving both repetitive and discrete activities, both network and LSM schedules can be combined and coordinated.

A significant advantage of LSM diagrams is the simplicity with which it can convey a detailed work schedule. LSM schedules are easily understood by management and field staff. LSM schedules' ease of preparation and flexibility also recommend their use as planning devices for comparing different scheduling alternatives of progress rates, equipment combinations, or sequence.

The major difficulty in LSM scheduling is determining minimum activity interval and/or buffers. Common sense prevents a contractor from conducting two operations at the same location at the same time or from starting an activity before it can be completed without interruption. Some buffer must be established to determine, for example, when to start the base course after starting the subbase in a specific location. The LSM can be used to calculate intervals or buffers once progress rates have been determined.

The start and finish dates for each activity or the entire project can also be determined from a LSM schedule. Just as with a network schedule, by applying the appropriate date of the start of an activity and adding an activity's duration, start and finish dates can be computed. Similarly, a labor histogram can be prepared by assigning craft codes to each activity line, much as with a network schedule. Resources may be balanced by adjusting progress. An earnings curve may also be produced by comparing the number of units forecasted to be completed on any progress payment date.

Disadvantages of LSM Schedules

Linear schedules are less effective when the project activities do not need to follow each other in the same order at every location or when repetitious activities are regularly interrupted. For example, a LSM schedule can be used to schedule construction of a road in a rural area, but in a city where many interferences, such as utility relocation, culverts, bridges, traffic, or temporary services may regularly be encountered, the repetition of activities is disrupted and LSM schedules are no longer effective as a scheduling tool. Even when repetitious sequential activities are not present, however, LSM schedules can be useful to evaluate the best combination of individual progress rates for least-cost or least-time calculations.

EXERCISES

1 An assessment of existing conditions at the Re-solve hazardous waste site is necessary in order to prepare a plan for cleaning the site. The 6-acre Re-solve hazardous waste site is an abandoned solvent-reclamation facility in a rural area. As a result of past disposal practices, contamination of soil, groundwater, and surface water has been documented at the site. In order to assess existing conditions, fifteen soil borings and six test pits are required. Each soil boring requires one drilling rig and is estimated to take $1\frac{1}{2}$ days to complete. Each test pit requires a backhoe and is estimated to take one $\frac{1}{2}$ day to dig and another $\frac{1}{2}$ day to obtain samples and study the exposed excavation. It is requested that all soil borings be completed and analyzed before the test pits begin, and that the data for each test pit be collected before a new pit is excavated. Soil borings require 5 days for analysis. Prepare a LSM schedule showing the completion of the site investigation using (a) one drilling rig and (b) three drilling rigs.

2 The following data have been developed for a small project:

Activity	Preceded by:	Duration a	Duration b	Duration c
A	—	2	5	8
B	A	5	6	7
C	A	3	5	7
D	B, C	8	10	11
E	B	6	9	12
F	C	4	9	11
G	D, E	3	5	7
H	F, G	6	6	7
I	E, F	3	11	15
J	H	10	13	16
K	I	4	6	9
L	J, K	3	5	8
M	J	6	10	14
N	K	5	7	7

(*a*) Determine the critical path.

(*b*) What is the probability that the project will be completed by day 50?

(*c*) What is the probability that the project will be completed by day 56?

(*d*) What is the duration that corresponds to an 80 percent likelihood of on-time completion?

(*e*) What is the probability that activity I will be completed by day 31?

3 Given the following project logic and estimates of activity duration, determine the mean project duration and project duration variance. How many days would be required to give a 90 percent chance of completing the project on time?

Activity	Followed by:	Duration a	Duration b	Duration c
A	—	5	6	9
B	J	5	8	13
C	A	3	4	9
D	A	8	9	10
E	D, J	10	12	15
F	H, G, K	8	11	12
G	D, J	2	3	3
H	B, C	5	7	11
I	E	3	5	9
J	A	8	9	10
K	A	12	14	18

4 Your company is responsible for installing a 24-inch pipeline. You have three crews, each of which performs a unique operation. Each crew is currently scheduled to work

5 days per week at 8 hours per day. The crews and their normal daily (8-hour) production rates are listed below. Assume that you would like to keep a minimum of 1 day or 200 linear feet, whichever is greater, between crews. The excavation crew started 10 working days ago. The pipe-splicing crew started 3 days ago, and the pipe-laying crew will start in 2 days.

Crew	Operation	Daily production (linear feet/day)
Excavation	Digs the trench	400
Pipe-splicing	Welds and coats the pipe	320
Pipe-laying	Sets pipe and backfills	450

(a) How long will it be until the pipe-splicing crew will control daily production at 320 linear feet/day?

(b) Assume that you will put the pipe-splicing crew on a 10-hour day. How long will it be until the pipe-laying crew catches the pipe-splicing crew?

(c) What would the ideal work schedules be to match production and have the minimum spacing between crews at completion?

5 Prepare a LSM schedule for a 1480-foot-long county road. The road will be a four-lane, undivided, paved highway. It has the following structure: a 10-inch layer of lime-stabilized earth with a 4-inch layer of ¾-inch crushed gravel and a 4-inch layer of asphaltic base with a 2-inch layer of asphalt topping. The entire length of the project will have concrete-formed curbs with storm sewer openings at approximately 200-foot intervals on each side. The curbs will have reinforcing bars and will be saw-cut at 50-foot intervals to allow for expansion. Site preparation will include clearing and grubbing. The last 280 feet of the project has major underbrush which will require extensive clearing and grubbing. There are four box culverts located at stations 240, 290, 730, and 1110. Only a moderate amount of cut-and-fill will be required, with all of the cut taking place between stations 420 and 700. Most of the fill will be needed between stations 150 and 370. An existing road that crosses the new road at station 570 will have to be demolished. No traffic will use either road until final completion. Final completion includes painting the stripes, highway signs, and fine grading and seeding work on both sides.

Prepare a complete LSM schedule meeting the above requirements. Determine the project duration using reasonable productivity rates.

6 Prepare a LSM schedule for a 3500-foot pipe rack that will consist of concrete-drilled pier foundations approximately 8 feet deep and 2 feet in diameter on which will be erected steel bents at 25-foot intervals with steel truss connection girders. The pipe rack will consist of two layers of piping: The upper level will have two 12-inch lines, two 8-inch lines, a 6-inch line, and four 4-inch lines; the lower level will have five 4-inch lines, two 6-inch lines, and twelve 3-inch lines.

Research the productivity and anticipated rate of progress for the erection of this pipe rack. Assume 40-foot lengths of pipe and a minimum number of turns. The entire steel frame will be sandblasted and painted after erection but prior to the installation of the piping. The piping systems will be insulated and tested prior to being placed in operation. Determine a reasonable time frame for the entire pipe rack project.

7 Assume that you are a contractor who has been awarded a contract to install 38 miles of two 16-inch natural gas pipelines to be routed through rural areas of North Dakota.

These lines will be interrupted by three pump stations spaced equally along the length of this project. The pump stations each require two high-speed compressors for the pumping of the natural gas. Each pump will be provided with a auxiliary pump in stand-by position. The pumps sit on concrete pads and must be piped into the pipeline system, complete with valving and control equipment. Each of the pumps will also be provided with a cooling system along with a cooling tower erected outside of the structure. The structure will be a typical metal building erected on perimeter footings to cover the natural gas pumps. The normal accessory equipment will be anticipated to include electrical panels, utility tie-in for electrical lighting systems, interior painting, phone lines, control panels, and automatic alarm systems.

Prepare a CPM diagram in I-J format for the three pump stations and a LSM format schedule for the pipeline. Integrate the two schedules.

8 Repeat the project in Exercise 7 but in PDM format with LSM scheduling for the pipeline installation.

9 Discuss whether a computer program could be developed for LSM schedules. Would it improve LSM scheduling? Would anyone buy it? Are there readily available substitutes?

CHAPTER
8

UPDATING
THE SCHEDULE

WHY UPDATE THE SCHEDULE?

Only in the rarest circumstances is a planned schedule followed precisely from the start of a project to completion. Considerable time and effort are required during the project to check actual progress against the planned schedule. Marking progress on the logic diagram and making sequencing adjustments to match actual construction is called "updating." Normally this process also includes transferring the project status and logic adjustments marked on the logic diagram to the computer database and recalculating the schedule.

Sometimes, because of lack of progress, changes in construction approach, or unanticipated problems, the project is delayed. To bring the work back on schedule, the schedule requires revision. These schedule revisions are part of what is called "rescheduling." Rescheduling, if required, is normally done at the same time as an update. Updating is concerned primarily with the effect that changes in the job plan have on the portions of the project yet to be completed. These changes may be sequential work changes or variations in progress on activities from the planned durations. Rescheduling is normally performed to revise the original schedule to the current plan in recognition of changes in the contractor's approach, absorption of poor progress, or mitigation of delays.

Updating is required frequently on projects facing high uncertainty, such as tunneling work. (Photo courtesy of Joy Manufacturing.)

There are many reasons why a contractor should update the schedule. In the construction industry, the manner in which the project is actually built in the field will often differ from the way the scheduler (or even the project manager) originally planned to build the project. The initial project schedule cannot anticipate every future circumstance. Unforeseen problems, adverse weather, change orders, more knowledge, and mistakes are discovered or experienced every day in the field. As a result it is highly improbable that the scheduled durations or sequences will match field durations and sequences. If the schedule is to continue to be used to predict project completion or manage construction, the initial schedule should be updated and revised to include these unanticipated events.

The updated schedule is important to more than just the contractor. Everyone involved in the construction project needs to understand how the project is progressing. Distribution of an updated schedule is a good way to keep participants informed. If the project is behind schedule, the designer needs to know to conform to the certification responsibilities of the contractor's progress payment requests. The designer needs to be made aware of changes in priorities for shop drawings and change-order reviews. The owner needs to know if the project is behind schedule so that necessary changes in move-in or owner-furnished equipment delivery can be made. The contractor needs to know to take appropriate actions to get the project back on schedule or request appropriate time extensions. Trade contractors and suppliers need to know how their work relates to the general contractor's work and to the work of other subcontractors. Work priorities can be determined by studying the updated schedule's critical path.

A project that is proceeding on schedule will not only hasten payment of contractor progress payment requests but also reduce more quickly the amount of retainage the owner holds to insure contractor completion of the work. Many contracts permit retainages to be eliminated or halved when the project is 50 percent completed and the project is on schedule. Prompt payment and retainage reduction keeps the contractor in working capital, and encourages the contractor to promptly pay subcontractors and suppliers, which further encourages effective construction progress. A financially healthy contractor has much less trouble borrowing money or bonding their projects.

A single activity that is delayed is generally so limited that lost time cannot be recovered by increasing the remaining effort and completing within the original duration of that activity. Corrective action, if possible, is usually based on logic changes or reduced durations to subsequent activities. Indispensable to corrective rescheduling is an updated schedule showing current status and how remaining activities are planned to occur.

If a project's scheduled completion date has been delayed, those involved will want to know the cause. Those responsible for the delay may face significant financial penalties. The contractor may be responsible for liquidated or other delay damages to the owner if the contractor's actions have caused the delay. The owner may be responsible for the contractor's delay costs if the owner has caused the delay. The designer may be responsible for either contractor or owner delay costs if the designer caused the delay. All potentially responsible or damaged will want to know the cause of delay and be able to explain why they are not responsible. A schedule regularly updated to show the effect of changes, errors, or delays can be used to determine and allocate responsibility for delay.

An updated schedule provides an important record of the project status at a specific time. These data will be extremely valuable later.

Without updates, the schedule loses its accuracy. The trade contractors, the designer, and the owner will stop relying on the schedule, and the contractor will lose a valuable project control tool which the schedule was intended to provide. If the schedule is not accurate, it cannot be used as a basis for time extension requests.

FREQUENCY AND DETAIL OF UPDATES

Updates may occur daily, weekly, bi-monthly, monthly, or less frequently depending on the size, complexity, and characteristics of the project. Monthly updates are the most common. Frequency of updating may be determined by the scheduling clause or by choice of the participants. If the progress of the work has followed the schedule closely, the updating process can be performed less frequently, although the authors recommend a regular interval of updating even when the project is progressing well. Do not, however, allow updating to stop completely. The updating process helps anticipate problem areas. If there have been a number of significant deviations from the schedule, then an updated schedule is probably necessary. Continuing to use an obsolete schedule can negate much of the advantage of formalized planning and scheduling.

Updating a project schedule for a project which is moving along as planned also gives the various contractors on the project a chance to meet and discuss upcoming work sequences. In addition, updating allows integration of subcontractors whose work is just starting (i.e., painting, flooring) into the construction process and fosters approaches for finishing the project early with the possibility of improved financial efficiency.

Updating reviews all parties' commitments to the use of the schedule and reminds all parties of their promises and responsibilities. The updated schedule can be used as an agenda for progress and planning meetings.

Turnaround time for an update is also important. Schedules need to be current to be effective as management tools. An incomplete update has no value. Since change can occur rapidly in construction, it is important that a minimum of time passes between data collection and processing and distribution of the updated scheduling documents. This assures that the update reflects a true picture of the project.

The degree of detail can also vary in an update. The update may identify actual start and finish dates for each activity or merely show that work was either completed or progressing if actual dates are unknown. An update may also attempt to estimate the percentage completion for each activity. If the schedule is used to compute progress payments, estimated percentage completion will be used to determine the value of progress payments. An update may also adjust the duration of activities to indicate no remaining duration (completion) or to reflect reevaluations of the duration of future activities based on actual field experience. An update for network schedules may revise the logic diagram to show the contractor's current plan for completion, reflect improved methods, or show the effect of changes.

UPDATING BAR CHARTS

Figure 2-4 is a bar chart used to schedule the completion of a 42-story office building. Figure 8-1 shows the same schedule updated in the middle of February to show the status of the work. The date of the update is shown by a vertical line of dots and dashes on the date of the update. At the time of the update, five activities had been completed. The actual finish dates are shown by solid triangles. Anticipated completion dates are indicated using open triangles. A dashed line has been drawn to show that the contract completion date has been revised to be 4/15/89 but does not give a reason for the delay of nearly 2 months beyond the revised completion date. Delay in completion may be due to additional work for which the contractor has not yet received time extensions, or to the contractor's failure to make adequate progress.

The update anticipates that six activities will complete within the next 2

FIGURE 8-1 Bar chart schedule.

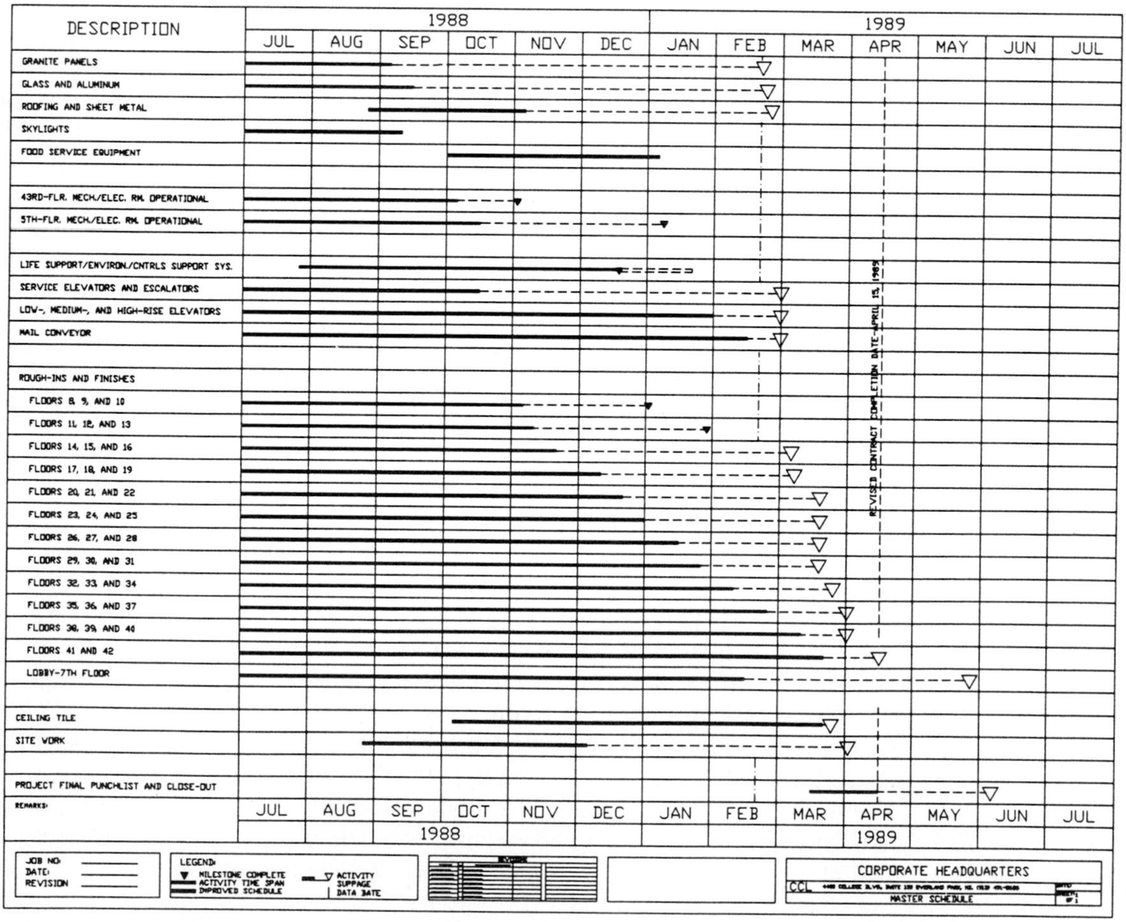

weeks, while remaining activities show completion during the next 3 to 4 months. Three activities are indicated not to complete until or after the revised contract completion date.

In contrast to the delayed completion of most activities on the February update, completion of systems for life support, environmental, and controls are shown to have been completed earlier than originally scheduled. The solid triangle used to indicate actual completion indicates that work was complete in November rather than January. The improved schedule is shown by the open outline of the bar representing the activity's original duration.

The original bar chart could show percentage completion or an S-curve. The update could also show actual percentage completed, revised anticipated future completion percentages, actual S-curve completion, or a revised S-curve. Figure 8-2 shows an updated bar chart with an S-curve.

FIGURE 8-2   Updated S-curve.

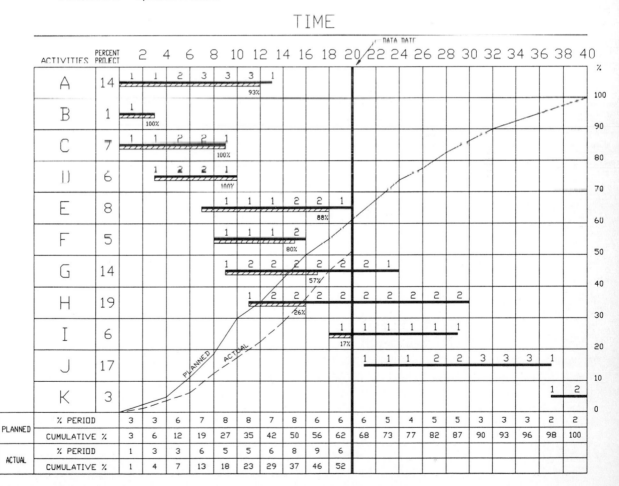

Activity bars have been updated by showing progress with a cross-hatched bar. To update the S-curve, additional information is necessary. Actual percentage completion for each activity for each period has to be determined. For the S-curve updated in Figure 8-2, the actual completion percentage has been determined and recorded below the activity bar for each period prior to the time of the update. Only this month's progress total is indicated as a percentage below the progress bar. Total project actual completion has also been calculated and recorded on the bottom line. Progress by period can be evaluated by comparing planned cumulative percentage completion to actual cumulative percentage completion.

Actual status may be shown on bar charts in a number of ways. The February update shown in Figure 8-1 used solid triangles to show actual completion dates and dotted lines from the original bars to show extended performance. Actual activity completion could also be shown as bars drawn beneath the scheduled completion in contrasting colors, shapes, or styles as indicated on Figure 8-2. Activities in progress but not completed can also be shown in a variety of ways. Bars may be colored or drawn beneath the scheduled completion.

Figure 8-3 shows two ways to update activities that have been started but not completed. The February update indicates anticipated revised completion dates later than originally scheduled. Some updates will not indicate revised completion dates for activities in progress as shown in Figure 8-1.

Information necessary to update the bar chart comes primarily from two sources: information supplied by the project's management team or the scheduler's observations made at the time of the update. The scheduler can use either or both sources in the updating process.

Commonly, a schedule is updated based on the scheduler's inspection and observation. A date for the update is selected; the scheduler appears and tours the project site. Usually the project manager or appropriate field superintendents will accompany the scheduler. Actual progress will be noted and antici-

FIGURE 8-3   Activities updated to show start but not completion.

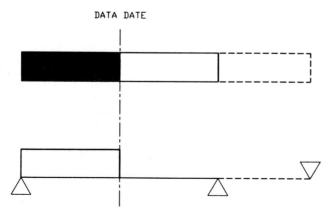

pated completion will be estimated. The scheduler discusses rates of progress, current delays or problems, and any adjustments to the planned sequence of work.

Based on the data observed and discussed during the project tour or supplied by the project's management team, the scheduler prepares an updated schedule within the next several days. The updated schedule is then distributed to appropriate individuals on the owner's, subcontractor's, designer's, and contractor's staffs. Whatever the source of status information, the basis of the update is the previously updated bar chart or the initial schedule. Bars are usually not rearranged even if the work sequence has changed.

If desired, an updated bar chart can illustrate if the activities are ahead or behind schedule and if the project will complete on time. Updated bar charts with percentage completion calculated can be used to support contractor requests for progress payments. An updated bar chart, however, does not show why activities have been delayed, or whether activity delay will result in project delay. Because a bar chart has no critical path, the observer cannot determine the anticipated project completion date based on the progress of the individual activities. Only the scheduler can insert a new anticipated completion date, which is normally just a good guess.

Time and money are not directly correlated in construction, thus the bar chart can easily be distorted to show the project at a more favorable status than it actually is. This distortion can be very cleverly hidden and made very difficult to detect.

Most projects scheduled with bar charts have a separate process for evaluating progress and payment. Contractors sometimes state, "We've used 60 percent of the time for this activity but only achieved 40 percent of the monetary value of the work, but we are on schedule because we have completed most of the time-consuming work." This statement can be true, as shown by the very nature of construction progress S-curves. Many times, however, this statement is an erroneous evaluation of the situation.

UPDATING NETWORK DIAGRAMS

One advantage of network schedules is their ability to easily incorporate schedule changes as the project is completed. Computer-stored data facilitate updates. Like updated bar charts, an updated network diagram is based on the initial network schedule or the latest update. The first step in updating network diagrams is to mark project status on the logic diagram. Figures 8-4a and 8-4b show updated logic diagrams in I-J and precedence formats. The scheduler has indicated revised activity durations by striking through the current estimated durations. These are marked to show either no duration remaining for activities that have been completed or a revised duration for those activities which have not yet been completed. Note that "ins" from the procurement subnetwork indicate that light fixtures, HVAC controls, and stone tile have been delivered and are ready when installation can begin.

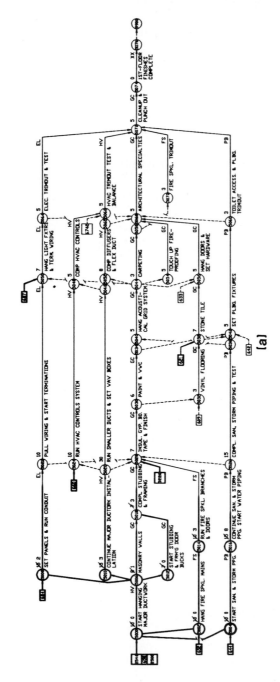

FIGURE 8-4a Updated rough-in and finish subnetwork in I-J format.

[a]

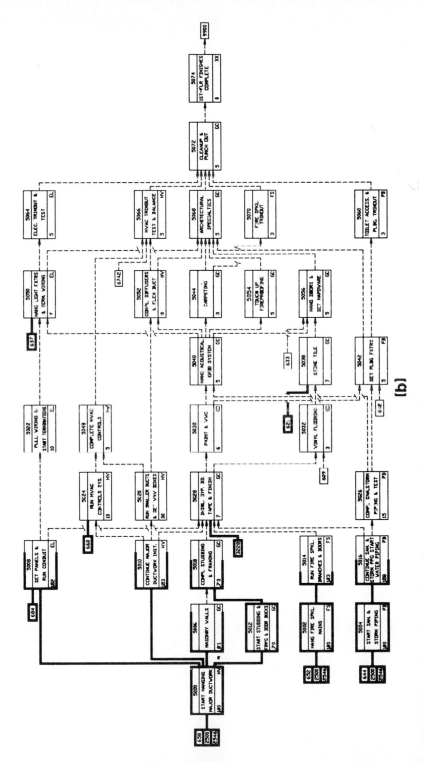

FIGURE 8-4b Updated rough-in and finish subnetwork in procedure format

Also note that "in" 3228-5028 is marked as complete. This restraint or relationship is from "Bldg. Essentially Watertight," indicating that the building is enclosed and drywalling can proceed.

Revised durations for incomplete activities can be determined in one of three ways. The remaining duration can be calculated by subtracting the time actually expired from the original durations; the remaining duration may be directly proportional to the original duration; and the percent of work complete or remaining duration can be determined from the scheduler's experience and actual rates of progress. For example, the scheduler or project manager may realize that, for any number of reasons, the originally estimated duration may have been incorrect. Actual remaining work may require that the original duration be increased even though some work on the activity has been performed. Alternatively, new and better information may indicate that remaining durations should be reduced. However the amount of remaining time is calculated, the durations on partially completed activities on the logic diagram are revised. "Informed judgment" durations are historically the most accurate.

As-built dates for start or finish of activities may also be recorded. If known, actual dates should be noted on the logic diagram and/or printout. Placing actual dates on the logic diagram in the field as work progresses is the best way to maintain accurate start and finish dates.

In addition to revising durations, network diagrams should show activity sequences updated. To the extent that completed sequences varied from scheduled sequences, the logic diagram may be revised to show actual sequences. Showing actual sequence may be important if the schedule is used as a time measurement tool although arguably not necessary if the schedule is to be used only to manage future work. Planned sequences of future work should be shown if the schedule is to be used as a time measurement or management tool. New activities may be added to the logic as indicated in Figures 4-5 and 4-6. Revised sequences may be indicated either by crossing out old logic and noting revised logic on the diagram or by redrafting or replotting the logic diagram.

Figure 8-4a shows some out-of-sequence progress at activities 5006-5018 and 5018-5028. Figure 8-4b shows the identical situation in precedence format. This situation is very common when updating CPM logic diagrams. Progress indicated does not mean that the contractor is not following the schedule. It also does not mean that the resultant logic is wrong. The original schedule indicated that all the masonry was going to be completed prior to "Complete Studding and Framing." The contractor has completed all but 1 day's worth of masonry wall installation. If the last day's work of masonry still must be completed prior to completion of the studding and framing, then the logic is still correct. Revising the logic each time minor out-of-sequence progress occurs is unnecessary and usually results in incorrect logic.

Major out-of-sequence work, however, does warrant a logic change. For example, if activity 5038-5056, "Stone Tile," was 100 percent complete with the update status on Figure 8-4a, then a logic adjustment should be made. Be aware, however, that except for major sequence changes, in most cases leaving the logic as is and merely updating the schedule status will result in accurate schedules. When the contractor's plan for completing the remainder of the project does not match the current schedule's logic sequence, then the logic must be adjusted.

Some scheduling software programs allow for different methods of calculating updates. Two calculation methods are currently used. They are known as "retained logic" and "progress override." Which method is selected can affect the projected project completion date. While both techniques have useful applications, most schedulers prefer to use the retained logic calculation method. The progress override calculation method allows the software to make sequencing changes to the logic based on status information on individual activities. Good scheduling practice dictates that scheduling decisions should be made by the scheduler, not the software.

When a schedule is updated using the retained logic method, the software calculates activity dates by using the current logic relationships and current durations to complete each activity. The software follows the sequencing regardless of partially or totally completed activities. The software does not modify the logic in any way. If the work on the project is proceeding as originally scheduled or with only minor differences, then no logic modifications are required. If the work sequence is significantly different from the scheduled sequence, then the logic should be adjusted. The retained logic calculation method is usually a more accurate projection of the current schedule and the completion date.

The progress override calculation method assumes that if an activity has started, then any preceding activity which is not 100 percent complete is now improperly sequenced. The progress override function resequences the activities so that every activity which reflects progress is calculated from the data date. Figure 8-5a indicates an original CPM schedule. Figure 8-5b indicates the updated status of the original schedule. The original project duration was 31 days. Using the retained logic calculation method, Figure 8-5b indicates a remaining project duration of 19 days. Figure 8-5c is essentially what the software would calculate using the progress override calculation method for the status information indicated in Figure 8-5a.

Either Figure 8-5b or Figure 8-5c could represent an accurate plan for completing the project. The scheduler, however—not the software—should determine the correct planned sequence and modify the logic accordingly.

Figures 8-5d and 8-5e indicate a different progress situation and the resulting progress override calculation.

As in updating a bar chart, information to update a network schedule may come from the project management team or the scheduler's observation of the site. Touring and observing the project while updating the schedule is commonly called a "site visit" update.

Once the update information has been gathered, the scheduler performs new

FIGURE 8-5 (a) Original sequence prior to updating. (b) Progress status information with retained logic. (c) Progress status information with progress override. (d) Updated logic using retained logic calculation method. (e) Updated logic using progress override calculation method.

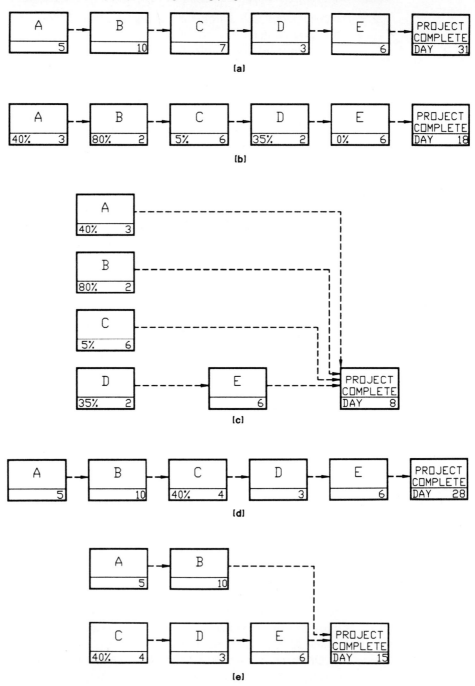

forward and backward passes, based on revised durations and sequences, calculates revised event times and float values, and determines the revised project completion dates and the critical path. The new forward and backward passes may be performed manually or with the aid of a computer. Whatever method the scheduler utilizes to perform the forward and backward passes, all revised calculations are based on the date the revised information was determined rather than the date the new calculations were performed. The date the status information was determined is usually called the "data date." Completed activities must show zero durations. As-built dates may be entered and printed on the computer printout. Anticipated start and finish dates that vary from scheduled dates due to known delays should be noted and reflected as increased durations on the schedule.

With large or complex schedules it is advisable to process the status information first. Running a printout based on the update date before revising the logic provides the contractor with important progress information. This step clarifies, even for the most experienced scheduler, what adjustments to the logic may be required to get the project back on schedule. The second running of the schedule with the same update data with adjusted logic provides the contractor with a measure of the effects of the logic changes. The contractor can then weigh the various methods of reducing remaining project time against costs.

One page of an updated printout showing new calculations and remaining durations for a typical logic diagram is shown in Figure 8-6. The updated printout permits an evaluation of the project's overall status when compared to the initial schedule. The updated network schedule calculates revised starting and finishing times for individual activities and indicates any changes in the schedule's critical path.

Network schedule updates can tell not only the revised project completion date but also why a project is behind schedule if actual dates and sequences have been used. If the start of activities has been delayed or performance has been too slow, a network schedule update will show these situations and indicate how they may have affected the project completion date.

I-J and AON schedules have been used to illustrate how a network schedule should be updated. Updating PDM and PERT schedules follows similar procedures. Both require the activity's duration and sequence to be reviewed and updated. PDM schedules are harder to update than I-J or AON schedules if lag, complex, or combined relationships have been used. Not only must duration and sequence of PDM schedules be reviewed, but the amount of lag and the validity of the complex or combined relationships must be checked and revised to match actual conditions or changed methods.

The updated logic diagram and printout should be accompanied by a "narrative report." A narrative report will describe in written form the meaning of the numbers, event times, and revised calendar dates presented in the printout. Although each author's style may differ, a narrative report will usually contain a description of progress since the last update:

A discussion of problem areas
Identification of alternate critical paths
Description of any logic revisions
Prediction of future problems
Possible recommendations for corrective action

The narrative report is easier and quicker to read than the computer printout and provides schedule information to those who cannot read or do not wish to take the time to read the CPM printout. When a narrative report is available, the printout may not be reviewed at all.

Because methods and details for network schedule updates may vary, construction specifications may mandate a particular procedure that must be followed. One agency that requires a detailed update procedure in its contracts is the U.S. Veterans Administration.

The Veterans Administration requires a date for the update that is agreeable to all parties to be identified in advance. Several days prior to the update, the

FIGURE 8-6   Primavera Project Planner/2nd update.

```
-------------------------------------------------------------------------------------------------------------------------
CCL CONSTRUCTION CONSULTANTS, INC.              PRIMAVERA PROJECT PLANNER

REPORT DATE 21MAY90  RUN NO.   12       SECOND UPDATE 16JUN89                    START DATE 10JUN88  FIN DATE 20FEB90
            12:03
NUMERIC SORT BY I-J NODE                                                         DATA DATE  16JUN89  PAGE NO.   27

          ORIG REM  -                                                           EARLY    EARLY    LATE     LATE    TOTAL
          DUR  DUR CAL  %   CODE            ACTIVITY DESCRIPTION                 START    FINISH   START    FINISH  FLOAT
PRED SUCC
3300 3310   0   0 1   0      GENC RESTRAINT                                      30JUN89  29JUN89   1AUG89  31JUL89    21
3300 3312   0   0 1   0      GENC RESTRAINT                                      30JUN89  29JUN89   1AUG89  31JUL89    21
3302 3312   0   0 1   0      GENC RESTRAINT                                      26JUN89  23JUN89   1AUG89  31JUL89    25
3302 8000   0   0 1   0      GENC RESTRAINT                                      26JUN89  23JUN89   6JUL89   5JUL89     7
3304 3312   0   0 1   0      GENC RESTRAINT                                      16JUN89  15JUN89   1AUG89  31JUL89    31
3304 3314   0   0 1   0      GENC RESTRAINT                                      16JUN89  15JUN89  14AUG89  11AUG89    40
3306 3328   5   5 1   0  GCSTGENC SET LIMESTONE THRU 3RD FLR EAST FACE            5JUL89  11JUL89  31JUL89   4AUG89    18
3308 3330  13   1 1  90  GCDWGENC EXT STUDDING & FRAMING @ 5TH FLR               16JUN89  16JUN89  24JUL89  24AUG89    25
3310 3322  13   4 1  20  GCDWGENC MISC INSUL & EXT GYP BD SHEATHING @ 4TH FLR    30JUN89   6JUL89   1AUG89   4AUG89    21
3312 3324   8   0 1 100  GCFPGENC SPRAY FIREPROOFING 3RD FLR
3314 3316   0   0 1   0      GENC RESTRAINT                                      16JUN89  15JUN89  14AUG89  11AUG89    40
3314 3318   0   0 1   0      GENC RESTRAINT                                      16JUN89  15JUN89  14AUG89  11AUG89    40
3314 3320   0   0 1   0      GENC RESTRAINT                                      16JUN89  15JUN89  14AUG89  11AUG89    40
3314 3326   6   0 1 100  VENTGENC SET ANCHORS FOR DUCTWORK 4TH FLR
3316 3326   4   0 1 100  HTACGENC SET ANCHORS FOR HTG & COOLING PPG 4TH FLR
3318 3326   4   0 1 100  FIREGENC SET ANCHORS FOR FIRE SPKL PPG 4TH FLR
3320 3326   4   0 1 100  PLUMGENC SET ANCHORS FOR PLBG PPG 4TH FLR
3322 3328   0   0 1   0      GENC RESTRAINT                                       7JUL89   6JUL89   7AUG89   4AUG89    21
3322 3332   0   0 1   0      GENC RESTRAINT                                       7JUL89   6JUL89   8AUG89   7AUG89    22
3322 3334   0   0 1   0      GENC RESTRAINT                                       7JUL89   6JUL89  14AUG89  11AUG89    26
3324 3334   0   0 1   0      GENC RESTRAINT                                      30JUN89  29JUN89  14AUG89  11AUG89    30
3324 8300   0   0 1   0      GENC RESTRAINT                                      30JUN89  29JUN89  14AUG89  31JUL89    21
3326 3334   0   0 1   0      GENC RESTRAINT                                      16JUN89  15JUN89  14AUG89  11AUG89    40
3326 3336   0   0 1   0      GENC RESTRAINT                                      16JUN89  15JUN89  17AUG89  16AUG89    43
3328 3350   5   5 1   0  GCSTGENC SET LIMESTONE THRU 4TH FLR EAST FACE           12JUL89  18JUL89   7AUG89  11AUG89    18
3330 3332   0   0 1   0      GENC RESTRAINT                                      19JUN89  16JUN89   8AUG89   7AUG89    35
3330 4002   0   0 1   0      GENC RESTRAINT                                      19JUN89  16JUN89  25JUL89  24JUL89    25
3332 3346  13   4 1  20  GCDWGENC MISC INSUL & EXT GYP BD SHEATHING @ 5TH FLR     7JUL89  12JUL89   8AUG89  11AUG89    22
3334 3344   8   0 1 100  GCFPGENC SPRAY FIREPROOFING 4TH FLR
3336 3338   0   0 1   0      GENC RESTRAINT                                      16JUN89  15JUN89  17AUG89  16AUG89    43
3336 3340   0   0 1   0      GENC RESTRAINT                                      16JUN89  15JUN89  17AUG89  16AUG89    43
3336 3342   0   0 1   0      GENC RESTRAINT                                      16JUN89  15JUN89  17AUG89  16AUG89    43
3336 3348   6   0 1 100  VENTGENC SET ANCHORS FOR DUCTWORK 5TH FLR
3338 3348   4   0 1 100  HTACGENC SET ANCHORS FOR HTG & COOLING PPG 5TH FLR
3340 3348   4   0 1 100  FIREGENC SET ANCHORS FOR FIRE SPKL PPG 5TH FLR
3342 3348   4   0 1 100  PLUMGENC SET ANCHORS FOR PLBG PPG 5TH FLR
3344 3352   0   0 1   0      GENC RESTRAINT                                       7JUL89   6JUL89  17AUG89  16AUG89    29
3344 9000   0   0 1   0      GENC RESTRAINT                                       7JUL89   6JUL89  14AUG89  11AUG89    26
3346 3350   0   0 1   0      GENC RESTRAINT                                      13JUL89  12JUL89  17AUG89  16AUG89    22
3346 3352   0   0 1   0      GENC RESTRAINT                                      13JUL89  12JUL89  17AUG89  16AUG89    25
3346 4038   0   0 1   0      GENC RESTRAINT                                      13JUL89  12JUL89  29AUG89  28AUG89    33
3348 3352   0   0 1   0      GENC RESTRAINT                                      16JUN89  15JUN89  17AUG89  16AUG89    43
3350 3354   5   5 1   0  GCSTGENC SET LIMESTONE TO ROOF EAST FACE                19JUL89  25JUL89  14AUG89  18AUG89    18
3352 3360   8   0 1 100  GCFPGENC SPRAY FIREPROOFING 5TH FLR
```

agency is given a copy of the printout marked to reflect progress since the last update. The agency verifies the revised information and notes disagreements or questions. The update meeting is devoted to discussing and resolving areas in which the agency and the contractor may differ.

The Veterans Administration requires progress on all activities, including procurement, to be shown by changing the original duration to the workday difference between the anticipated completion date and the date of the update. Incomplete activities are updated for both time and money, since Veterans Administration schedules are dollar-loaded. Durations for completed work are changed to zero, percentage is changed to 100, and as-built dates must be recorded. Activities for material stored on site are created and valued. Any changed work must be added to the schedule by approved additional activities or logic revisions.

The Veterans Administration itself produces part of the update information by summarizing contract information such as change order value and the durations of approved time extensions. The agency also updates those activities for which they are responsible, such as approvals of shop drawings and other submittals or government-furnished equipment. Presently, the VA produces the computer-generated reports while the contractor maintains accurately marked diagrams.

Most construction contracts do not specify as detailed an update procedure as the Veterans Administration does. As experience with network scheduling grows, more contracts will specify updating procedures.

UPDATING LINEAR SCHEDULING METHOD DIAGRAMS

Methods for updating LSM diagrams are similar to methods described for updating bar charts. The date of the update is marked with a vertical line corresponding to the appropriate time on the horizontal time scale. On LSM schedules, however, progress on individual activities is indicated by location. Progress can be marked in contrasting colors or shading which parallels the activity's performance line or by symbols located on the line. Those activities with intervals diagrammed should have indicated both the most advanced location started and the most advanced location completed. Figures 8-7 and 8-8 show two ways to update LSM schedules.

As long as the project is reasonably on schedule, the LSM diagram need not be redrawn. Delayed or accelerated projects, however, require a new LSM diagram to be prepared. An LSM diagram is easily redrawn.

USING AN UPDATED SCHEDULE

The updated CPM schedule is used much the same way the original CPM schedule is used. Distribution of revised printouts and plots to all appropriate parties allows more accurate planning for material and equipment deliveries, manpower requirements, and interfacing considerations. The update should be

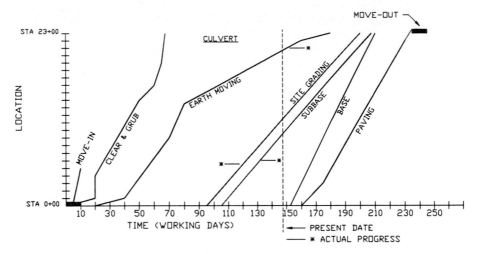

FIGURE 8-7   Updated linear scheduling method schedule.

viewed as maintaining the accuracy of the original schedule, as a status check, and as an adjustment to the completion plan for the project. The update should not be viewed as a new schedule.

An updated schedule shows a revised critical path, recalculated project and activity completion dates, and perhaps revised logic for subsequent activities. Most often an updated schedule is used to predict future completion. An updated schedule can also be used to expose and evaluate past delays. To identify and evaluate past delays, one must understand how the delays occurred and how the contractor proposes to overcome the delays.

FIGURE 8-8   Updated and redrawn LSM schedule.

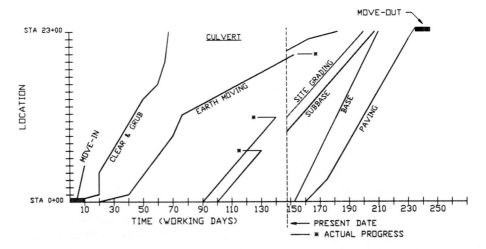

Performance of the following steps is necessary to identify and evaluate past delays in an updated schedule: (1) list completed activities; (2) compare the original duration with the actual duration of completed activities; (3) select those activities which had delays; (4) attempt to identify responsibility for the delay(s); and (5) project future performance from past performance.

Identifying responsibility for past delays will assist in evaluating time extension requests. Projecting future performance from actual performance will help the scheduler produce a realistic current schedule. Using an update in this way will also build a project history for use in future planning and in resolving disputes or claims efficiently.

EXERCISES

1. Data for a small project are given below:

Activity	Preceded by·	Duration
A	—	10
B	A	6
C	A	18
D	E, F	8
E	B	17
F	B	21
G	D	11
H	C, F	10
I	D, H	6
J	H	9
K	G, I, J	4

(*a*) Draw a network diagram (your choice of methods) and calculate event times and floats.

(*b*) On day 18, the following information is obtained:

Activity C will be starting in 3 days.

Activity F has been in progress 5 days but will take an additional 3 days more than the time originally allotted.

Delivery problems have delayed the early start of activity I by 10 days.

Activity G must start at the same time activity I begins but cannot be completed until 2 days after the completion of I.

Activity E is on schedule.

Update your network and calculate the revised event times, free floats, and total floats and critical path for the remaining activities.

2. Update the schedule in Figure 8-9 with the data provided. What is the new completion date? Assume that the schedule was statused as of day 15.

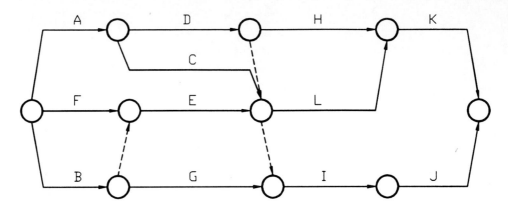

FIGURE 8-9 Updating exercise logic.

Activity	Original duration	Remaining duration
A	9	— (finished)
B	13	— (finished)
C	12	4
D	8	3
E	11	8
F	8	— (finished)
G	5	2
H	7	7
I	17	15
J	4	4
K	10	10
L	13	12

3. Update logic diagrams in Figures 6-9a, 6-9b, and 6-9c with the following information:

			%	Days remaining
(a)	Erect skin system	N, E, S, W	100	0
(b)	Brick lintels and miscellaneous frames	N, E	100	0
(c)	Brick lintels and miscellaneous frames	South	60	2
(d)	Brick lintels and miscellaneous frames	West	40	3
(e)	W.R. drywall	North	100	0
(f)	W.R. drywall	East	80	2
(g)	W.R. drywall	South	40	4
(h)	Insulation and brick veneer	North	100	0
(i)	Insulation and brick veneer	East	10	11
(j)	Glazing and mullions	North	5	4
(k)	Construct inside parapet wall		100	0

	%	Days remaining
(*l*) Lay roof insulation (but not roof)	100	?
(*m*) Interior roof drain piping	100	0
(*n*) Erect office front entrance	100	0

4. Perform manual forward passes and backward passes on diagrams updated in Exercise 3. Use the retained logic method of calculation.
5. How do the results of Exercise 4 vary from each other? What adjustments must be made to the logic from updated Figure 6-9c to make it match the other forward passes? Which updated schedules are more accurate?
6. Discuss how the progress override calculation method would change the forward passes on each diagram.
7. Would progress override calculations be correct?

CHAPTER
9

EFFECTIVE
USE OF
SCHEDULES

Between the good intentions at the start of a project and the realities of workload as construction progresses, the schedule is often ignored. As it continues to be ignored, it may even become ridiculed—considered inaccurate and useless by the trade contractors, designer, and owner.

If the schedule has been developed with the involvement of the subcontractors, architect, owner, and superintendent as discussed in Chapter 6, then the groundwork for accepted and used quality schedules will have been set. Maintaining use and commitment to the CPM schedule and its updates takes far less effort than the original schedule development.

An old business adage indicates that "attitude flows from the top." The same is true of construction. If the project manager uses and believes in the CPM, so will all the other players. If the project manager uses the CPM when discussing the project plan and progress in the weekly subcontractor meetings, if the project manager uses the CPM when discussing shop drawings or change orders with the architect, if the project manager uses the CPM when discussing pay applications and move-in dates with the owner, then these players will read and use the CPM. When all the players use and follow the CPM, the many values of the CPM that have been discussed will come to fruition.

The real benefit of schedules comes in their use to analyze progress and replan the project.

Simply using the CPM, however, is not enough to maintain confidence in the schedule. The authors have compiled guidelines for effective use and coordination of the many different schedules which can be generated during a project. Chapter 6 gave many tips on good scheduling techniques. This chapter expands on those techniques and discusses additional methods for effective use of schedules. The schedule can also be a valuable control tool for the project management team. This chapter also explains a control approach known as "earned value," which can be implemented with network-based schedules.

SUMMARY SCHEDULES

When negotiating a project, many times a general contractor or construction manager will develop and submit a summary schedule. The summary schedule portrays the general strategy for executing the project. A summary schedule also provides an opportunity for the contractor to demonstrate scheduling abilities, software, and graphics.

An effective summary schedule should be simple. It should show key relationships and milestones on the project, including designer and owner commitments. A summary schedule will show anticipated notice to proceed and project completion dates. A summary schedule should show major assumptions made or representations that have been relied on concerning how the project will be constructed.

If the project is fast-tracked, each of the promised release dates for design document packages should be indicated on the schedule. All durations and sequences must be realistic and complete. Often, most of the activities on a summary schedule will be critical.

A summary schedule is the owner's and designer's first indication of how the contractor intends to build the project and how the owner's and designer's responsibilities interface with that intention. Most owners and designers underestimate their roles in the construction project. A summary schedule is a contractor's first opportunity to alert them that they are a key element in the project team. The summary schedule serves to identify roles and responsibilities of the key team members. This schedule offers the first opportunity to plan the project's execution mutually.

When a more detailed project schedule is developed, it should be based on what was presented in the summary schedule. If the detailed project schedule does not resemble the summary, an explanation for the changes should be provided.

DEVELOPING THE ORIGINAL SCHEDULE ON FAST-TRACK PROJECTS

The traditional project delivery sequence of design, bid, and construction lends itself to the schedule development procedure outlined in Chapter 6. When developing the original schedule on a "fast-track" project, some adjustments to procedure are required.

On fast-track projects, as the specifics of the project develop, the CPM schedule should be refined and expanded. It is essential that the schedule be as detailed as practical, and prepared as quickly as possible after the release of construction documents. Detailed schedules for only one or two months ahead with summary activities shown through the end of the project do not provide sufficient information for trade contractors to plan their work. Lack of detail also leads to a lack of credibility for the schedule.

Developing an entire network schedule early in the project allows the scheduler and project manager to study the entire project and discover problems long before they occur on the job site. Various approaches to completing the project can be studied. A bar chart or summary schedule cannot provide this type of planning or problem avoidance and provides no basis for time extensions if the scope of the project increases or the project experiences delay. A summary schedule which contains an activity, "Construct Garage, 140 Days," cannot be used to justify a time extension for a new masonry wall within the garage.

Even if all design information has not been completed, a detailed schedule should be prepared based on knowledge to date. Although such a schedule may have to be revised, it will still provide a basis for determining time extensions if the scope of the work changes.

Some contractors expand the fast-track schedule only when work for the detailed portion of the logic is nearing completion. This approach is incorrect.

It causes problems in interfacing the work among trade contractors when the scope is increased or omissions are discovered. When the expanded planned schedule is finally distributed, it may be met with disbelief, rejection, or allegations that the schedule has been manipulated to create or prove a delay that became apparent before the expanded planned schedule was completed. Trade contractors may have anticipated more time or a different sequence for their work which was not apparent in the summary portion of the previous schedule.

Waiting for trade contractors to bid and sign contracts before developing a detailed schedule for their work is inappropriate. If the detailed information is available, as in construction documents, then the CPM schedule should include detailed activities with proper restraints to show the work plan accurately.

On fast-track projects a construction schedule can also be distributed with the bid packages for the trade contractors bidding the project. By this distribution the trade contractors become aware of how they are to interface with each other and the construction manager. Trade contractor bids will be based on the schedule, eliminating the construction manager's need to interface their work after the trade contract is signed. This reduces the problems inherent with integrating the trade contractors' plans for executing the work into the project schedule at various times during the project. The commitment of the trade contractors to the detailed master schedule is made as they join the project and reduces later possible disruption to the schedule.

If the trade contractor cannot match the sequence and timing indicated in the schedule, the construction manager will be informed immediately. The schedule may be modified in exchange for a more favorable price from the trade contractor. Early warning of potential problems allows the construction manager greater time and consequently greater flexibility in resolving the problem. The opportunity to negotiate the scheduling and completion criteria for the trade contract while the construction manager is still in a strong bargaining position enhances cooperation from the trade contractor.

Fast-track projects are typically under severe time constraints. Quality and detailed CPM schedules and timely updates are essential to maintain project progress and minimize delays.

MULTIPLE SCHEDULES

Do not prepare multiple schedules on the same project. Some projects have procurement schedules, bid package schedules, project summary schedules, construction schedules, tie-in schedules, and owner move-in schedules. It is very difficult to interface multiple project schedules accurately. Adjustments in one schedule may affect other schedules, but the other schedules may not be adjusted. The schedule's credibility is lost if errors because of lack of interface are discovered. Anyone refuting a claimed time extension may choose the favorable, noninterfaced schedule to show that no time extension is due. The construction schedule should be the master schedule and integrate all activities necessary to complete the project. No matter how detailed a specific work package may be, it should be included in the construction schedule. The construction schedule is the place where late deliveries or poor performance manifest themselves in real costs.

If necessary, a partial printout or plot can be produced from the master schedule that shows only the area of interest. For example, a plot and printout can be produced that shows only the activities related to owner move-in or procurement.

If confronted with a situation where a new CPM schedule is developed or expanded from the previous CPM, the scheduler should indicate that the old schedule is no longer to be used. The scheduler should provide completely new plots (or hand-drafted logic) and printouts to all parties. The scheduler should offer to explain the changes and/or new schedule.

MULTIPLE PRIME SCHEDULES

Projects with multiple prime contracts present some unique scheduling problems. Normally the prime contractor awarded the general construction contract is specified as having responsibility for scheduling and coordinating the work, while the remaining prime contractors (typically, electrical, HVAC, and plumbing) are required to cooperate with the general contractor's scheduling efforts.

While conceptually developing a schedule for a multiple prime project works the same as a general contractor developing a CPM schedule with subcontractors as detailed in Chapter 6, the reality is different. The other primes have no contractual responsibility to the general, and the general contractor prime has no financial control over the other primes. The owner is obligated to provide the impetus for the other primes to cooperate with the general contractor prime in executing the project. Most owners lack the skill or desire to provide this continuous pressure to maintain scheduling cooperation.

Multiple prime contracts usually are best scheduled using an owner- or architect-furnished scheduling service. This procedure places the responsibility of producing the physical portions of the CPM (plots, printouts, etc.) on the architect or owner and provides unbiased scheduling information. The following section explains owner/architect-furnished scheduling services.

OWNER/ARCHITECT-FURNISHED SCHEDULING SERVICES

There are a number of reasons an architect or owner may wish to furnish scheduling services on a project. Contractually, the progress of the project may be the responsibility of the owner, as is possible in some multiple prime contract situations. The architect or owner may desire or need a schedule of greater quality than is typically provided by general contractors. The owner may also wish to have unbiased evaluations of delays that may occur on the project. Finally, the architect or owner may want the schedule maintained and updated regularly, an event that may not have occurred with past general contractors.

Accurate predictions of the project completion date can be crucial to the owner. Further, if the owner becomes involved in a delay claim, all the updates and a CPM consultant already familiar with the project and the delay issues will be available. This process can save substantial money over attempting to re-create the original project schedule and updates after the project is completed.

Typically, the owner may specify that the services of a construction scheduling consultant have been retained and the consultant has responsibility for managing the general contractor's (or all primes') schedule development efforts. The consultant is provided free of charge to the general contractor. The consultant develops a detailed CPM schedule based on information and guidance from the general contractor and the subcontractors (or primes) and performs all the technical functions of a scheduler.

The owner must be careful not to dictate "methods, means, or materials" to the contractor. The scheduling consultant must develop the contractor's schedule, not the owner's schedule. This works much the same way as a scheduler works with a superintendent as described in Chapter 6. The owner, however, is in control of the production of the CPM and can dictate to the scheduling consultant when updates will be performed and the subsequent production of plots (or logic diagrams), printouts, and bar charts. The owner then also has an unbiased source for advice concerning delays, impacts, and progress.

If the schedule sequence and plan are the owner's, not the contractor's, the owner assumes many of the responsibilities of the contractor, resulting in increased risk to the owner. To the extent that a contractor cannot complete the work according to the owner's sequence, the contractor may demand that the contract price be increased. The contractor may make the same demand if the actual work for any activity cannot be accomplished within the scheduled duration.

Owners go to great lengths to avoid this added responsibility that increases the owner's risk. Chapter 14 includes a number of scheduling clauses which should be carefully considered when drafting a contract or developing a scheduling specification.

TECHNICAL CONSIDERATIONS

Level of Detail

The construction schedule should include every major part of the project. Missing portions of the project indicate that the schedule is incomplete and cannot be trusted. Adequate detail includes the logic necessary to show accurate relationships among the various trades. Commonly, trade relationships are shown unrealistically. Figure 9-1 shows a traditional subnetwork describing mechanical, electrical, and general construction interface completion of a con-

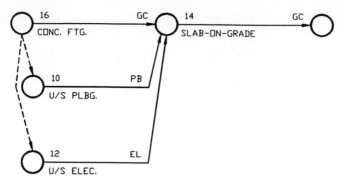

FIGURE 9-1   Traditional subnetwork trade interface.

crete slab-on-grade. It shows that underslab plumbing and electrical will be performed concurrently with footing construction. When all three preceding activities have been completed, slab-on-grade can begin.

The true sequence of underslab work is indicated in Figure 9-2. When concrete footings are approximately 50 percent complete, underslab plumbing will begin. After underslab plumbing has been 50 percent completed, underslab electric will begin. After underslab electric is 50 percent complete, slab-on-grade can begin. The sequence in Figure 9-2 shows the realistic overlapping of trades.

Figure 9-2 more accurately portrays the sequential relationship among the trades. The subnetwork in Figure 9-2 will also more accurately portray delays to any work in the sequence, because it more accurately portrays the sequence.

FIGURE 9-2   True sequence of underslab work.

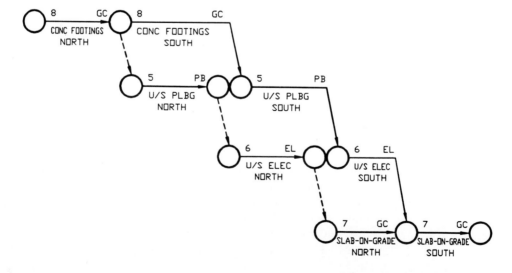

The construction schedule must have sufficient detail to permit anyone to understand the scope of an activity. An activity for "Site Concrete" is not adequate for a project which has curbs, sidewalks, underground utilities, and plaza retaining walls. Sufficient detail will allow more accurate control of project progress and provides a location for change orders or additions to the scope of the work. Without sufficient detail, the schedule cannot measure the impact of changes or variations.

Imposed Finish Dates

When a construction contract is signed, a specific completion date is usually included. Occasionally the contract will include intermediate completion dates. When developing the schedule, however, only contractual completion dates (mandatory finishes, imposed finishes, etc.) should be included. Self-imposed late finish dates, such as "Top Out Structure," "Building Essentially Watertight," etc., make the schedule difficult to read, produce deceptive critical paths, and may hinder chances of receiving time extensions when due. Imposed finish dates do not provide the physical relationships necessary for quality scheduling.

Self-imposed late finish dates may be desired for a variety of valid reasons. Topping out may be necessary in order to avoid winter weather, for example. However, these self-imposed controls can be put into the schedule without using imposed finish dates. Figures 9-3a and 9-3b compare the use of imposed start and finish dates and sequenced activities for "Winter Shutdown Concrete Operations" and "Building Essentially Watertight." The sequenced activities are superior because they are tied into the schedule, identified on the printout, and can show delay if previous activities' late completions prevent the occurrence of the desired late finish dates.

Some software programs permit the scheduler to "block out" time for self-imposed dates. Although they are better than imposed late finish dates, these "block outs" do not appear on the printout, preventing communication of the reserved time. Hidden information does not support the objective of good scheduling: good communication. Avoid the use of "blocked-out" time.

Enclosure prior to winter is often crucial for interior work. This critical relationship between construction work and the weather needs to be shown in the logic. Clarity in the logic will assist the contractor in obtaining a time extension if enclosure is delayed because of acts or omissions of the designer. Clarity in the logic will also assist the contractor in collecting additional funds for temporary enclosure. The submitted schedule provided notification to the owner of the contractor's intent. The placement of an imposed finish date on enclosure of the building does not communicate the reason the contractor desires to finish by the imposed date. Failure to communicate intentions makes recovery of delay damages very difficult.

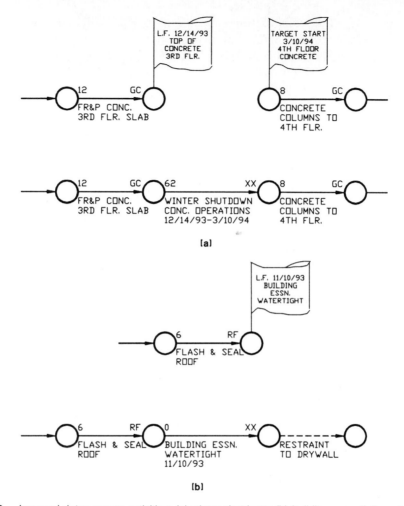

FIGURE 9-3 Imposed dates versus activities: (a) winter shutdown; (b) building essentially watertight.

Demonstration of Intentions

Show the intentions of the planned approach to the project. The CPM schedule should show all the contractor's major intentions for executing the project. If it is planned to erect curtain wall after two levels of steel are up and plumb, show that in the logic. If concrete is planned to be shut down over the winter, show that in the logic. If the structure is intended to be enclosed temporarily in order to heat interior work, show that in the logic. These intentions do not necessarily reflect productive construction work, but they are crucial to informing the owner and designer that the project is planned in a certain manner. For example, if the owner is to provide natural gas for the contractor's use and the contractor plans to heat the structure using the owner's natural gas and permanent equipment, this intention should be shown in the schedule. It will be difficult to ask for a time extension or additional costs if the owner fails to provide the gas if there has been no demonstration of the contractor's reliance on it.

Other items which may be included in the schedule are durations a crane will be on the job site, durations for temporary safety enclosures, scaffolding, road closures, shoring, temporary enclosure, temporary heat, joint occupancy, utility system tie-ins, and shutdowns. Always remember that the CPM is a great communications tool: Use it as such and it will reward you with smoother running projects and fewer disputes.

Demonstration of Commitments

The most important commitments to include in the schedule are those of parties over which you have no control. If the owner commits to vacate a portion of the structure for renovation at a certain date, that date should be specifically indicated in the schedule. If the architect has agreed to return shop drawings in 2 weeks, then that duration should be included in the procurement portion of the schedule. If the designer has agreed to produce design documents on the skin system by a certain date, then that date needs to be included in the construction schedule. Moreover, the dates need to be correctly sequenced into the schedule. By including them in the schedule, the owner is given the opportunity to clarify any commitment shown in the schedule. Failure to comment can be interpreted as approval. If the commitments are critical or nearly critical and the owner or designer fails to deliver as promised, the schedule will provide a clear indication of the necessity for a time extension.

Demonstrate commitment to the schedule by using it to plan both the contractors' and trade contractors' work. Use the schedule at job meetings as an agenda for discussions. The schedule should reflect the current plans. If the schedule does not show the way the project is to be completed, then change the schedule. The schedule's users need to be confident that the schedule is an accurate representation of the current project plan.

Do not manipulate the schedule by adjusting the project completion or other contractual completions in the updating process unless a time extension has been granted. Do not skip updates even if the project is progressing as scheduled. By regular updates, confidence in the schedule as an accurate plan for the project is communicated to trade contractors. By updating, one can discover hidden delays that may not be apparent on the job site. By updating, a record of progress is produced which also provides notice to the owner and designer of the status, problems, and delays to the completion date.

During an update, the logic should be adjusted for changes in the planned approach. If the sequencing change adjusts the critical path, then it should be done in a separate process. The schedule should first be updated to determine project status, and a printout should be produced and filed. Second, the necessary sequence adjustments should be made and another printout produced.

The second printout should be distributed to all subcontractors, the architect, and the owner, and the first printout should be retained as evidence of the project status prior to the logic change. The first printout will be used to justify claimed time extensions.

The monthly update sent to the owner or owner's representative should provide notice of delay even if the delay started several months earlier. The contractor should point it out and provide a change order request.

The above procedures are just a few of the ways the scheduler (and project manager) can maintain the commitments made to the project schedule.

FIELD USE OF CPM SCHEDULES

Teaching CPM Use

If the procedure detailed in Chapter 6 has been followed, especially as it relates to meetings and discussions with the various parties, the scheduler will have established the groundwork to provide an accurate and usable construction management tool. This CPM schedule, used with confidence by the superintendent, will be a major asset in the planning and management of the project. The completed quality CPM schedule can then be used to run the project.

Although the scheduler has explained how the CPM is created, how to read the diagram, and how to read the various printouts produced by the CPM software, it is advisable to both review and expand on this teaching which started in the developmental process of the master CPM schedule. This additional teaching can easily be accomplished during or immediately following the final presentation of the CPM schedule to all the parties involved. "Refresher courses" can also be provided each time the CPM schedule is updated. The purpose of the training sessions is to encourage use of the CPM schedule. The CPM schedule is an essential communication tool for management of the project. However, without its continuous use and reference by the majority of parties involved in the construction process, it will not fully serve its intended purpose.

The scheduler should prepare for the final presentation meeting by producing plots and/or hand-drafted logic diagrams, numeric-sort printouts with relationships, responsibility-by-early-start printouts, and detailed responsibility-by-early-start bar charts from the software package. (Different schedulers may elect to produce different presentation materials.) The bar charts need only be for the first three or four months of the project. With these diagrams and printouts the CPM scheduler can both present the final schedule and thoroughly teach those who will be expected to use the schedule the ease with which it can be read and interpreted. The first explanation should be a review of the concept of the CPM scheduling method used (I-J or PDM). This quick review should take only a few minutes but should remind the users how the CPM was put together and how the different parts of the logic diagram function. If the logic diagram has been plotted by computer, it is best to defer its use until after explanation of the printouts. If the logic diagram has been hand-drafted, however, explanation of how the work activities are shown on the diagram and how they are related is a concise way to start the learning process. Cross-referencing between the printouts and the diagrams reinforces that there is only one plan and the documents portray the same information in different formats.

After explaining the hand-drafted logic diagram, the scheduler should show the various information presented in the numeric-sort listing. Point out the I and J node columns (or the activity number in PDM), the duration, remaining duration, code and responsibility columns, activity description column, etc. The scheduler should explain each column and how it relates to the diagram. The scheduler should then explain to the users the meaning of early start, early finish, late start, and late finish, and the use of the float calculation column. Normally contractors recommend to their subcontractors that they try to meet all early start and early finish dates. This is normally good scheduling practice and good construction procedure.

Some general contractors go so far as to produce printouts with no late start or late finish dates and no float column. This procedure, while encouraging the use of early start-early finish dates, does not give the subcontractor the ability to evaluate the priority of the different activities. If there are problems later, it will be difficult for the general contractor to accuse the subcontractor of not handling things in a prioritized manner. It is usually best to provide the sub-contractor with all important information.

The scheduler should explain that the float listed in the float column for each activity is the total float available, and that float is not for the exclusive use of the activity indicated. Make sure the schedule readers understand that once that float is consumed, it is gone forever and never regained unless project progress exceeds the anticipated rate of the schedule. At this point it is advisable to ask each of the various subcontractor superintendents to give you some answers to some specific questions about the project schedule. This will force them to look through the logic printout and find various items and read the CPM schedule from the printout so that you are assured they are capable of using this document. Continue this process until you are confident that they have a clear understanding of their commitments and the meaning of the sched-ule documents.

Because of the ease of reading a hand-drafted logic diagram, questions should be asked relative to specific work areas within that diagram. For example, ask a question related to the structural subnetwork of the subcontractor who is responsible for installing structural decking. Ask the subcontractor when the first decking will be installed on the project. The subcontractor should look to the CPM diagram to find the activity which first indicates structural steel deck-ing erection. The subcontractor can then identify the I and J node of the activity, refer to the printout, and locate its early start date. This would be a good guideline date to use for the steel decking start on the project. Warn that absolute dates for when this will occur need to be coordinated with the general contractor's field superintendent, as progress may be ahead or behind schedule when time to install the decking arrives.

The next subcontractor may be the plumbing subcontractor. Ask a question relative to the underslab plumbing installation. The plumbing subcontractor should follow the same procedure to find the activity on the logic diagram and then refer to the I-J numeric-listing printout, find the activity, and identify its early start date. Once you have asked a specific question of each of the various superintendents, you can establish whether or not they have the savvy to understand the CPM schedule. Some superintendents may require additional one-on-one assistance to read the CPM schedule. Other subcontractors superintendents may totally refuse to make any effort or attempt to read the CPM schedule. The scheduler must make his or her best attempt to "convert" these people into believers and users of CPM schedules.

After completing the previous step, have the superintendents review the responsibility-by-early-start printout. This printout greatly eases the process of finding their individual activities among all the activities that occur in the numeric listing. Most of the superintendents will be very happy to see this printout. It lists their work in chronological order. This printout will be easy to read compared to the full numeric listing. It is important to indicate that the information on this printout is identical to the information on the numeric-listing printout. However, they cannot follow the restraints (or relationships) or determine how their work activities relate with the other work without referencing either the numeric listing or the logic diagram.

Plotted logic diagrams for large projects can be difficult if not impossible to refer to for specific relationships. While a sophisticated scheduling individual can work through a plotted logic diagram, most trade people should not be expected to make that effort. Showing them how to read the diagram may assist in their understanding of the CPM schedule; however, do not expect that they will continue referring to a plotted logic diagram.

The plotted logic diagram, however, if it is time-scaled, can be a great asset in the field. It will produce a diagram similar to a bar chart. The field superintendents will generally ignore the restraints and use the plot very much like a detailed bar chart. This logic diagram should be placed on the wall of the trailer for reference by all parties. The superintendent may elect to darken in or color the bars as work is completed. This makes the plot even easier to read.

The final printout to be presented is a bar-chart printout produced by the computer software of the CPM schedule activities. Normally a bar chart in responsibility-by-early-start format is used. Each individual activity is represented by a different bar on the printout. This printout is a clear bar chart for each subcontractor's work in chronological order. This promotes the execution of the project in the most optimum manner possible. This printout will be most readily accepted even by the "noncomputer types."

After fully explaining the technique for reading and using the CPM schedule, the scheduler needs to caution all of the superintendents that the CPM schedule is not an absolute follow document. The CPM schedule serves only as a guide for the execution of the construction project. Major changes in sequence should not be attempted without full concurrence of all the construction parties. However, minor adjustments to the sequencing can be made after discussions with the general contractor's superintendent and the subcontractor's superintendent. The schedule should never be used as a tool for telling contractors they cannot do certain work. That approach would defeat the purpose of the construction schedule, which is to complete the project in a timely manner. Completing work as early as possible without risking damage or causing undo inconvenience for other contractors is good construction practice.

Rolling Bar Charts

If the CPM schedule has been developed with sufficient detail to produce quality-information bar charts for trade foremen to use in the direction of their work, then further bar charts will not be required. However, typically on a large project the CPM schedule does not identify and name every step of the construction process. On these projects it may be desirable to produce bar charts which are derived from the CPM schedule and normally cover a period of 1 to 2 weeks. These bar charts break into finer detail the work required to complete the activities listed in the CPM schedule. They may also indicate the specific direction of work within an area. For example, the CPM schedule may indicate drywall in the northwestern corner of the first floor of a major office building. The 2-week rolling schedule will indicate by room when the drywall will be done and in what order. This is to allow a follow-on contractor to start in a timely manner. The original CPM schedule may not have been developed to provide this level of detail. Whether to use rolling bar charts or CPM bar charts or both is a management decision. Always coordinate the bar charts with the CPM schedule. Do not risk having multiple schedules for the same project that do not agree.

Continuous Dedication and Use of the CPM Schedule

The most important individual to maintain the accuracy and use of the CPM schedule is the general contractor's superintendent. He needs to use the CPM schedule to direct and schedule the individual work items occurring on the project. The superintendent should use the CPM schedule to run the project.

Whenever schedule discussions arise on the job site, the superintendent should refer to the CPM schedule for answers and directives on sequences of work. Whenever the owner may consider a change order, the CPM schedule should be referred to. The superintendent cannot be expected to determine delays and impacts from the CPM schedule on the job site. If potential delays develop, the scheduler should be contacted for advice and information. If the project is substantial enough in size or time is extremely important, an on-site scheduler is a worthwhile addition.

An on-site scheduler can provide updates much more timely than an off-site scheduler. The on-site scheduler can provide advice immediately as to the effects of change orders or other delays on the construction process. This type of immediate schedule information response provides a quicker and more accurate directive for construction management.

The schedule on a major project should be updated at least monthly. Updating the schedule provides an opportunity for all the subcontractors, contractor, engineer, and owner to discuss both the positive and negative progress aspects of the project. If the project is falling behind, it gives the contractors time to discuss various options for bringing the project back on schedule. This updating process allows everyone to review the project in detail and discuss ways to improve the management and efficiency of the project.

The scheduler who developed the master project schedule should be the individual on site updating the CPM schedule. While this is not always possible, it is desirable because the CPM scheduler can recall all the reasons and commitments made by the various parties in the development of the original schedule. When one party states that it cannot meet the schedule the way it is currently indicated, the scheduler can remind that individual of the original commitment which was made. The updates provide the opportunity for all contractors to recommit to their efforts to complete the project on schedule. The processing of an update allows the construction manager to determine if and who is holding up project progress. The project can then be brought back on schedule before excessive delays occur by taking corrective action in a timely manner.

Field use of the CPM schedule for planning and communication is the most important asset of CPM scheduling. The CPM's ability to identify, prevent, and assist in resolving construction project delays before they expand in magnitude to the point where they cannot be easily resolved is a secondary yet important asset of the CPM process.

Project Control

The greatest payback for scheduling comes about in the use of schedules in project management for project control. Project control requires three elements: (1) a baseline, (2) measurement of performance, and (3) effective corrective actions. The schedule is a valuable tool to be used in this process to represent the baseline and provide a means to measure performance. Earlier chapters have discussed ways to schedule project activities and discussed the importance of updating schedules to portray the most current plan. This section addresses approaches to measuring progress on individual activities and an approach to calculating the project's performance at intermediate times in the execution of construction.

Measuring Work Progress

There are many different types of activities required for projects. There is not a single "best" way to measure the progress on an activity. There are at least five approaches which can be used to determine the percent complete for an activity:

1 Units completed
2 Incremental milestone
3 Start/finish
4 Opinion
5 Cost ratio

The "units completed" approach applies to activities which involve many identical repetitive tasks or units of work. It is assumed that these are not interrelated subtasks and that each unit will represent the same amount of time for completion. Excavating a trench, laying block in a wall, and pulling wire in a cable-tray are all examples of activities which might be appropriate for this method. The percent complete is determined by dividing the number of units completed by the total number of units. For example, if 1200 blocks are required and 300 blocks have been laid, the percent complete for block laying is 300/1200 = 25 percent.

The "incremental milestone" approach applies to activities which have multiple units of work with several sequential tasks. An example activity would be form reinforce and pour concrete footings. There might be many footings, with each one requiring the tasks of forming, reinforcing, and concrete placement. Completion of each task for each unit is considered a milestone. A percentage of the effort is normally assigned to the completion of each milestone based on the relative manhours required for the task compared to the total. The total percent complete for the activity is determined by summing the product of the incremental milestone percentage times the percent of total units for each unit for which the milestone has been achieved. For example, if the tasks below were assigned the following incremental percentages,

Thru forming	30%
Thru reinforcing	65%
Thru placement	100%

and there were a total of 120 footings with completion of milestones as follows:

Thru forming	20
Thru reinforcing	15
Thru placement	12

the percent complete is determined as:

Thru forming	$0.30 \times 20/120$	$= 0.050$
Thru reinforcing	$0.65 \times 15/120$	$= 0.081$
Thru placement	$1.00 \times 12/120$	$= \underline{0.100}$
	Total	$= 0.231$

or

23.1%

The "start/finish" approach is used for activities of relatively short duration where the only recognizable milestones are the start and finish. Examples of activities of this type include hydrotesting and aligning large rotating equipment. The approach involves performing the same type of calculation as for the incremental milestones: start and finish. The finishing of a task is assigned 100 percent, while the start is assigned a value between 0 and 100 percent. Typically a percentage of 20 percent to 30 percent is used for the start of a task. This tends to give a conservative (underreported) progress assessment when several units are involved in the activity.

The "opinion" approach is simply an informed judgment concerning completion status. This approach relies on the experience and perceptive abilities of the supervisor to judge the relative completion status of the activity based on a subjective assessment.

The "cost ratio" is an approach which is used for evaluating tasks that are continuous and bear a strong relationship to the total effort required in the project. Examples include activities such as inspection, safety, and project control. The progress is measured by the ratio of the costs to date divided by the total forecast cost.

EARNED VALUE

A method has been developed to determine the progress on a project basis from the progress assessments of various activities. The method has been termed "earned value." The earned value method can be used on projects where the budgets change and still provide meaningful information. With the earned value approach, a relationship exists between the current budget for an activity and the measurement of percent complete.

The earned value equals the percent complete times the activity budget in either dollars or manhours. By summing all the earned values for the various activities and dividing this by the sum of the activity budgets, one can arrive at a measure of the project percent complete. As an example, consider the following project data:

Activity	Budget	Actual or expended	% complete	Earned value
A	400 mh	300 mh	60	240
B	600 mh	500 mh	90	540
C	100 mh	50 mh	60	60
D	1000 mh	125 mh	10	100
E	400 mh	200 mh	60	240
F	300 mh	240 mh	75	225
G	200 mh	180 mh	80	160
Totals	3000 mh	1595 mh	—	1565

$$\text{Percent complete} = \frac{1565}{3000} = 52.2\%$$

The earned value for activity A is calculated by multiplying the percent complete, 60 percent (determined by one of the methods listed above) times the current budget, 400 mh, to arrive at the earned value expressed as 240 earned work-hours. Similar calculations are made for each activity. The project status is determined by dividing the cumulative earned work-hours by the budget. The earned value approach provides a meaningful measure of project status even when budgets change as a project progresses due to changing quantities.

COST AND SCHEDULE PERFORMANCE

It is often desired to analyze a project and obtain a measure of the performance of the project compared to the plan. The earned value approach can be used for this purpose. To develop these measures, several pieces of information, identified in the examples above, concerning the activity or project are used. These are:

1 The budgeted manhours or budgeted cost for work scheduled (BCWS)
2 The earned manhours or budgeted cost for work performed (BCWP)
3 The actual manhours or actual cost for work performed (ACWP)

By using combinations of these three pieces of information it is possible to obtain cost and schedule performance measures.

The performance measures include:

1 Schedule variance (SV) – earned manhours – budgeted manhours

$$SV = BCWP - BCWS$$

2 Schedule performance index (SPI) = earned manhours/budgeted manhours

$$SPI = \frac{BCWP}{BCWS}$$

3 Cost variance (CV) = earned manhours − actual manhours

$$CV = BCWP - ACWP$$

4 Cost performance index (CPI) = earned manhours/actual manhours

$$CPI = \frac{BCWP}{ACWP}$$

Under favorable conditions of performance the variances will be positive and the indices will be greater than 1.0.

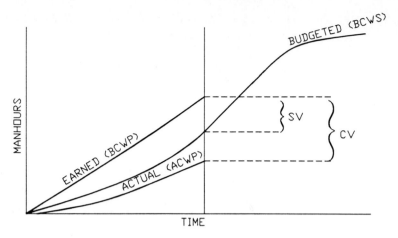

FIGURE 9-4 Cost and schedule variance.

For example, assume that a total of 2000 mh of work has been scheduled up to a particular point in a project, with 1900 mh actually expended and 2100 mh earned (based on percent complete). The performance measures would be as follows:

$$\text{BCWP} = 2100 \text{ mh} \qquad \text{BCWS} = 2000 \text{ mh} \qquad \text{ACWP} = 1900 \text{ mh}$$

$$\text{Schedule:} \qquad \text{SV} = 2100 - 2000 = +100 \text{ mh} \qquad \text{(favorable)}$$

$$\text{SPI} = \frac{2100}{2000} = 1.05 \qquad \text{(favorable)}$$

FIGURE 9-5 Cost and schedule performance indices.

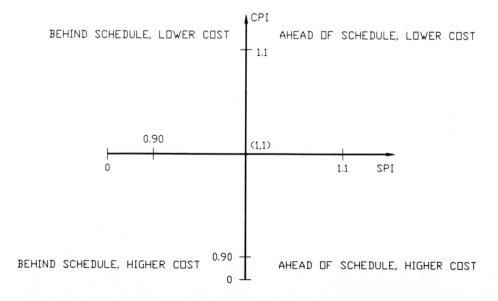

$$\text{Cost:} \quad CV = 2100 - 1900 = +200 \quad \text{(favorable)}$$

$$CPI = \frac{2100}{1900} = 1.11 \quad \text{(favorable)}$$

These measures can be shown graphically in two forms, as illustrated in Figures 9-4 and 9-5.

These performance measures provide a useful tool for focusing management attention on problem areas, giving the needed feedback to the project management team on activities which are not being performed according to the plan.

EXERCISES

1 List three situations other than "Winter Work Shutdown" where a construction schedule would need to indicate a break in the work.
2 List three situations where "Target Start Activities" might be used.
3 Create a summary CPM schedule (I-J or PDM) for a fast-track grade-school construction project. The school will house three classrooms for each grade (K-6) and has the typical support functions of cafeteria, administration offices, library, two gymnasiums, and a playground. The contractor will be given notice to proceed when the foundation drawings are issued. The following groups of construction documents will be issued thereafter.

Superstructure	1 month	After	N.T.P.
Enclosure and roofing	$2\frac{1}{2}$ months	After	N.T.P.
Sitework and civil	$3\frac{1}{4}$ months	After	N.T.P.
Mechanical and electrical	$3\frac{1}{2}$ months	After	N.T.P.
Architectural finishes	$4\frac{1}{2}$ months	After	N.T.P.

Create the summary schedule with approximately 100 activities.

4 Discuss the difficulties a scheduler might encounter on:
(a) A multiple-prime project
(b) An owner-furnished scheduler project
(c) An architect-furnished scheduler project
(d) A design/build project
Discuss ways to resolve or minimize those difficulties.
5 Discuss the level of detail for different CPM schedules and develop guidelines for determining the detail level required: What is insufficient detail? What is adequate detail? What is excessive detail?
6 Calculate the percent complete for the project based on the project data provided using the earned value approach.

Activity	Budget	% completed
1	10,000	98
2	5,000	66
3	3,100	43
4	4,650	51
5	1,220	90
6	6,250	11
7	5,100	70
8	7,100	16
9	1,900	30
10	2,300	15

7 Calculate the schedule variance, cost variance, schedule performance index, and cost performance index given the following:

2100 mh expended to date

1900 mh earned to date

2000 mh budget to date

Explain the status.

CHAPTER 10

DURATION-COST TRADE-OFF

The previous chapters described the general process used to develop a schedule and the mechanics of scheduling. At this point it is important to note that the first schedule developed may not be a suitable schedule for beginning the project. Refinements to the schedule may be necessary to make the schedule compatible with the objectives and the constraints imposed on the project. The scheduler must be able to modify the schedule in a manner consistent with the objectives to satisfy the contractual time requirements which have been established.

In many cases the contractual time requirements have been established by the engineer or owner without regard to a reasonable assessment of activity durations. It is then imperative that the scheduler attempt to adjust activity durations and sequences to fit within the contractual requirements, even if inefficient adjustments such as multiple shifts, overtime, or large crew sizes are required.

The usual case with most first schedules is a need to shorten completion of all or part of the project. There can be many reasons for this need. In some cases the contractor may have to shorten durations to avoid contractually imposed liquidated damages. In other cases the contractor and the scheduler want to take advantage of bonuses that have been offered. The scheduler may want to take advantage of seasonal weather variations that may affect productivity. Yet another reason is the need to either avoid or take advantage of holiday periods or other imposed calendar restrictions such as industrial plant

There are many alternatives for equipment and resources, each of which has a unique cost. (Photo courtesy of Manitowoc Engineering Co.)

shutdowns or turnarounds. The contractor may want to free resources from the immediate project for future projects. The owner may wish to shorten the overall schedule to improve project economics by accelerating the cash flow derived from the project. Market conditions may dictate that an owner must make changes in the completion date to maintain a viable competitive product line. This may require the contractor to expedite the overall completion of a project or a portion of the project.

The easiest way to reduce the duration of a project, or a portion of a project, is to eliminate unnecessary restraints between the various activities. Another way is to resequence activities, dividing them into smaller activities that can be performed concurrently and thus reducing overall duration by overlapping various operations. The final method, the topic of this chapter, is to adjust durations. In most cases, reduction of duration is from the normal estimated value. This chapter will explain a system that expedites activities but still minimizes the overall cost of the project.

The process of expediting a project is often called ''crashing'' the project. The term ''crashing'' refers to the reduction of activity durations with the overall effect of reducing the project duration. The crashing process is a deliberate, systematic, analytical process that involves examining all the activities in the project and focusing on those activities on the critical path. The crashing process uses an assessment of activity variable cost with time to determine which durations to reduce to economically minimize the overall project duration. The process seems very simple in concept, but in reality it is very complex. There are several ways to reduce activity durations, and many combinations of activity durations and costs that must be considered in a complete analysis. This chapter will explain the process.

CRASHING ACTIVITIES

Activities on a project can be crashed in one of the following ways:[1]

Multiple-shift work
Extended workdays
Using larger or more productive equipment
Increasing the number of craftspeople
Using materials with faster installation methods
Using alternate construction methods or sequences

The use of multiple shifts permits the performance of particular activities in fewer calendar days. The appropriateness of shift work must be judged for each activity in terms of lighting and other work environment constraints, support services, and safety. Typically, shift work will cost more on a unit-price basis due to the wage differential for shift work, the additional costs of support services, and the reduction in productivity. Therefore, it can be assumed that activity cost will increase to some degree when shift work is used to reduce activity duration. However, shift work allows for dramatic reductions (50 percent or more) in activity durations.

Extending workdays helps reduce the total time required for an activity. The craftsperson works 10 or 12 hours per day rather than the usual 8-hour day. This can reduce an activity duration by up to 33 percent. Typically, additional costs for support services and loss of productivity are encountered. Additionally, most employers pay an overtime rate of 1.5 times the normal wage for work beyond 8 hours per day.

Using larger or faster equipment, increasing crew size and using materials with faster installation methods, or using alternate construction methods or sequences are also ways to reduce time, but each will typically increase the cost above the normal standard approach originally chosen (because of its efficiency and economic advantage) over these expediting alternatives.

DURATION-COST ANALYSIS

This section illustrates the analysis that must be done to determine the alternative durations and costs for an activity used in the crashing process.

[1]Edward M. Willis, Scheduling Construction Projects, New York: Wiley, 1986, pp. 299–300.

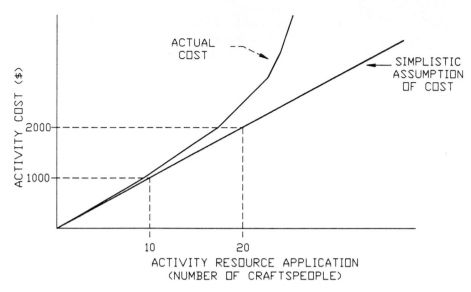

FIGURE 10-1 Resource-cost relationship.

Before going into a detailed discussion of the duration-cost relationship, it is important to understand several fundamental concepts and assumptions related to the model which will be developed. The first of these is the relationship between the application of resources such as craftspcople and the activity cost. The resource-cost relationship illustrated in Figure 10-1 shows a simplistic assumption of cost and its relationship to the application of resources. This simplistic assumption of cost illustrates the typical view that if the number of resources is doubled, the activity cost is also doubled. However, as can be seen from the actual cost line, when the number of resources is doubled, the cost is more than doubled.[2] This is due to the fact that typically the least expensive resources or most productive are first applied, but any resources applied later are increasingly expensive or less productive. The cost is not simply doubled when the resources are doubled, but more than doubled due to the use of more expensive resources.

A second concept related to understanding the duration-cost relationship can be seen in Figure 10-2, which shows the resource-duration relationship. The typical assumption here is that the total amount of work accomplished (a product of the rate of resource application times the number of days) remains constant. In Figure 10-2 the typical assumption is shown: that an activity could be completed in either 1 day by eight craftspeople or in 8 days by one craftsperson. The total amount of work, 8 work days, remains constant. Other com-

[2]R.B. Harris, *Precedence and Arrow Networking Techniques for Construction,* New York: Wiley, 1978, p. 187.

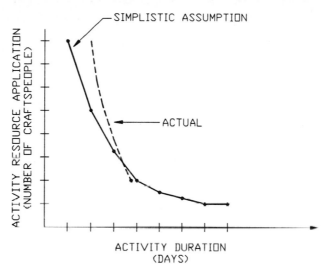

FIGURE 10-2 Resource-duration relationship.

binations could be illustrated, such as two craftspeople for 4 days or four craftspeople for 2 days.

The line labeled "actual" illustrates the deviations from this typical assumption. Among the causes of the differences between the typical and actual is that it may not be possible to perform the activity with fewer than two craftspeople: The physical limitations of the work may require a minimum of two craftspeople to perform the activity. A second cause of the difference from the simplistic assumption is that applying more resources will not result in the same level of productivity by all of those resources. This concept was illustrated in Figure 10-1, which shows less productive resources being applied as the number of resources is increased. Additionally, productivity declines as a result of the potential crowding effect of putting too many resources in a limited space.

With these concepts in mind, the relationship between duration and cost can be determined for any given activity. Example 10-1 illustrates the way in which basic planning and estimating information is used to develop estimates of the duration and cost for an activity. In this example, the planner has evaluated the options available for determining the activity duration to be used in the original schedule. This example illustrates how basic cost data and productivity information are utilized to develop each of the durations used in a schedule. As will be illustrated later, this information can be utilized again to adjust the schedule by adjusting activity durations.

EXAMPLE 10-1. A subcontractor has the task of erecting 84,000 square feet of metal siding. He can use several sizes of crew for its erection, with various crew costs (including scaffolding). The subcontractor expects production to vary with crew size and has prepared the following estimates of production and cost:

Estimated daily production (square feet)	Crew size/makeup
1300	4 (1 scaffolding set, 2 laborers, 1 carpenter, 1 carpenter foreman)
1660	5 (1 scaffolding set, 2 laborers, 2 carpenters, 1 carpenter foreman)
2040	6 (2 scaffolding sets, 3 laborers, 2 carpenters, 1 carpenter foreman)
2300	7 (2 scaffolding sets, 3 laborers, 3 carpenters, 1 carpenter foreman)
	Laborer $12.00/hour (8-hour day)
	Carpenter $16.00/hour (8-hour day)
	Carpenter foreman $18.00/hour (8-hour day)
	Scaffolding set $60/day

To determine the duration for installing the metal siding with each crew, the total quantity is divided by the estimated daily production:

$$\frac{84,000}{1,300} = 64.6 \text{ days}$$

The cost can be determined by summing the daily costs of labor and scaffolding and multiplying by the number of days:

Crew size	Duration (days)	Cost	
4	64.6 (use 65)	$33,850	$34,060
5	50.6 (use 51)	32,991	33,252
6	41.2 (use 42)	33,289	33,936
7	36.5 (use 37)	34,169	34,632

Daily crew cost:

2 laborers $12.00/hour × 8 hours =		$192.00
1 carpenter $16.00/hour × 8 hours =		128.00
1 foreman $18.00/hour × 8 hours =		144.00
Scaffolding $60/day =		60.00
		$524.00/day

$524/day × 65 days = $34,060

This example illustrates the option which the planner develops as he or she establishes the normal duration for a project choosing the least-cost alternative and assigning the duration which was calculated for the activity for the original schedule. As can be seen, there are several possible combinations of duration and cost which could be used for this activity. The relevant portions of this example which could be used for crashing a project include the crew-size options of 5, 6, and 7. The crew size of 4 would not be used if the objective were to minimize duration or cost but would be used only in a case where the objective was to minimize the number of craftspeople working on the project. Thus we would get a curve or a relationship for the cost and duration which would be of the form shown in Figure 10-3.

Figure 10-3 illustrates the basic duration-cost relationship in a general form. This curve was developed by a method similar to that used in Example 10-1, the points illustrated as points I, II, III, and IV being developed through a similar process. The solid line linking point I and point III and the solid line linking point II and point IV illustrate that there are a continuous set of optional costs between those points. The lack of a line between point III and point IV illustrates that there is a discontinuity between these points and therefore there are no options between those two for duration and cost. Point I represents the point of minimum activity duration and maximum activity cost; this point is a limiting point for the duration. Point II represents the minimum activity cost and the associated activity duration corresponding to that cost. Point I is often referred to as the "crash cost," and point II is referred to as the "normal duration." Similarly, point I also represents the "crash duration" and point II represents the "normal cost." At this point it should be mentioned that while these lines connecting the various points approximate a straight line, they are normally arcs connecting the points. Figure 10-4 illustrates the way in which these points are connected and an approximation which can be used to connect

FIGURE 10-3   Duration-cost relationship.

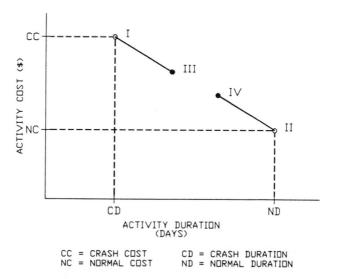

CC = CRASH COST   CD = CRASH DURATION
NC = NORMAL COST   ND = NORMAL DURATION

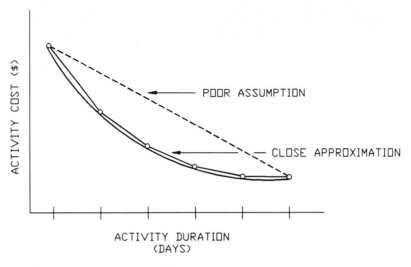

FIGURE 10-4 Complex duration-cost relationship.

various points. For some activities, simply connecting the normal duration with the crash cost point will yield a suitable approximation. In other cases, as can be seen in Figure 10-4, it is necessary to connect the points with a series of segments or to calculate intermediate points in order to approximate the actual curve with a series of segments and thus provide a better analytical solution. As we will see later in this chapter, we use the slope of these segments to determine which activities or at what point we select the activity duration for use in crashing the overall project schedule. It is the slope of this line that is used to determine the impact on project cost of reducing project duration. We can calculate the slope of this line or the slope of a segment mathematically by using the coordinates on the duration-cost graph to calculate this value. For the simplest case, where we have made a close approximation by a straight line between the crash cost and the normal duration points, we can calculate the cost-per-day slope by the following formula:

$$S = \frac{CC - NC}{ND - CD}$$

where

CC = crash cost
NC = normal cost
ND = normal duration
CD = crash duration
S = slope

Example 10-2 illustrates the calculations for cost slope between various segments for the activity described in Example 10-1.

EXAMPLE 10-2. Using the information provided in Example 10-1, we can graph the duration-cost relationships and determine the cost slopes between the points (which could be done by changing the size of crews for increments of time while working on the activity).

The duration-cost relationships are shown in Figure 10-5. The slopes are found from:

$$S_1 = \frac{\$33,936 - \$33,250}{51 - 42} = \$76.22/\text{day}$$

$$S_2 = \frac{34,632 - 33,936}{42 - 37} = \$139.20/\text{day}$$

Figure 10-6 illustrates in general form how this relationship might appear. Again, the direct cost for the project increases as we reduce or crash the overall project duration. Crashing is accomplished by crashing activity durations, so the curve for a project is made up of segments which represent the slopes of either individual activities or groups of activities as will be discussed later in this section. A curve like this is developed by analyzing and determining the cost of crashing various activities in the network. Figure 10-7, a typical construction network, will be used to illustrate this process. Example 10-3 shows the duration-cost relationships for the activities identified in Figure 10-7 and presents a stepwise solution for crashing this network from an original duration of 140 days to a final duration of 110 days.

FIGURE 10-5   Example 10-2 duration-cost relationship.

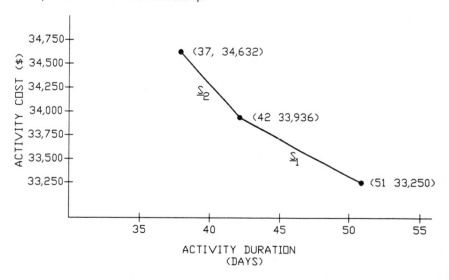

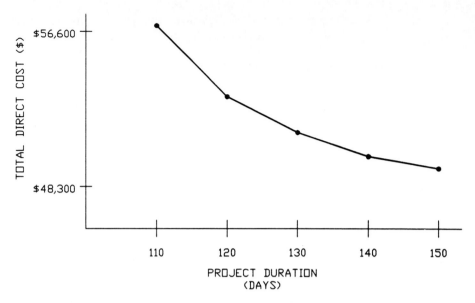

FIGURE 10-6 Project direct cost-duration relationship.

EXAMPLE 10-3

Activity	Normal duration (days)	Crash duration (days)	Normal cost	Crash cost
A	120	100	$12,000	$14,000
B*	20	15	1,800	2,800
C*	40	30	16,000	22,000
D*	30	20	1,400	2,000
E*	50	40	3,600	4,800
F	60	45	13,500	18,000

*Critical path activity.

FIGURE 10-7 Simple project network for compression.

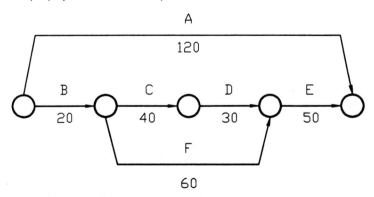

Assume that the duration-cost relationship for each activity is a single linear, continuous function between the crash duration and normal duration points. Using the normal duration (ND), crash duration (CD), normal cost (NC), and crash cost (CC), the crash cost slope for each activity can be determined as follows:

$$S_A = \frac{CC - NC}{ND - CD}$$

$$= \frac{\$14,000 - \$12,000}{120 - 100}$$

$$= \$100/\text{day}$$

$$S_B = \$200/\text{day}$$

$$S_C = \$600/\text{day}$$

$$S_D = \$60/\text{day}$$

$$S_E = \$120/\text{day}$$

$$S_F = \$300/\text{day}$$

where the subscripts refer to activities.

The normal cost for the project is the sum of a normal cost for each activity. In this example, the normal cost for the project is $48,300 and the normal duration is 140 days. If the duration of the project is to be crashed, then the activities will need to be crashed. The first activity which should be crashed is the one on the critical path which will add the least amount to the overall project cost. This will be the activity with the flattest or least-cost slope. The duration can be reduced as long as the critical path is not changed or a new critical path is created. In addition, the activity duration cannot be less than the crash duration. For Example 10-3 this is activity D, which will add only $60 per day for each day of reduction in activity duration. A maximum of 10

FIGURE 10-8   Network after one step of compression.

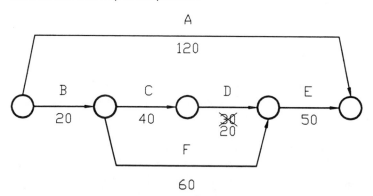

More scaffolding and equipment are required to speed progress. This additional use of resources will contribute to a higher average unit cost.

days can be cut from this schedule by reducing the duration of activity D to the crash duration (minimum duration) of 20 days. Figure 10-8 shows the revised network. Note that the overall duration is now 130 days and there are multiple critical paths (B-F-E and B-C-D-E). The total project cost at this duration is the normal cost of $48,300 plus the cost of crashing activity D by 10 days ($60 per day times 10 days), or $600, for a total of $48,900.

Applying the same approach, the next activity to be crashed would be activity E, since it has the least-cost slope ($120 per day) of any of the activities on the critical path. Activity E can be crashed by a total of 10 days. Crashing activity E by 10 days will cost an additional $120 per day or $1200. Figure 10-9 illustrates the simple project network after the second activity has been crashed. The project duration is now 120 days and the total project cost is $50,100. There are now three critical paths (A, B-C-D-E, and B-F-E).

The next stage of crashing requires a more thorough analysis since it is impossible to crash one activity alone and achieve a reduction in the overall project duration. It is necessary to crash activity A and one of the other activities along the other two critical paths to achieve the desired reduction of 10 days in overall project duration. In this case the combined cost slopes are used and a minimum total value is chosen. Activity A is paired with each of the other activities to determine which has the least overall cost slope for those activities which have remaining days to be crashed. This subset of the critical

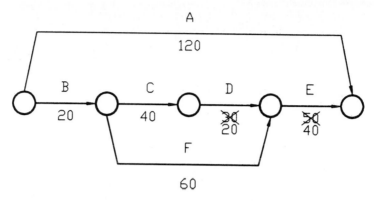

FIGURE 10-9 Network after two steps of compression.

path activities includes activity B, activity C, and activity F. It can also be seen from Figure 10-9 that activity C cannot be crashed unless activity F is also crashed. The choices for activities to be crashed are therefore as follows:

Activity A ($100) + activity B ($200)
Activity A ($100) + activity C ($600) + activity F ($300)

The least-cost-combination cost slope would be activity A + activity B for a cost increase of $300 per day. Reducing the project duration by 5 days would require crashing activity A by 5 days to 115 days and crashing activity B by 5 days to 15 days. This would add $300 per day × 5 days or $1500 to the total project cost. The project duration would be 115 days with a total project cost of $51,600. The final step in crashing the project to 110 days would be accomplished by reducing the duration of activity A by 5 days to 110 days, reducing activity C by 5 days to 35 days, and reducing activity F by 5 days to 55 days. The combined cost slope for the simultaneous reduction of activity A, activity C, and activity F would be $1000 per day. For 5 days of reduction this would

FIGURE 10-10 Network after three steps of compression.

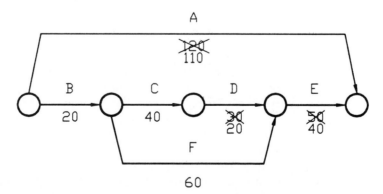

be an additional $5000 in total project cost. The total project cost for the crashed schedule to 110 days of duration would be $56,600. Figure 10-10 illustrates the duration-cost relationship for the project showing the changes in duration and the associated total costs for the project.

TOTAL PROJECT COST ANALYSIS

Total project costs include both direct costs and indirect costs of performing the activities in a network. Total project costs are related to the duration, as will be shown in this section. Direct costs vary inversely with the duration, as was seen earlier. Indirect costs, on the other hand, increase with project duration.

Direct Costs

Direct costs for a project include the direct costs for materials, labor, and equipment which can be related specifically to a task or an activity. The direct costs for a project are the sum of all of the activity direct costs. Examples earlier in this chapter have shown how direct costs can be computed.

Indirect Costs

Indirect costs are the necessary costs of doing business which cannot be related to a particular activity and in some cases cannot be related to a specific project. Indirect costs can be classified into two subsets: general overhead and project overhead. General overhead items include those items that are a part of the cost of doing business but cannot be related to a specific project. Examples of such costs include home office operations such as utilities, rent, accounting, purchasing, and payroll. Job or project overhead includes those items of cost that can be related to a particular project but cannot be assigned to a specific activity. Such items include site supervision, job site trailers, site utilities, project insurance, and scheduling costs. Indirect costs tend to increase as project duration increases. Indirect costs are often a function of time. As one example, site offices are often rented by the month. The cost of site offices, an indirect cost, increases as the number of months of project activity increases. Figure 10-11 illustrates two types of indirect costs. The first type is constant indirect costs such as those items identified above as general overhead items. Also in this category of constant indirect costs might be project overhead costs such as an office trailer. This curve illustrates the concept that indirect costs grow very rapidly at the beginning of the project and then remain constant over the course of the project. The second curve in Figure 10-11 is the variable indirect costs. Variable indirect costs are related to the level of activity on the project but cannot be identified with any specific work tasks. Variable indirect costs include such items as site cleanup, site security, insurance on equipment used on this site, and temporary utilities used on this site. Variable indirect

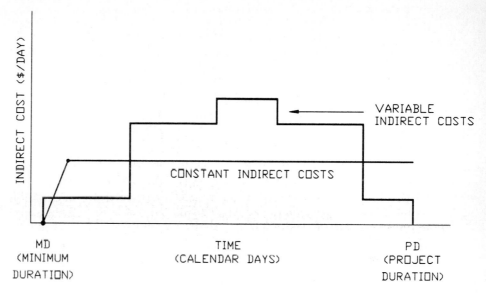

FIGURE 10-11   Indirect cost-time relationships.

costs have the feature of being related to some level of activity on the site, such as the number of craftspeople employed on the site or the volume of work put in place during a particular period. The usual assumption is that there is a relatively constant indirect cost profile for a project. Figure 10-12 illustrates how we utilize this knowledge of indirect costs and total activity direct costs to develop a total project duration-cost relationship curve.

FIGURE 10-12   Total project duration cost relationship.

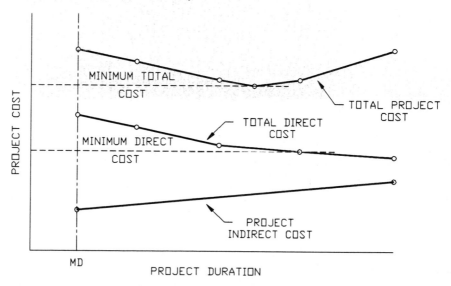

Assuming a constant indirect cost profile, there is a straight-line increase in indirect costs as the project duration increases. The slope of this line would be the value of the constant indirect cost profile. On the same graph, as shown in Figure 10-12, we can plot the duration-cost relationship for direct costs for project activities. The curve at the top represents the total cost. The total cost is the sum of the indirect costs plus the direct costs of activities at various durations. There is a minimum value for total costs, which may be different from the minimum direct cost value. This curve could be further complicated if we imposed variable indirect costs on the project, which would introduce a curve rather than a straight line for indirect costs. We could also evaluate a situation where liquidated damages or a bonus clause was included. This would have the effect of creating a breakpoint or a two-segment line for the indirect cost, and would have implications on shifting the duration for the minimum total project cost. Example 10-4 will illustrate development of a total duration-cost relationship. The example includes a situation which has a bonus/penalty involved, as well as variable indirect costs.

EXAMPLE 10-4. We will use the network shown in Figure 10-7 and the activity direct-cost data from Example 10-3. The indirect costs for the project include job overhead of $250 per day throughout the project. There are also additional support services needed from day 20 to day 90, which increase the indirect cost by $100 per day during this period. General overhead support related to the staffing size of the project and duration can be assumed to be

FIGURE 10-13   Indirect cost profiles for Example 10-4.

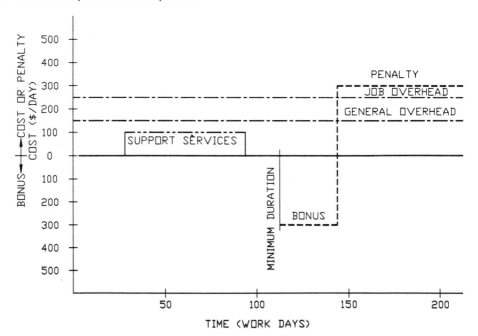

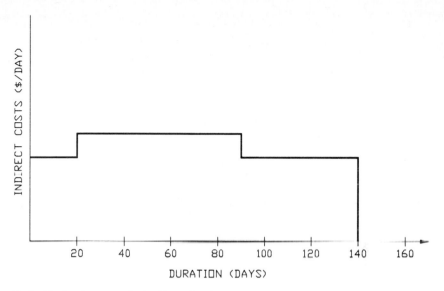

FIGURE 10-14   Project indirect cost profile for Example 10-4.

$150 per day. There is a bonus/penalty clause in the contract of $300 per day with a target duration of 140 days. It is desired to determine the optimal duration for the project based on the least cost to the contractor.

The indirect cost profile includes two parts. The first part is the constant portion of indirect cost, which is the sum of the job overhead ($250/day) plus the general overhead ($150/day), for a total of $400/day. There is also a variable indirect cost from day 20 to day 90; Figure 10-13 illustrates the indirect cost profile for various durations. The total indirect cost curve, Figure 10-14, is developed by adding the indirect costs for each possible duration. The total cost curve, Figure 10-15, shows the relationship between duration total costs

FIGURE 10-15   Total cost profile for Example 10-4.

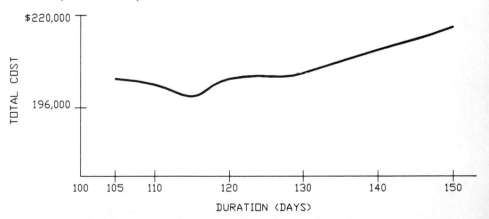

TABLE 10-1   PROJECT COST ANALYSIS

Project duration (days)	Project direct cost	Project indirect cost	Subtotal	Bonus/ penalty	Total cost
150	$148,300	$67,000	$215,300	$ 3,000	$218,300
145	148,300	65,000	214,800	1,500	214,800
140	148,300	63,000	211,300	0	211,300
130	148,900	59,000	207,900	− 3,000	203,900
120	150,100	55,000	205,100	− 6,000	199,600
115	151,100	53,000	204,900	− 4,500	196,600
110	156,100	51,000	207,100	− 9,000	198,100
105	161,100	49,000	210,100	−10,500	199,600

for the project with all items considered. The minimum-cost point represents the optimum duration-cost schedule. Table 10-1 lists the direct costs, indirect costs, and bonuses and penalties for several possible durations.

EXPEDITING LSM SCHEDULES

The procedures used to expedite an LSM diagram are similar to the procedures used to expedite network diagrams. The amount by which activities may be expedited is identified, the critical activities are determined, and the cost per day to expedite each activity is calculated. The critical activity with the least expediting cost per day is selected and that activity is reduced as far as possible.

FIGURE 10-16   LSM schedule to be expedited.

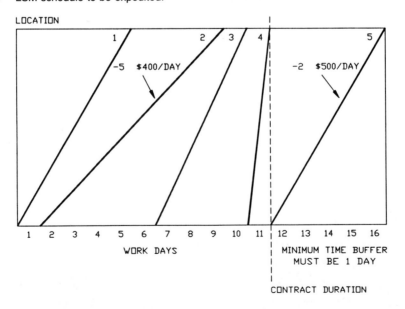

The revised schedule is evaluated to determine if further reductions are necessary. If further reduction is necessary, the next-least-cost activities may be expedited in turn until the project duration cannot be reduced further.

Figure 10-16 shows an LSM schedule to be expedited. Critical activities can be determined to be activities 2, 4, and 5 because each has the specified minimum time buffer of 1 day at the start and finish of the activity. It is determined that activity 2 can be expedited 5 days at a cost of $400 per day, activity 4 cannot be expedited, and activity 5 may be expedited 2 days at a cost of $500 per day. It is necessary to reduce the scheduled project duration by 5 days to complete within the time required by contract.

Because activity 2 has the least cost per day to expedite, its duration is reduced first. However, although activity 2 may be reduced a total of 5 days, only 3 days may be deleted if the 1-day minimum buffer is to be maintained.

Figure 10-17 shows the revised LSM schedule with activity 2's duration reduced 3 days. The originally scheduled slope for the unchanged activities has been maintained, while the slope of the expedited activity is adjusted to match the increased speed of progress. The expedited cost for 3 days has been recorded. The revised schedule shows that the project duration must be further reduced to meet contract completion requirements.

Figure 10-18 shows the LSM schedule further revised by reducing activity 5's duration by 2 days. The expedited cost has been recorded. The schedule shows completion within the contractually required duration. No further expediting is necessary.

FIGURE 10-17   LSM schedule after first expediting.

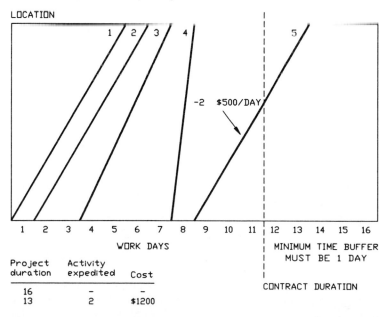

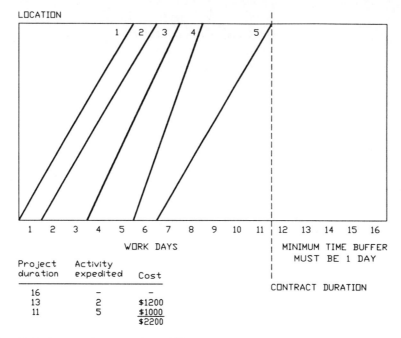

FIGURE 10-18 LSM diagram after second expediting.

USE OF SCHEDULES TO EXPEDITE PROJECTS

The difficulties involved in expediting a project are not learning how to utilize the schedule as an analytical procedure to guide the evaluation, but rather determining how much individual activities may be reduced and the cost of the activities' reduction. Knowing how much an activity may be reduced and the cost of the reduction requires experience in estimating and methods of construction. If the scheduler does not have this experience, others must be included in the expediting process who do.

The scheduler may also know intuitively where least-cost project-duration reduction time is available from the effort of scheduling the project, determining durations, and making the forward and backward passes. The analytical procedure described here may be unnecessary in such cases.

The ultimate decision about how much expediting a project should experience is a management decision. The decision must be based on how much money management is willing to spend for analysis. The consequences of later completion must be weighed against the cost of shortening the project.

EXERCISES

1 Given the following information about a project and the network diagram in Figure 10-19 (drawn using the normal durations), compress the network from its normal duration of 9 days to its expedited duration. List those activities that are expedited

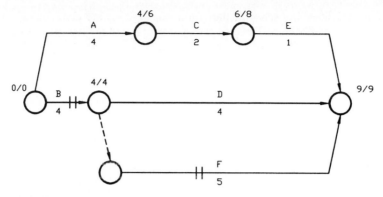

FIGURE 10-19 Expediting exercise.

and the costs associated with each compression. Plot the duration-cost curve for this project.

Activity	Normal duration (days)	Expedited duration (days)	Normal cost	Expedited cost per day
A	4	3	$100	$ 25
B	4	3	250	150
C	2	1	150	150
D	4	1	450	150
E	1	0.5	200	400
F	5	2	200	50

2 Given the following data, draw the network and plot the least-cost curve for going from a crash duration to a normal duration.

Activity	Duration (days)		Preceded by:	Crash cost	Cost rate
	Crash	Normal			
A	15	20	H	$ 400	$6000
B	5	7	H	550	2750
C	6	12	I	500	3000
D	3	9	B, C	1500	4500
E	6	10	C	700	4200
F	3	7	A, D, E	200	600
G	2	4	—	200	400
H	10	14	G	400	4000
I	7	9	G	400	3800

3 Cost and schedule data for a small project are given below. Assume an indirect cost of $200/day. Develop the least-cost curve for the project.

Activity	Preceded by:	Cost Crash	Cost Normal	Duration (days) Crash	Duration (days) Normal
A	—	$ 3,900	$3,600	6	7
B	A	6,500	5,500	3	5
C	B	7,200	6,350	7	9
D	B	4,900	4,700	18	19
E	B	2,200	2,050	9	10
F	C	1,700	1,200	6	8
G	F	7,200	7,200	5	5
H	E	10,000	9,450	10	11
I	D, G, H	4,700	4,500	6	7

4 Draw the precedence diagram for the following data:

Activity	Followed by:	Duration (days) Normal	Duration (days) Minimum	S
A	B, E, F	7	5	$200/day
B	K	9	5	$450/day
C	H, D	8	7	$400/day
D	I, N	11	4	$100/day
E	G, M	9	6	$400/day
F	L	8	7	$500/day
G	C	7	5	$200/day
H	I, N	6	2	$200/day
I	—	12	9	$200/day
J	E, F	10	8	$600/day
K	G	14	10	$350/day
L	M	18	16	$700/day
M	C	9	8	$550/day
N	—	12	9	$200/day

(*a*) Calculate the ES, LS, EF, LF, FF, and TF for each activity and identify the critical path.

(*b*) Compress the schedule to a 65-day duration. How much more would the project cost? (Minimize the additional cost.)

CHAPTER
11

RESOURCE SCHEDULING

INTRODUCTION

Putting together a workable schedule that satisfies all constraints is not an easy task. After contractors have evaluated the work to be performed and the most logical and cost-effective sequence of performing that work, there remains further analysis to produce a workable and efficient construction schedule. Often contractors find that labor, equipment, or materials may be in short supply. Shortages of these essential resources can significantly affect the initiation, performance, and completion of activities on the schedule and can cause the project to be extended beyond the scheduled duration.

Limits imposed on resources can be dealt with in several ways. One such approach is to attempt to level the use of resources in such a way that the scheduled completion date can be met and activity early and late start and early and late finish dates can be respected. This process is known as "resource leveling."

In some cases it may not be possible to schedule activities so that the maximum number of resources does not exceed the available resources and also respect the early start, late start, early finish, and late finish schedule dates for activities. In these cases it may be necessary to extend the overall duration to meet the limits. This type of allocation of inadequate resources is called "resource-restrained scheduling."

As will be seen in the examples in this chapter, there are often wide fluctuations in the daily need for various types of resources based on either the early start or late start dates. These wide fluctuations, if permitted, could affect

Draglines are very specialized pieces of equipment.

the contractors' productivity, completion, and the ability to attract needed resources such as craftspeople. Carpenters, welders, and electricians do not want to work a schedule which requires 3 days of work, being idle for 2 days, working another 2 days, then being idle for 1 day and working another 3 days, and so on. Resource scheduling can help avoid such fluctuations. Contractors attempt to create smooth, balanced resource utilization within the various crafts and for the various types of equipment. When this resource balancing or leveling is applied to labor, it is often termed "manpower leveling." The term used in this chapter will be "resource leveling" and applies to all types of resources: materials, equipment, and manpower.

Resource leveling does have certain limitations, and these should be noted. First, differences in productivity between resources of the same type are not considered. Second, within certain ranges, the differences between productivity of different-size crews are also not taken into account. It is common not to split activities for resource leveling once the activity has started. This may be a real-world solution even though it is not considered in traditional resource leveling, since the impact on productivity of starting and stopping cannot be accounted for adequately. The other main limitation of resource leveling is that of consideration of moving the same crew from one activity to the next. While this can have a significant impact on cost and productivity, the scheduler is unable to track a single crew or group of craftspeople through an entire project while performing resource leveling. Where this is crucial, one can use logic ties to assure continuity of crews. The normal goal in resource leveling is simply to provide a smoothing and leveling process to the resources.

Manual resource leveling can be extremely complex and time-consuming. Even with the considerable effort required, it can be difficult to determine if the best solution has been found. Before making an attempt to schedule re-strained resources, the scheduler should determine whether this effort will provide the benefits that would justify the time and cost associated with it. In most cases, where contractors anticipate scarce resources in advance, there may be an opportunity to contract or stockpile the needed resources in some manner to meet the demands of the project. Manpower fluctuations cannot be so easily overcome as equipment or material limits. An attempt to level all resources would be a monumental task. In most cases it would be impossible and certainly not cost-effective. Most schedulers will attempt to level one principal resource and possibly a few unrelated resources, such as plumbers, electricians, and steel-erecting equipment.

The procedures that are shown in this book, along with the examples, are kept simple to illustrate the concepts. It should be remembered that the process is extremely complex and time-consuming for multiple resources. This chapter does not attempt to address the issue of multiple-project resource leveling. Leveling resources on multiple projects is an extended application of the concepts introduced in this chapter.

Experienced schedulers tend to build their initial schedule with some resource restraints already in it. This is accomplished because the sequences of activities are predetermined based on experience. The approach presented in this chapter assumes that this process has not taken place and that all resource restraints will need to be addressed as a separate step after the initial logic has been developed.

RESOURCE LEVELING

With most projects there are calendar dates for beginning and ending the project that are fixed and cannot be changed. These milestones become goalposts for measuring the success or failure of the project. It is the responsibility of project management to find the necessary resources and utilize them to achieve the goals of the project, including these schedule dates.

Using the logic and durations developed in the initial development of the schedule, it is possible to identify the resources that will be used in each activity and to develop an assessment of the total resources on a daily basis that will be required based on the activities beginning on their early start dates and similarly the same information based on activities beginning on their late start dates. This process is known as "resource loading." In its simplest form, the network is reduced to a time-scaled bar chart format where the resources of interest are identified for each activity where they will be utilized. The daily quantities of resources for each activity are then identified on this time-scaled bar chart.

The next step in the resource-loading process is to sum the total of resources of interest for each day to determine the total project resource requirements on a daily basis. An example of this resource-loading process is presented in Figures 11-1 and 11-2.

For this example, the first step has been to draw the network diagram shown in Figure 11-1. Based on this network diagram, a time-scaled bar chart is

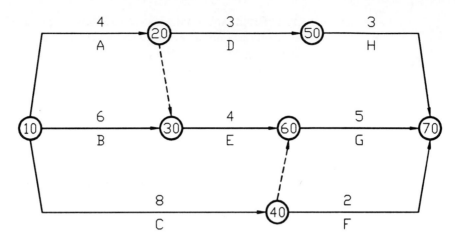

FIGURE 11-1 Network diagram example: resource loading.

constructed using the early start and early finish dates for each activity. By applying the number of craft labor per day to each day on the time-scaled bar chart for each activity it is possible to determine the total craft labor for each day of the project.

For this particular example the resources appear to be imbalanced, varying from three craft labor required up to a maximum of 13 craft labor required. Resource leveling attempts to arrive at the same number of craft labor to be utilized during each day of the project. To determine this number the total craft labor on a daily basis is summed and divided by the number of workdays available. For this example the total number of craft labor workdays is 112.

FIGURE 11-2 Time-scaled bar chart: resource loading example.

ACTIVITY	NUMBER OF CRAFT LABOR/DAY	0	1	2	3	4	5	6	7	8	9	10	11	12	13	14	15
A	2																
B	3																
C	2																
D	5																
E	4																
F	3																
G	3																
H	6																
TOTAL CRAFT LABOR		7	7	7	7	10	10	11	12	13	13	3	3	3	3	3	

TIME (ALL SHOWN AS EARLY START TIMES)

Thus, 112 divided by 15 equals approximately 7.5, indicating that the leveled resources for this particular project will be ideal when there are seven or eight craft utilized each day.

This network will be resource-leveled by utilizing the available float from noncritical activities. It is not possible to alter the critical activities and their associated resource consumption rates without changing the completion date. The first approach used will be to move those activities that have float to different positions so that the total resources on a daily basis come closer to the calculated average value of 7.5. Several approaches for leveling these resources will be illustrated later in this chapter with examples.

The manpower requirements shown in Figure 11-2 are usually prepared by estimators who determine the bid price. If manpower loading is to appear on a construction schedule, the scheduler must consult with the estimator or have access to the estimator's bid calculations. It is also useful to get input from field superintendents and foremen who will supervise these craftspeople. The scheduler can use the data to determine the day-by-day requirements for particular types of resources on activities.

Network schedules must be time-scaled in order to provide a format that will identify the activities planned to be in progress on particular days. A time scale for a 3000-activity, 2-year schedule will involve a large piece of paper and show too many activities performed on a particular day to permit analysis of all craftspeople for the entire project. Thus, most schedulers will use computer programs that have been designed for this purpose. If he or she is performing this process manually, the scheduler will actually work with one trade at a time to level the resources. The following sections will discuss two manual methods for resource leveling and then later discuss the use of the computer and computerized approaches.

TRIAL-AND-ERROR APPROACH

The trial-and-error approach begins by drawing a time-scaled bar chart of the activities, starting with the critical path activities at the top. The resources of interest are summed for these activities. Next, bars representing the remaining activities, grouped into individual paths where possible, are added to the bar chart diagram. Resources for these activities are now included and totaled along with the critical activities on a separate line immediately below the bar chart, as shown in Figure 11-3 for the example network in Figure 11-1.

The first step in the trial-and-error approach for moving activities involves identifying the float available for the latest activity in the project and for previous activities in the path leading to that activity. It is useful to identify logical constraints on the diagram so that when activities are moved, the corresponding changes in float can be noted on the remaining activities.

On Figure 11-3 are shown the various steps taken with the trial-and-error approach to improve the leveling of resources. The first trial is made by moving activity F 5 days to the right (move, F-5R), using all of its float and moving it

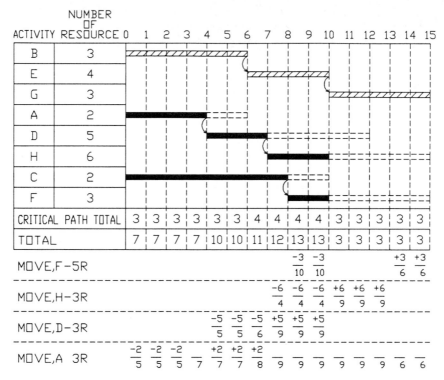

ACTIVITY	NUMBER OF RESOURCE	0	1	2	3	4	5	6	7	8	9	10	11	12	13	14	15
B	3																
E	4																
G	3																
A	2																
D	5																
H	6																
C	2																
F	3																
CRITICAL PATH TOTAL			3	3	3	3	3	3	4	4	4	4	3	3	3	3	3
TOTAL			7	7	7	7	10	10	11	12	13	13	3	3	3	3	3

	1	2	3	4	5	6	7	8	9	10	11	12	13	14	15
MOVE, F −5R								$\frac{-3}{10}$	$\frac{-3}{10}$					$\frac{+3}{6}$	$\frac{+3}{6}$
MOVE, H −3R								$\frac{-6}{4}$	$\frac{-6}{4}$	$\frac{-6}{4}$	$\frac{+6}{9}$	$\frac{+6}{9}$	$\frac{+6}{9}$		
MOVE, D −3R					$\frac{-5}{5}$	$\frac{-5}{5}$	$\frac{-5}{6}$	$\frac{+5}{9}$	$\frac{+5}{9}$	$\frac{+5}{9}$					
MOVE, A 3R	$\frac{-2}{5}$	$\frac{-2}{5}$	$\frac{-2}{5}$	$\frac{+2}{7}$	$\frac{+2}{7}$	$\frac{+2}{7}$	8	9	9	9	9	9	9	6	6

FIGURE 11-3 Trial-and-error resource leveling.

Specialized "Holegater" is used to bore horizontal holes in rock. (Photo courtesy of Aeker Manufacturing.)

out to its late start time. In the calculations at the bottom the total resources are adjusted by subtracting the three resources per day for each of the 2 days from the previous location in the project duration and adding three resources per day to the 2 days now occupied by the activity and the project duration. This same approach is followed for activity H by moving it 3 days later, then making the adjustments in the total resources, in this case six resources per day.

This same process can be utilized over and over again until an acceptable total resource histogram is achieved. The four steps illustrated in Figure 11-3 can possibly be improved upon by further revision, but this example illustrates the trial-and-error process. This type of approach is somewhat haphazard and can become time-consuming without producing any appreciable improvement after initial changes have been made.

MINIMUM MOMENT ALGORITHM

The minimum moment algorithm represents a systematic process for resource leveling that allows the scheduler to measure the degree of improvement in the leveled resources. The goal of the minimum moment algorithm is eventually to achieve a uniform distribution of resources for each day of the project, thus achieving a histogram with a minimum moment about the x axis, or time. The proof of the minimum moment algorithm and the derivation of the improvement factor are given in R.B. Harris, *Precedence and Arrow Networking Techniques for Construction* (New York: John Wiley, 1978). This section will simply introduce the terms and illustrate its application.

The minimum moment algorithm is a sequential process that examines the potential improvement in resource distribution for the project. When followed properly, the approach considers all possibilities for the assignment of scheduled times for activities and their effects on the resource histogram. Several terms are used in the approach that need to be defined before the process can be explained.

An "improvement factor" (IF) is used to successively determine the schedule with the best resource leveling. The improvement factor is calculated as[1]

$$ IF_{A,5} = r \left(\sum_1^m x_i - \sum_1^m w_i - mr \right) $$

where

$IF_{A,5}$ = improvement factor for shifting activity A 5 days out in time

r = daily resource rate for the activity

[1]Adapted from R.B. Harris, *Precedence and Arrow Networking Techniques for Construction*, New York: Wiley, 1978. See pages 272–273 for complete derivation.

m = minimum number of days that the activity is shifted or the duration of the activity

x_i = daily resource sum for the current time frame over which resources will be deducted

w_i = daily resource sum for the time frame over which resources will be added

An activity can be shifted up to the number of days of free float for the activity. As long as a nonzero positive number is obtained for the improvement factor, the resource histogram is improving. The step-by-step process described in the next section continues until no additional improvement is achieved. Figure 11-4 illustrates several of the terms.

Another term used in the minimum moment algorithm approach is the sequence step. The "sequence step" is the order in which a dependent chain of activities is started. The first activity in the chain is step 1, and those activities which can start following completion of step 1 are all step 2. For the network shown in Figure 11-5, activity A is sequence step 1, activities B, C, and D are sequence step 2, activities E, F, and H are sequence step 3, and activity G is sequence step 4.[2]

To begin the minimum moment algorithm approach, a bar chart is prepared

[2]R.B. Harris, *Precedence and Arrow Networking Techniques for Construction*, New York: Wiley, 1978, pp. 50, 279.

FIGURE 11-4   Resource leveling: time-scaled bar chart.

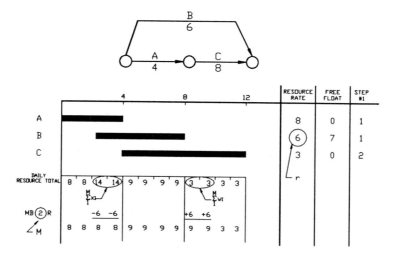

$$\text{IF}_{B,\,2} = 6\left((14+14)-(3+3)-2(6)\right) = +60$$

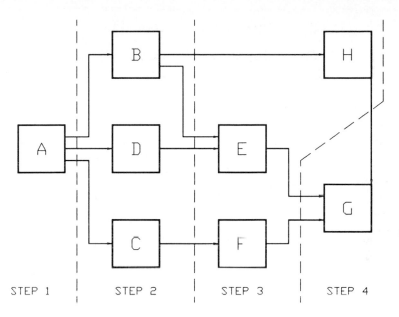

FIGURE 11-5   Resource leveling: steps and counting.

with activities drawn at their early start time. The critical activities are plotted at the top, followed by activities grouped by step number (step 1 first, then step 2, and so on). To the right of the bar chart, columns are labeled resource rate, free float, and step number. These columns are completed for each activity. Beneath the bar chart are totaled the daily resources which would be used based on the time placement of the activities in the early start position.

The process begins by sorting and evaluating the activities in the highest-numbered sequence step. Activities with no free float are not considered. Activities requiring none of the resources being leveled are moved to the right, to the limit of the free float. For all other activities, the improvement factors for each day the activity can be moved are calculated. The activity and shift with the highest positive improvement factor is moved. If a tie exists among the choices, the activity with the largest resource rate is selected. If a tie still exists, the activity is chosen which creates the largest free float for preceding activities. Where a tie continues, the activity with the latest start date on the first activity in the chain is used.

Upon completion of the shifting, the daily resource totals are recalculated. This process is repeated for the remaining activities in the highest-numbered sequence step until all improvement factors are negative. At this point the process can begin again on the next earlier sequence step. When all the sequence steps have been processed in a similar manner, the first cycle is completed: All of the activities have been shifted to later starting positions.

The next cycle involves using any increased float that results from the difference between an activity's early finish and its current scheduled finish. The

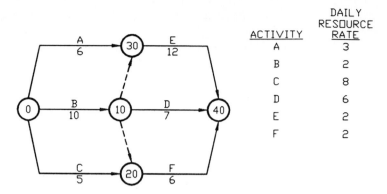

ACTIVITY	DAILY RESOURCE RATE
A	3
B	2
C	8
D	6
E	2
F	2

FIGURE 11-6 Minimum moment method: network example.

cycle is repeated by going through each sequence step beginning with the first. This should consider all possible positioning of the activities and produce a schedule with the minimum moment on resources. To illustrate the minimum moment algorithm, a simple network example is shown in Figure 11-6.

EXAMPLE PROCEDURE

The first step involves determining the critical path. The critical path is B-E for this network (nodes 0-10-30-40). A bar chart is then drawn with the critical activities first and the noncritical activities ordered by sequence step and free float as shown in Figure 11-7.

FIGURE 11-7 Bar chart for network example.

ACTIVITY	RESOURCE RATE	FREE FLOAT	SEQUENCE STEP
B	2	–	1
E	2	–	2
A	3	4	1
C	8	5̶,11	1
D	2	5	2
F	2	6	2

TOTAL DAILY RESOURCES: 13 13 13 13 13 5 2 2 2 2 6 6 6 6 6 6 4 2 2 2 2 2
-2 -2 -2 -2 -2 -2 -2 -2 -2 -2 -2
4 4 4 4 4 4 6 4 4 4 4 4

MOVE F 6 DAYS: -8 -8 -8 -8 -8   -8 -8 -8 -8 -8

MOVE C 6 DAYS: 5 5 5 5 5 5 10 10 10 10 10 12 4 4 4 4 4 6 4 4 4 4 4

For this example, all noncritical activities have some free float. All activities have resources which require leveling associated with them. Two activities are on the last sequence step: activities D and F. For each of these activities the possible improvement factors are calculated as shown below.

For activity D,

$$IF_{D,1} = 2[6 - 2 - (1 \cdot 2)] = 4$$
$$IF_{D,2} = 2[12 - 4 - (2 \cdot 2)] = 8$$
$$IF_{D,3} = 2[18 - 6 - (3 \cdot 2)] = 12$$
$$IF_{D,4} = 2[24 - 8 - (4 \cdot 2)] = 16$$
$$IF_{D,5} = 2[30 - 10 - (5 \cdot 2)] = 20$$

For activity F,

$$IF_{F,1} = 2[6 - 4 - (1 \cdot 2)] = 0$$
$$IF_{F,2} = 2[12 - 6 - (2 \cdot 2)] = 4$$
$$IF_{F,3} = 2[18 - 8 - (3 \cdot 2)] = 8$$
$$IF_{F,4} = 2[24 - 10 - (4 \cdot 2)] = 12$$
$$IF_{F,5} = 2[30 - 12 - (5 \cdot 2)] = 16$$
$$IF_{F,6} = 2[36 - 14 - (6 \cdot 2)] = 20$$

For this step, shifting activity D by 5 days or activity F by 6 days gives the same improvement factor. Shifting activity F by 6 days will give the largest free float for activity C. Adjustments are made as shown in Figure 11-7.

The process is now repeated for activity D, the remaining activity on the last sequence step.

For activity D,

$$IF_{D,1} = 2[4 - 4 - (1 \cdot 2)] = -4$$
$$IF_{D,2} = 2[8 - 8 - (2 \cdot 2)] = -8$$
$$IF_{D,3} = 2[12 - 12 - (3 \cdot 2)] = -12$$
$$IF_{D,4} = 2[16 - 16 - (4 \cdot 2)] = -16$$
$$IF_{D,5} = 2[20 - 20 - (5 \cdot 2)] = -20$$

No shifting takes place! (All are negative.)

The next step is to repeat the process for the activities on the previous sequence step: activities A and C.

For activity A,

$$IF_{A,1} = 3[13 - 2 - (1 \cdot 3)] = 24$$
$$IF_{A,2} = 3[26 - 4 - (2 \cdot 3)] = 48$$

$$IF_{A,3} = 3[39 - 6 - (3 \cdot 3)] = 72$$
$$IF_{A,4} = 3[52 - 8 - (4 \cdot 3)] = 96$$

For activity C,

$$IF_{C,1} = 8[13 - 5 - (1 \cdot 8)] = 0$$
$$IF_{C,2} = 8[26 - 7 - (2 \cdot 8)] = 24$$
$$IF_{C,3} = 8[39 - 9 - (3 \cdot 8)] = 48$$
$$IF_{C,4} = 8[52 - 11 - (4 \cdot 8)] = 72$$
$$IF_{C,5} = 8[65 - 13 - (5 \cdot 8)] = 96$$
$$IF_{C,6} = 8[65 - 12 - (5 \cdot 8)] = 104$$
$$IF_{C,7} = 8[65 - 14 - (5 \cdot 8)] = 88$$
$$IF_{C,8} = 8[65 - 16 - (5 \cdot 8)] = 72$$
$$IF_{C,9} = 8[65 - 18 - (5 \cdot 8)] = 56$$
$$IF_{C,10} = 8[65 - 20 - (5 \cdot 8)] = 40$$
$$IF_{C,11} = 8[65 - 20 - (5 \cdot 8)] = 40$$

Activity C is shifted 6 days.
Activity A is again calculated.
For activity A,

$$IF_{A,1} = 3[5 - 10 - (1 \cdot 3)] = -24$$
$$IF_{A,2} = 3[10 - 20 - (2 \cdot 3)] = -48$$
$$IF_{A,3} = 3[15 - 30 - (3 \cdot 3)] = -72$$
$$IF_{A,4} = 3[20 - 40 - (4 \cdot 3)] = -96$$

No shifting takes place! (All IFs are negative.)
Further shifting using backfloat will not produce any positive improvement factors. Therefore, the leveling process is complete.
The reader should recognize that the process illustrated above can be extremely rigorous and time-consuming when performed by hand for large numbers of activities. This is an area where the speed and accuracy of microcomputers can and should be put to work to develop a resource-leveled schedule. Several of the popular scheduling programs will perform these calculations and make the adjustments.

EXERCISES

1 Data for a project are given below. Use the minimum moment algorithm to level the resources.

Activity	Preceded by:	Duration	Resource rate
A	B, C	7	7
B	—	6	5
C	—	5	6
D	C	4	3
E	A	9	2
F	D	6	4
G	D	11	4
H	F	8	6
I	E	4	3
J	I	6	2
K	H, I	9	5

2 Given the following network, level the total manpower to reduce the peak manpower requirements to a minimum.

Task code	Description	Duration	Total manpower	Preceded by:
10	Clear site	1	1	—
20	Strip and stockpile topsoil	2	2	10
30	Excavate trench for sewer	2	1	20
40	Grade and compact subgrade	4	2	20
50	Install site sewer	10	3	30
60	Excavate for footings	2	2	20
70	Form footings	5	4	60
80	Reinforce footings	5	4	70
90	Pour footings	3	5	80
100	Form foundation walls	10	5	90
110	Reinforce foundation walls	10	5	100
120	Pour foundation walls	6	6	110
130	Backfill	3	2	140
140	Waterproof foundations	5	2	120
150	Erect steel columns and beams	10	6	90
160	Erect steel joists	3	4	150,130
170	Fireproof structural steel	5	3	160
180	Install metal deck	3	4	170
190	Erect precast concrete wall panels	15	5	160
200	Rough-in plumbing	20	2	50, 140
210	Rough-in electrical	15	2	90
220	Rough-in heating, ventilation, and air conditioning	10	3	190
230	Rough carpentry work at roof	15	2	180
240	Form, reinforce, and pour concrete topping on metal deck	5	4	230

Task code	Description	Duration	Total manpower	Preceded by:
250	Install built-up roofing	20	6	240
260	Install metal finish	5	2	250
270	Form slab-on-grade	3	4	180
280	Reinforce slab-on-grade	3	4	270
290	Pour slab-on-grade	2	5	280
300	Install windows and glazing	5	4	260
310	Install exterior doors and hardware	2	3	300
320	Install caulking and insulation	3	2	310
330	Install ornamental metal	5	2	300
340	Install roof hatches	2	2	260
350	Install curbs and gutters	3	4	310
360	Pave parking lot, including sidewalk	18	6	350
370	Site landscaping	12	4	350

3 Assume that a large earth-moving project will be undertaken to clean up several hazardous waste sites. It is desired to minimize the number of operations required. The current schedule and resource assignments are shown in Figure 11-8. Adjust the activities and schedule to obtain the desired allocation.

FIGURE 11-8 Diagram for Exercise 3.

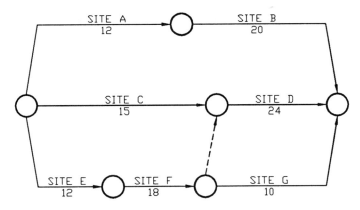

SITE	CURRENT PLANNED NUMBER	DESIRED ALLOCATION
A	6	4
B	6	5
C	6	4
D	6	8
E	6	9
F	5	4

4 Given the following information, draw a network diagram, perform a forward and a backward pass, and draw a time-scaled diagram. Assign the manpower and equipment requirements for this project and draw an S-curve (cost versus time) for each of the project resources. Assume a linear distribution of resources through each activity. Adjust the schedule to the maximum number of resources available.

Activity	Resource or craft type	Number of resource units	Preceded by:
A	Bulldozer	2	—
B	Laborer	4	A
C	Laborer	4	A
D	Bulldozer	2	C
E	Welder	4	C
F	Bulldozer	2	B
G	Laborer	4	B
H	Welder	3	F
I	Laborer	7	E
J	Crane	1	H
K	Loader	2	G
L	Crane	1	N
M	Welder	2	N
N	Loader	1	D
O	Laborer	2	N
P	Crane	1	J
Q	Welder	4	I, M
R	Welder	1	I, M
S	Laborer	3	P
T	Laborer	5	O
U	Welder	3	L, Q
V	Laborer	4	R
W	Laborer	6	S, T, U, V

RESOURCE LIMITS

Type of resource	Maximum number available	Resource unit	Unit cost
Laborers	8	Manhour	$ 8.50
Crane	1	Equip-hour	62.50
Bulldozers	2	Equip-hour	37.50
Welders	5	Manhour	18.00
Loaders	1	Equip-hour	28.00

CHAPTER
12

USING
A SCHEDULE
TO PROVE
DELAY CLAIMS

WHAT DELAY IS

Definitions

In construction claims, a "delay" is the time during which some part of the construction project has been extended or not performed due to an unanticipated circumstance. An incident of delay may be caused by the contractor or by any of the other factors that influence the construction project. Delay may also be caused by the owner, the designer, other prime contractors, subcontractors, suppliers, labor unions, utility companies, nature, or any number of other organizations and entities which participate in the construction process.

Many things may occur on the construction project to increase the time of performance of any given activity or the overall project. Most common causes are differing site conditions; changes in requirements or design; inclement weather; unavailability of labor, material, or equipment; defective plans or specifications; and owner interference. These and other delays not only increase the time required to perform the contract work, but may also increase the costs for many of the parties involved.

Schedules play an important role in construction delay claims. Delays can be identified, defined, and explained by schedules. By rerunning the schedule through the computer with the delays, the effect of delays on the project com-

Paving is one of the last activities associated with a building project.

pletion date can be shown and the effect of future delays can be anticipated. Delay defined by a schedule is difficult to deny.

Excusable and Nonexcusable Delays

Construction delays fall into two major categories: excusable and nonexcusable. An "excusable delay" is one that will serve to justify an extension of the contract performance time. It excuses the party from meeting a contractual deadline. Common excusable delays for a contractor include design problems, employer-initiated changes, unanticipated weather, labor disputes, and acts of God. A "nonexcusable delay" is one for which the party assumes the risk of delayed performance and its consequences—not only its own performance, but possibly the impact upon others as well. Common nonexcusable delays for a contractor include failure to perform work within the allotted time frame, unavailability of manpower, subcontractor failures, and improperly installed work that must be replaced.

Generally, whether a delay is excusable or nonexcusable is a matter of contract. Many standard construction contracts specifically enumerate various types of delay that will entitle the contractor to an extension of time. The American Institute of Architects (AIA), in its Document A201, *General Conditions of the Construction Contract*, permits an extension of time for the following delays:

8.3   Delays and Extensions of Time

8.3.1   If the Contractor is delayed at any time in the progress of the work by an act or neglect of the Owner or Architect, or of an employee of either, or of a separate contractor employed by the Owner, or by changes ordered in the work, or by labor disputes, fire, unusual delay in deliveries, unavoidable casualties or other causes beyond the Contractor's control, or by delay authorized by the Owner pending arbitration, or by any other causes which the Architect determines may justify delay, then the Contract Time shall be extended by Change Order for such reasonable time as the Architect may determine.[1]

However, standard construction contracts vary as to which delays will be excusable. One must review the relevant contract clauses to determine if a particular delay is excusable.

Most standard contracts also contain a catch-all phrase that sets out a general standard of excusable delay, such as "causes beyond the contractor's control" or "unforeseeable causes beyond the control and without the fault or negligence of the contractor." FIDIC (*Federation Internationale des Ingenieurs-Conseils*) *Conditions of Contract (International)* provides the following "catch-all" phrase in Clause 44—Extension of Time: "or other special circumstances of any kind whatsoever which may occur, other than through a default of the contractor, be such as fairly to entitle the contractor to an extension of time. . . ."[2] In the AIA's Clause 8.3.1, the catch-all phrase is "or other causes beyond the Contractor's control, . . . or any other causes which the Architect determines may justify delay. . . ."

Most contracts require an excusable delay to be unforeseeable. For example, in the United Kingdom, the Institution of Civil Engineers Clause 14(6) permits a time extension if the delay "could not reasonably be foreseen by an experienced contractor at the time of tender." Determining foreseeability is often difficult. Even though a delay occurs that may apparently fall within the language of the time extension clause, it may not be excusable because of contractor control or foreseeability. For example, it was held that the late delivery of steel due to a strike was not excusable because at the time the parties signed the contract, the strike was clearly foreseeable, and the contractor assumed the risk of delay by not specifically providing for it in the contract.[3] If a contractor "causes" a strike by pursuing an unfair labor practice, such a strike would not be an excusable delay. However, if the strike that delayed the project was "caused" by the unfair labor practices of a subcontractor, the subcontractor's actions might be deemed to be beyond the control of the contractor, and the delay might be excusable. If a contractor negligently set fire to its own plant, the delay resulting from the fire would not be excusable.[4] Other examples of foreseeable delays which are not excusable are a shortage of capital or an

[1] American Institute of Architects, *General Conditions of the Construction Contract*, 14th ed., Document A201, AIA, Washington, D.C., 1987.

[2] Federation Internationale Des Ingenieurs-Conseils, *Conditions of Contract for Works of Civil Engineering Construction*, 4th ed. Chailly, Switzerland (1987).

[3] *John F. Miller, Inc., v. George Fichera Constr. Corp.*, 7 Mass. App., Ct. 494, 388 N.E.2d 1201 (1979).

[4] R. Nash, *Government Contract Changes*, Federal Publications, Washington, D.C., 1975, p. 327.

inability to obtain money that may affect the contractor's performance, failure to provide adequate equipment or labor, failure to order materials in a timely manner, failure of the general contractor to coordinate the work of its subcontractors, delays due to normal weather conditions, failure to evaluate the project site, inadequate supervision of the projection, and delays due to the removal and replacement of nonconforming work.

A nonexcusable delay may also be a breach of the construction contract and may justify the termination of the contract. If there is a liquidated damages clause, the owner may assess liquidated damages for nonexcusable delays. Normally, nonexcusable delays are not granted time extensions, and the contractor is expected to make up or absorb the delay into the schedule.

If a contractor encounters excusable delays, it is entitled to an extension of time and should not be assessed liquidated damages for such delays, nor should the contract be terminated for default. Whether the contractor may recover its delay costs for an excusable delay depends on whether the excusable delay is also a compensable delay and whether it is concurrent with other delays.

The concept of excusable delays also applies to the performance of the owner and designer. For example, the contractor must excuse the owner or designer for the time necessary to secure a permit if the time was reasonable. The contractor may deserve a time extension if time necessary to obtain a permit was unreasonable and the late permit extended the project completion date. In this situation, the unreasonable late permit is an unexcusable delay. The Armed Services Board of Contract Appeals discussed reasonable time to respond to contractor's request for information in *Fishbach & Moore International Corp.*, as reprinted at the end of this chapter. Interestingly, the designer may not be responsible to the owner for delays resulting from actions that are consistent with "professional skill and care" or that are the result of "reasonable cause" even if the owner is responsible to the contractor.

Types of delay that extend the project completion date that will be nonexcusable for the construction contractor are numerous. These include failure to provide access to the site, failure to remove obstructions, failure to make decisions within a reasonable time, and failure to coordinate multiple prime contractors.

Compensable and Noncompensable Delays

Excusable delays may be further classified as "compensable" or "noncompensable." If the delay is deemed compensable, the party will be entitled to additional compensation for the costs of the delay, as well as additional time for contract performance. However, in some special circumstances a compensable delay does not always mean that additional time is due. Sometimes only additional costs will be compensable. Generally speaking, a delay that could have been avoided by due care of one party is compensable to the innocent party suffering injury or damage as a result of the delay. Taking too long to secure a permit may make the unreasonable portion of the time compensable, even though the contract completion date has not been extended, if the unreasonably late permit caused the contractor's costs to increase.

What additional compensation for delays may be recovered often is limited by the terms of the contract. Many delay claims are brought under the auspices of the changes clause. Recovery under the changes clause may be limited to the delay costs related to the direct performance of the change-order work. Such recovery would not encompass the delay impact on the other aspects of the contract work. Delay resulting from changed conditions may be compensable under the differing-site-conditions clause. Some contracts have a specific clause to compensate the contractor for delay, often known as the "suspensions clause." FIDIC Conditions of Contract (International), Clause 40—Suspension of Work, permits part or all of a project to be suspended by the designer, but provides all additional performance costs as a result of the suspension to be reimbursed to the contractor. The FIDIC clause also permits the contractor to treat the contract as abandoned if the suspension lasts more than 3 months. If the contractor abandons the contract, the contractor may recover costs defined in the change-order (variation) clause or full damages.[5]

AIA Document A201, *General Conditions of the Construction Contract*, similarly provides for delays as a result of suspension and payment for any additional costs which may result:

14.3   Suspension by the Owner for Convenience.

14.3.1.   The Owner may, without cause, order the contractor in writing to suspend, delay or interrupt the work in whole or in part for such period as the Owner may determine.

14.3.2   An adjustment shall be made for increases of the contract, including profit on the increased cost of performance, caused by suspension, delay or interruption. No adjustment shall be made to the extent:

.1   that performance is, was or would have been so suspended, delayed or interrupted by another cause for which the contractor is responsible; or

.2   that an equitable adjustment is made or denied under another provision or this contract.[6]

Suspension of the contractor's work is also discussed in the *Fischbach & Moore International Corp.* decision at the end of this chapter.

Even without appropriate contractual authority, delays may constitute breaches of the contract, allowing for recovery of damages. When the delay is a breach of contract, limitations on cost recovery that are included in the changes or variation clause do not apply. For example, an owner's delay in furnishing the property for construction of massive concrete column support for the concrete

---

[5] FIDIC, *Conditions of Contract (International)*, op. cit.
[6] AIA, *General Conditions of the Construction Contract*, op. cit.

guideway on a light rail system in Vancouver and delay in removing utilities that obstructed the construction of the concrete columns was determined to be a breach of contract entitling the contractor to delay damages, extra work, and lost productivity costs.[7]

Some contracts attempt to preclude contractors from recovering compensation for delays by including a no-damage-for-delay clause in the contract. However, such exculpatory clauses will not always be successful in preventing the contractor from recovering employer-caused delay costs. Courts have recognized many exceptions to the enforceability of no-damages-for-delay clauses.[8]

Although a no-damage-for-delay clause provides the owner with a substantial benefit, the benefit often is achieved at the expense of the contractor. No individual familiar with the construction process believes that a contractor has the ability to foresee all the possible delays, disruptions, and interferences possible on a project. If a contractor included money in its proposal for the potential delays and disruptions that typically occur on a project, it would have almost no chance of being low bidder on that project. Further, the no-damage-for-delay clause invites unreasonable behavior by the architect, the owner, or the owner's other agents against the contractor. The contractor cannot exclude the no-damage-for-delay clauses in its bid, because this would change the terms offered by the owner. Any contractor including such an exclusion in its bid will normally be disqualified from the bidding process.

Instead of the no-damage-for-delay clause, a much more reasonable and fair approach to limiting owner responsibility for delay damages is to include a specific clause defining permitted contractor delay costs, much as a liquidated damage clause defines the owner's delay costs. A contractor could be required at bid time to state a daily overhead rate for both its field and home office. The amount the contractor includes can be used to evaluate bids. A specific procedure for analyzing and evaluating delays would also assist in resolving the problem of project delays and their resolution (see Chapter 14).

The owner is commonly compensated for delay costs if the delay is attributable to the contractor. The contract may provide for liquidated damages—a stated dollar amount that may be withheld from payments otherwise due the contractor for each day of contractor-caused delay which results in late project completion. Stipulated liquidated damages facilitates the employer's recovery of its delay costs by providing an easy calculation and allowing the employer to retain the liquidated damages from progress payments. However, if the contractor's delays are excusable, the owner will not be entitled to assess liquidated damages. A liquidated damage clause not only fixes the amount of delay damages, but also limits the delay damage to the liquidated amount. Without a liquidated damage clause, an owner can recover its actual costs of delay.

---

[7] *W. A. Stevenson Construction (Western) Limited v. Metro Canada Limited* (1987), 27 C.L.R. 113 (B.C.S.C.).

[8] B. Bramble and M. Callahan, *Construction Delay Claims*, New York: Wiley, 1987, p. 54.

Critical and Noncritical Delays

Not all delays result in delay to the overall project completion. For example, a change in the type of electrical switchplates may not delay completion of the project—assuming that the substituted switchplates are readily available, other substantial work is proceeding, and the installation of the switchplates will not delay other construction activities that may delay completion. On the other hand, changing the type of structural steel members while the contractor is erecting structural steel will likely delay the contractor's overall completion of the project. The change order must be prepared; shop drawings for the changed members have to be prepared and approved; and the steel has to be fabricated and delivered to the project site. In the meantime, most other substantial contract work cannot proceed. Delays that result in extended project completion are known as "critical delays." Concomitantly, delays that do not extend the project completion date are called "noncritical delays."

It is generally held that a contractor will not be entitled to a time extension for an excusable delay unless the delay extends the overall project completion. However, a delay need not delay the overall contract performance time for a contractor to recover its costs of delay. The delay need only affect the contractor's cost of performance. For example, on a wastewater-treatment facility, a 2-year suspension to the construction of a maintenance garage may not delay the overall completion of the operating plant but may increase the contractor's cost of building the maintenance garage. In such cases, the contractor may be entitled to its additional performance costs, despite the fact that it completed the overall project within the required contract performance period.

THE IMPORTANCE OF NETWORK SCHEDULES

Schedules are an important part of the need to measure the effect of delays. Once a delay has been experienced and it has been determined whether the delay was compensable or critical, the length of the delay and the effect on the remaining work need to be defined. An accurate evaluation will help the designer and contractor calculate both the value of any required change order (direct costs, overheads, indirect costs, impact, lost efficiency) and the appropriate length of time extension. Schedules used to measure the effect of delay can avoid serious disputes.

The best way to measure the length of a delay and effect on the uncompleted work is by a network schedule. Bar charts are less effective than network schedules for measuring delays. Both the construction and the legal community recognize the limitations of bar charts for measuring delays. Justin Sweet, in his book, *Legal Aspects of Architecture, Engineering and the Construction Process*, says: "Requiring. . . a CPM schedule has at least three advantages. . . . Third, from a litigation standpoint, requiring the contractor to maintain a CPM schedule helps prove or disprove the impact of the owner-caused delay."[9]

U.S. courts have consistently held bar charts to be less effective than network diagrams in defining delays. A U.S. federal court found that bar charts are "not designed to afford an overall coordinated schedule of the total work covered by the contract."[10] The U.S. General Services Board of Contract Appeals took a similar position when it refused to accept a bar chart to prove a delay because the bar chart could not depict the effect of changes on the interrelationship of job activities as a CPM schedule could.[11]

Because of the high visibility of bar charts, however, they may be used to demonstrate the process or results of a critical path analysis. Figure 12-1 shows a bar chart used to explain a network delay analysis. The bar chart is used to illustrate the network analysis because of the bar chart's visual simplicity, but it is not a substitute for the network diagram.

Network schedules are also important for showing interrelationships among multiple causes of project delay. If the owner and contractor both contribute to delayed project completion, neither can recover monetary damages unless the delay can be apportioned between them. If both parties contribute to a delay and neither is able to apportion the delay, neither party can collect

[9] J. Sweet, *Legal Aspects of Architecture, Engineering, and the Construction Process*, 4th ed., St. Paul, MN: West Publishing, p. 579.
[10] *Natkin & Co. v. George A. Fuller Co.*, 347 F. Supp. 17 [W.D. Mo. (1972)].
[11] *Haas and Haynie Corp.*, GSBCA No. 5530, 84-2 BCA 17,446 (1984).

FIGURE 12-1   Use of a bar chart to illustrate CPM analysis.

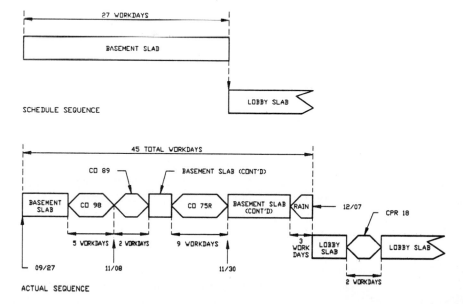

additional costs or liquidated damages from the other. Bar chart schedules cannot show the interrelationships among multiple causes of delay and therefore have not been accepted as clear proof of the apportionment of delay; network analysis systems have.[12]

Schedules are an important part of providing (or refuting) delay claims because they provide a detailed medium for comparing and measuring time and intent. A detailed method of presenting a time claim is important to carry the claimant's burden of proof in order to establish its entitlement to whatever additional time or costs a delay may have caused. Since their rise in popularity, CPM schedules have been grasped as the medium and method to prove a delay in the United States and other countries. In Canada, a contractor's CPM schedule showing as-planned sequence for construction of concrete ''T'' columns to support an elevated concrete guideway for the Vancouver light rail transit system was compared to the as-built sequence shown in another schedule to award $2,348,765 in lost productivity damages.[13]

Schedules have every potential for fulfilling this key role, but it is important to recognize that since schedules are prepared by people, schedules are subject to the same good and bad, achievement and failure, legal and criminal behavior that people are. A schedule can be incorrect—if not plainly wrong—despite an impressive stack of computer printouts full of detailed scheduling information. Intentional deception is possible, and, yes, the computer can be made to lie. As a result, before a schedule can be used to measure the effect of delay, certain basic standards must be demonstrated.

DETERMINING THE EFFECT OF DELAY

Reliability

A schedule used to prove a delay must be prepared and used as correctly as any other tool necessary for project completion. Thus, before acceptance to measure a delay, the schedule should be demonstrated to be reliable. To be able to use a CPM to measure a delay properly, initial data must be sensible, accurate, and reasonable. Correct scheduling technique and methodology must have been used, and no open ends are permitted. All activities must be accurately and logically sequenced. If the input was introduced carelessly, or in error, it makes the original schedule incorrect and the analysis of the delay distorted, if based on the incorrect original schedule.

To be used to prove a delay, the schedule and the underlying information must be reliable. Various U.S. court decisions have accepted challenges to the reliability of particular schedules offered to prove delays and have rejected the

---

[12] *Pathman Construction Co. v. Hi-Way Electric Co.*, 65 Ill. App. 3rd 480, 382 N.E.2d 453 (Ill. 1978); *Fischbach & Moore International Corp.*, ASBCA No. 18146, 77-1 BCA 12, 300 (1976).

[13] *W. A. Stevenson Construction (Western) Limited v. Metro Canada Limited* (1987), 27 C.L.R. 113 (B.C.S.C.). [In addition, see *Thiess Properties Pty. Ltd. v. Ipswich Hospitals Board* (1985), 2 Qd.R. 318, aff'd (1985) 2 Qd.R. 323, a decision from Australia.]

schedules. The judicial challenges discussed herein do not limit future challenges or detail the presentations necessary to demonstrate a schedule's reliability. U.S. courts and boards respond to situations and defects in individual schedules. Much more can go wrong with a schedule than the errors these courts have discussed.

Schedule Must Be Complete

A New Jersey court[14] discussed the importance of a schedule being complete before the schedule may be used to prove or refute a construction delay. As often occurs on construction projects, the schedule presented to the New Jersey court had not been completed at the start of the project. As the project progressed, completion of the schedule also progressed. At the time of the first update, the schedule included the mechanical, electrical, and general prime contractors' intent to construct the building. At the time of the first update, the schedule was complete through close-in or until the building was tight, but omitted finishing activities such as toilet and laboratory equipment installation and equipment hookup. Both the mechanical and electrical prime contractors stated that the schedule at the time of the first update demonstrated their intent to complete the project. The schedule at the time of the second update had evolved to indicate installation of laboratory equipment and owner-furnished equipment, finish details, and the electrical contractor's procurement. But it was not until the third update, when the schedule reflected the general contractor's procurement schedule, that the court found the schedule sufficiently complete to be used as a reference to demonstrate delays.

Equipment procurement is a vital, although often overlooked and disdained part of the construction schedule. Contractors regularly ignore procurement in developing their schedules. Without procurement, however, a network schedule is incomplete. An incomplete schedule should not be used to measure delay, because it does not reflect the interaction of all the activities necessary to complete the entire project. Delays must be measured from their effect on the entire, not partial, project. If they are necessary to make the schedule complete, procurement activities can be added to the project schedule before delay analysis is started.

Schedule Reliability Not Dependent on Field Use

A court may accept a schedule as a reliable framework to demonstrate delay even though the schedule was not used by the contractor to complete the project. The U.S. Board of Contract Appeals[15] utilized an outside scheduling

[14]*Dobson v. Rutgers*, 157 N.J. Super 357, 384 A.2d 1121 (1978), aff'd sub no. *Broadway Maintenance Corp. v. Rutgers*, 180 N.J. Super 350, 434 A.2d 1125 (App. Div. 1981), aff'd 90 N.J. 253, 447 A.2d 906 (1982).

[15]*Blackhawk Heating & Plumbing Co.*, 75-1 BCA 11,261 (1975).

consultant's schedule, prepared after project completion, rather than the contractor's construction schedule, because the contractor's schedule did not break down the activities necessary to analyze a delay into appropriate subparts. The scheduling consultant's schedule divided the contractor's schedule into more detail, permitting more meaningful analysis of project delays. In adding activities, the board found that the consultant had made the contractor's schedule more useful while maintaining the basic construction plan.

Information in the Schedule Must Be Reliable

The information on which the schedule is based must be demonstrated to be reliable. One U.S. board[16] found the CPM schedule could not be accepted to substantiate the contractor's delay claim because no satisfactory evidence of the origin of the data that went into the preparation of the schedule was presented. Another board[17] refused to find that a CPM schedule proved government responsibility for delay, because the contractor presented no testimony or documentation to show what information was furnished or the accuracy or validity of the information furnished to prepare the CPM.

One way to demonstrate the reliability of the schedule analysis is to show that the unaffected part of the planned schedule matched the actual progress. This suggests that the planned schedule could have been achieved if the delays had not occurred. A contractor[18] demonstrated the reliability of its as-planned CPM schedule to show the effect of delay and out-of-sequence work by showing that those portions of the project that were *not* affected by the delays were completed within the originally anticipated sequence and duration.

However, it is not always possible to demonstrate reliability by showing that planned and actual sequence matched when they were not affected by delays. Many construction projects are so seriously delayed at the outset that no portion of the planned schedule will be unaffected. In these cases, reliability may be demonstrated either by showing how past similar projects were completed as originally scheduled or that the original sequences or durations match industry standards.

A schedule may also be found to be not reliable if the logic does not consider practical field restraints that control the construction sequence. One board[19] rejected the contractor's CPM analysis purporting to show delay because the schedule ignored foreseeable winter weather conditions and showed compacted fill to be installed in December, January, February, and March. During trial, the contractor admitted that compacted fill should not be placed on frozen ground. Because the schedule did not present a reliable framework for mea-

---

[16] *Chaney & James Construction Co.*, 66-2 BCA 6066 (1966).
[17] *Lane-Vergudo*, 73-2 BCA 10,271 (1973).
[18] In *W. A. Stevenson Construction (Western) Limited v. Metro Canada Limited* (1987), 27 C.L.R. 113 (B.C.S.C.).
[19] *Joseph E. Bennett Co.*, 72-1 BCA 9364 (1972).

suring the delay, the court did not permit the CPM analysis to be used to measure the delay.

Schedule Must Reflect the Intended Plan

A schedule may be found to be not reliable if the logic intentionally deviates from the manner in which the contractor actually intends to complete the project. In *E. C. Ernst, Inc. v. Manhattan Construction Co.*,[20] the contractor attempted to use a schedule to measure delay that had its duration reduced 1 year from the actual time it had intended to take to construct the project. The contractor attempted to measure alleged subcontractor delays by the reduced-duration schedule. The court refused to accept the reduced-duration schedule because it did not accurately reflect the contractor's intent to complete the project. The U.S. Corps of Engineers, in its *Modification Impact Evaluation Guide*,[21] requires a contractor's schedule that is used to measure delay to be first modified to match the contractor's actual intended plan.

In order to be accepted as a demonstration or measurement of a delay, a schedule must accurately reflect both the contractor's intent to construct the project and the practical field restraints that apply to the activities. The schedule must be a logical analysis of the individual tasks which comprise the project. This applies to schedules prepared to complete the project and to postcompletion scheduling analysis prepared to prove a delay.

Approved Schedules

A schedule also does not have to be formally accepted by the owner or designer to be accepted by a court to prove a delay. If the schedule is submitted with no responsive comments from the owner or designer, it can reasonably be assumed that the submitted schedule has been accepted, if not approved. A schedule approved by the owner will of course help facilitate the proof that the schedule is reasonable.

The British Columbia Supreme Court used a more detailed as-planned schedule to measure delay even though it had not been approved, rejecting the less detailed but approved schedule that the defendant-owner had argued should be used but that did not show any particular sequence. The defendant argued that the contractor had not effectively planned its original construction sequence sufficiently for out-of-sequence damages to be measured because the approved schedule did not show any particular sequence. The Canadian court used the more detailed but unapproved schedule to measure delay because it

---

[20] *E. C. Ernst, Inc. v. Manhattan Construction Co. of Texas*, 387 F. Supp. 1001 (S.D. Ala. 1974), modified, 551 F.2d 1026 (5th Cir. 1977), cert. denied *sub nom. Providence Hospital v. Manhattan Construction Co. of Texas*, 434 U.S. 1967, 98 S. Ct. 1246, 55 L. Ed. 2d 769 (1978).

[21] Department of the Army, Office of the Chief of Engineers, *Modification Impact Evaluation Guide*, Washington, D.C., 1979.

provided the necessary detail and reasonably reflected the contractor's intent.[22] U.S. courts have agreed that a schedule need not be approved to be used to measure delay.[23]

Mistakes

Mistakes in the schedule may also cause a court to reject a CPM analysis. In one case,[24] the Board of Contract Appeals refused to accept a CPM delay analysis that used duplicate subnetworks which had been modified to demonstrate a delay and inserted into a schedule without removing the original subnetworks. No explanation was given of the effect on the schedule of the multiple subnetworks for the same activities. The contractor had also claimed delay periods which ran concurrently with initial performance estimates. For these and other reasons, the board concluded that the contractor did not prove the government's responsibility for the claimed job delay.

In another case[25] a board also found the scheduling analysis invalid because of mistakes in the schedule. The inaccuracies were so substantial that they rendered the analysis suspect. Mistakes made in the schedule analysis included the following:

1 The change analysis showed a 223-day delay as a result of the first change order. The analysis erroneously showed only 10 days taken for submission of the contractor's proposal and 179 days taken by the government for review and approval of the proposal. Actually, the contractor took 131 days for proposal submission, and government review took 58 days.

2 A clerical error existed in the change-analysis study of change request no. 118. The 77-day delay indicated should have been 42 days.

3 Change request no. 116 involved replacing an installed manhole with a larger one. The board found the 65-day duration for the change excessive, but the contractor could offer no explanation why so much time was taken.

4 Change request nos. 158, 163, and 165 involved adding small sidewalk areas and lowering an isolated manhole that took 1 or 2 days. Their direct cost totaled less than $500.00. All changes were issued after government acceptance of the project. The board could not understand how those small changes could have delayed the project from 129 to 164 days as indicated in the change-study analysis merely because they were issued after the originally scheduled completion date.

5 The change-study analysis indicated a 276-day delay for change request no. 10 although the work was completed February 21 and the late start date

[22](1987) 27 C.L.R. 113 (B.C.S.C.).

[23] See, for example, *Dobson v. Rutgers*, 157 N.J. Super 357, 384 A.2d 1121 (1978), aff'd *sub nom. Broadway Maintenance Corp. v. Rutgers*, 180 N.J. Super d350, 434 A.2d 1125 (App. Div. 1981), aff'd 90 N.J. 253, 447 A.2d 906 (1982).

[24] *Lane-Verdugo*, 73-2 BCA 10,271 (1973).

[25] *William Passalacqua Builders Inc.*, 77-1 BCA 12,406 (1977).

for the same activity on the original schedule was June 13. In other words, the work was completed earlier than scheduled. The board characterized this as "grossly erroneous."

These cases indicate that, in order to be accepted as a demonstration or measure of a delay, a schedule must not include any mistakes that would affect the mathematical calculations that define the additional performance time. In contrast, inconsequential errors in the schedule that do not affect logic or durations may be excused. For example, when a schedule assigned responsibility for embedded plumbing inserts to the wrong trade contractor, the court found that this had no effect on the schedule.[26] An error in labeling that identified certain activities as "dummies" that were intended to carry time durations was also found not to have any effect on the validity of the schedule analysis.[27]

Sometimes a schedule may fail as a logical analysis of the project not because of mistakes in the schedule but because of failures in the scheduling process. Contractors often fail to sequence all work. Activities that theoretically may be accomplished at any time during the construction may be shown with open ends, able to be completed at any time during the performance period. While perhaps realistic, this is not good scheduling practice and prevents the schedule from accurately measuring the effect of a delay in some situations. Contractors also often fail to incoporate the shop-drawing procedure of submission, approval, fabrication, and delivery of material tied to construction activities within the schedule. As discussed earlier in this chapter, schedules without procurement activities are incomplete. Further, general contractors often fail to recognize and schedule other trade work. For example, many schedules do not contain sufficient mechanical and electrical activities. On a given floor in a building consisting of ductwork, piping, coils, VAV boxes, and temperature controls, one activity, "Install Mechanical Services," is typically provided. This type of gross scheduling does not allow for project control or analysis of the interaction necessary to prove a delay involving electrical or mechanical activities. Schedules that do not reliably reflect the work should not be used to prove or refute a delay without correction.

Manipulation of the Schedule

Much of a schedule's failing can be attributed to simple human error or unintentional oversight. However, some failings are intentionally committed to manipulate a schedule so that it will support a particular party's position.

For example, one can emphasize or reduce the effects of a particular delay or time extension by increasing or decreasing durations. As durations increase, float decreases, which means that less time is required before a delay becomes

---

[26] *Dobson v. Rutgers*, 157 N.J. Super 357, 384 A.2d 1121 (1978), aff'd *sub nom. Broadway Maintenance Corp. v. Rutgers*, 180 N.J. Super 350, 434 A.2d 1125 (App. Div. 1981), aff'd 190 N.J. 253, 447 A.2d 906 (1982).

[27] *Blackhawk Heating & Plumbing Co.*, 75-1 BCA 11,261 (1975).

critical. In one case,[28] the contractor had allocated 285 days for mechanical/electrical rough-in. The U.S. government contended that 120 days was a more reasonable duration. Mechanical/electrical rough-in was delayed. With the 285-day duration, the delay to the activity delayed project completion, but with a 120-day duration, the activity maintained sufficient float to be unaffected. The board concluded that the delay could not be measured with the original lengthy activity duration for mechanical/electrical rough-in and made its own judgment of what the duration should be. Other decisions have identified and rejected schedule analyses that have manipulated restraints[29] to emphasize lack of performance and that have manipulated float to change the critical path.[30]

Another technique that is often used to emphasize or reduce the effect of a delay is to manipulate the status of a job at the time a delay or change occurred. Some delay-measurement systems require a delay to be measured at the time it occurred. If the delay was not contemporaneously measured and inserted into the schedule, the status of the project at the time the delay occurred must be re-created from project records. A considerable amount of judgment is often necessary to arrive at project status at any particular time, because adequate records are not maintained for many projects. By exercising that judgment in favor of one particular disputant and increasing or decreasing actual durations of significant activities, the critical path and thus the outcome of a dispute may be influenced.

By performing monthly updates to the CPM schedule, the contractor automatically produces contemporaneous information as to correct project status and the current plan for executing the project. If schedule analysis is required after the project is complete, accurate records will be immediately available through the monthly updates.

Occasionally, the computerized schedule's software program will itself contain algorithms that may result in manipulation of the schedule. Some scheduling software assigns performance of noncritical activities that are linked concurrently with critical activities to the noncritical activities' late start position rather than to their early start position. This moves the noncritical activities' float time to before performance of the activity rather than after. Figure 12-2 describes the differences in positioning of float. Activity A, with a hypothetical duration of 5 days, has a start-to-start and finish-to-finish relationship with activity B. Activity B has a hypothetical duration of 25 days. With instructions to show activities in early start position, the computer would show activity A to be performed between day 1 and day 5 with 20 days of float. With instructions to show activities in late start position, the computer would show activity A to be performed between day 20 and day 25 with 20 days back float. The position of the float will result in different calculations of the dates on which activity

[28] *C. H. Leavell & Co.*, 70-2 BCA 8437 (1970).

[29] *E. C. Ernst, Inc. v. Koppers Co.*, 476 F.Supp. 729 (W.D. Pa. 1979), aff'd in part, rev'd in part, remanded in part, 626 F.2d 324 (3rd. Cir. 1980), 520 F.Supp. 830 (1981).

[30] *Joseph E. Bennett Co.*, 72-1 BCA 9364 (1972).

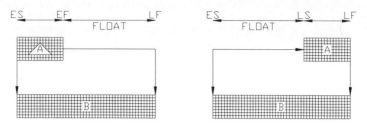

FIGURE 12-2 Position of float.

A should start and finish. Any succeeding activity that has a finish-to-start relationship with activity A will also show different performance dates depending on how the float was positioned.

Positioning an activity's float prior to the activity's performance is not common in critical path method scheduling. Late start position is not common because it results in a ''no later than'' date for the start of a concurrent activity, rather than identifying the earliest date on which work could proceed. Standard scheduling theory suggests that start dates should be shown as early as possible to encourage beginning work and to minimize possible delays. Showing later start dates may create a false sense of security.

Positioning float at the beginning of an activity has several other undesirable results. The printout (which normally indicates calendar dates, in contrast to the logic diagram, which only indicates relationships) will not show that an activity may start earlier but restricts the indicated start until the later date. Further, it builds in potential delay because any delayed start will necessarily impact its successor activity, since some or all float has been consumed. In delay analysis, the position of float can manipulate the calculation of the length of delay because positioning float at the beginning of an activity may alter the critical path. For example, see Figure 12-3.

On one recent schedule evaluation, project completion dates on a 2-year project were compared by first positioning float at the beginning of performance and then in its more common position at the end of an activity. Project completion changed from July 1 when float was positioned at the beginning of performance to May 19 when float was moved to the end of the activity. This 42-day difference in completion dates can result in substantial cost as well as time differences. If on-site and home-office overhead is $1000 a day, this change in the computer's calculation methods can add $42,000 to a delay claim.

As will be discussed in Chapter 13, schedule software can also manipulate a schedule's delay calculations through the algorithms used when out-of-sequence work is performed or when as-built dates are not used in the update process.

Some software packages permit a scheduler to impose a date on a schedule and manipulate both the critical path and interim activities' completion dates. Start and finish dates for any network schedule are calculated by adding the

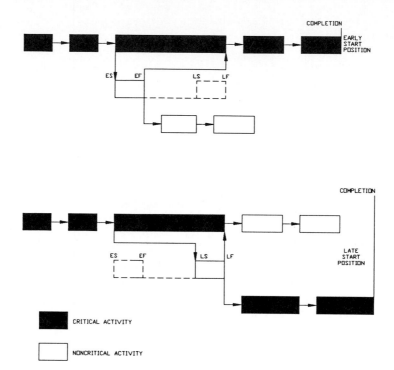

FIGURE 12-3 How float positioning affects critical path.

activities' durations and the durations of preceding activities. An imposed date on a specific activity may change the calculated date. For example, imposing a start date of November 12 on an activity with a calculated start date of October 19 will delay the activity's start until the imposed date. Performance dates for all succeeding activities in the network would be determined from the imposed rather than calculated date. An imposed date may also make an otherwise noncritical activity critical or reduce its float. Similarly, an imposed finish date later than the network-calculated finish date will impose float on preceding activities if no other controlling restraint exists.

Making an otherwise noncritical activity critical may be important in justifying a claimed delay, because current legal requirements demand that the delayed activity be critical before extensions are awarded. By changing the calculated schedule dates, imposed dates can also extend or reduce a claimed delay. Increasing the length of delay by imposing a later-than-otherwise-calculated start date may also increase the claimed amount of time-related damages.

Although imposed dates may serve a legitimate planning purpose (for example, when used to record major milestone dates), because of their potential for manipulating a schedule analysis, they should be removed from the network when evaluating the effect of a delay. Contractually imposed dates, however,

```
+----------------------------------+----------------------------------+
|          SPRING-LOADED           |             NORMAL               |
|                                  |                                  |
|             NO. OF               |             NO. OF               |
|             WORK                 |             WORK                 |
|           ACTIVITIES    FLOAT    |           ACTIVITIES    FLOAT    |
|  CRITICAL PATH   10       0      |  CRITICAL PATH   10       0      |
|  OTHER PATHS     90  X   10 = 900|  OTHER PATHS     90  X   20 = 1800|
|                 100              |                 100              |
|                                  |                                  |
|  900 = 9 AVG FLOAT               |  1800 = 18 AVG FLOAT             |
|  100   PER WORK ACTIVITY         |  100     PER WORK ACTIVITY       |
+----------------------------------+----------------------------------+
```

FIGURE 12-4   Comparison of average float.

should remain in the CPM during the delay analysis. Contract dates establish the period from which delay should be measured. More than one contractual completion date, however, makes schedule analysis more difficult.

Restraints may also be added or sequences changed to make critical a particular chain of activities anticipated to be delayed. Sometimes referred to as being "spring-loaded," these schedules have been manipulated to reroute the calculated critical path to a path on which delays are anticipated. One can identify a "spring-loaded" schedule in one of several ways. First, durations may be in even multiples of a 5-day or 7-day week, indicating that no real effort has been made to estimate activity durations. Instead, durations have been used to make a particular chain of activities critical. Second, a higher portion of activities will be critical than usual. Usually, approximately 10 percent of the schedule's activities are critical. More than that may indicate manipulation. Third, if restraints have been manipulated to make otherwise noncritical activities critical, unexpected or unusual activities will be critical. Are normally flexible activities critical? Finally, one can identify a "spring-loaded" schedule by a higher-than-usual average float per activity; see Figure 12-4.

For example, a "spring-loaded" schedule may sequence site work before finishes to "avoid delivery of finish material through mud" (really to spring-load it). If a "spring-loaded" schedule has been used to schedule the project, each unnecessary restraint must be removed and activities must be resequenced to reflect normal sequences before the schedule can be used to measure delays. By resequencing site work to later in the network, the manipulated schedule can be neutralized.

CORRECTING MISTAKES, MANIPULATION, AND OMISSIONS

If an original schedule to be used to measure delay or impact has errors, omissions, or other deficiencies, the problems should be corrected. Once revised to eliminate any problems with the schedule, the corrected or revised original schedule can be used to measure delay.

Errors in the original schedule do not necessarily mean that the original schedule must be abandoned and cannot be used to measure delay. However, if the original schedule contains too many errors, the delay analysis may be more reliably completed by abandoning the original and preparing a new, model construction schedule with which to measure the delay.

Manipulated schedules should not be accepted to demonstrate a delay. However, these types of manipulation are sometimes difficult to discover in such an intricate, complicated medium as a network diagram. Often only an expert scheduler can discover these changes. Even when manipulation has been discovered, it is often hard to convince a judge or jury that stacks of computer calculations do not possess an innate accuracy that automatically discovers and corrects errors without human assistance. Do computers lie?

METHODS OF SCHEDULE ANALYSIS

As-Built Method

The traditional method used to determine amount of delay was to compare the contractor's planned schedule with calculated dates to an "as-built" schedule that had substituted actual completion dates for all activities. The traditional as-built method has many handicaps, however. As-built schedules are costly to prepare because of the research necessary to determine actual dates. Considerable judgment is also necessary, since detailed records are not always available and even if they are available, work in the field does not match the theoretical divisions of a network schedule. Some commentators have suggested that comparing planned to actual progress should be insufficient to prove delay unless the planned schedule can be demonstrated to be reliable and reasonable.

Creation of an accurate as-built schedule from daily logs or other contract documentation is also difficult, since the sequencing or relationships of work activities may have changed from the as-planned schedule. The relationships and sequences are important because they determine the length of the delay. Reestablishing the actual sequence from project records is extremely difficult, if not impossible. Merely using dates and lines for the as-built schedule without using relationships produces only an elaborate bar chart. While an as-built approach can be very accurate for some types of delays, because of these difficulties it can also easily be manipulated and distorted to reflect the position of the claimant.

An as-built schedule may be prepared in several different ways. Thus, there is no one as-built schedule methodology. Some research the actual start and finish dates for planned activities, record them, and calculate extended project duration based on the originally planned sequence. Some may compare the last monthly update to the initial, planned schedule if the schedule has been updated regularly. Some may prepare an as-built schedule that includes both actual dates and sequences all activities whether they were included on the planned schedule or not. Which of these ways will be chosen to prepare an as-built schedule may be influenced by which activities have been delayed or what kind of delay must be measured, what information is available, and how economically it can be used.

Equally important in using an as-built analysis is to demonstrate that the as-built sequence and duration were reasonable. The fact that the project was completed in a certain manner does not automatically mean that manner was reasonable. Often during a delay, manpower, equipment, and materials may be diverted to other activities, serving to prolong completion of the delayed activities. Whether the actual delay was also a reasonable delay is a question that should be answered in all as-built delay analyses. Measuring the duration of a time extension by the length of the delayed activities is inappropriate unless the contractor can demonstrate that the equipment, manpower, or material was promptly redirected to the activity when the delay ended, or that prompt redirection was not in the best interest of the project.

Modified As-Built Method—U.S. Corps of Engineers

There are many other ways in which a schedule can be analyzed. The U.S. Army Corps of Engineers uses a "modified as-built" approach which uses as-built dates only to the point in time when the delay started. The CPM is updated to just before the delay started. This update becomes the "original" or base schedule that is then compared to the impacted completion date on another CPM schedule on which a delay is included. Others refer to the Corps method as the "contemporaneous impact analysis method," "update-impact method," "time impact analysis," and "snapshot method."

The Corps requires a step-by-step procedure to implement its "modified as-built" approach. The Corps first requires the current status of the job to be determined without influence from the contractor's formal project schedule, because the contractor's real plan for pursuing the work may not be the same as indicated on the formal project schedule, or the schedule may not have been revised to reflect the effects of previous modifications. Second, the Corps suggests that the scope of the modification or whatever else has initiated the analysis be studied to determine what subsequent activities will be directly affected and how the schedule should be revised to accommodate the modification. If all or part of the work to be performed under a modification does not fit an existing activity, new activities may be created and inserted into the schedule. Third, the schedule, as revised, is used for new calculations to determine new critical paths and project completion dates. From the newly determined critical path and revised completion dates, time extensions and other effects of the delay can be granted.

The Corps suggests that no time extensions should be granted unless the project's critical path has changed. A schedule evaluation is required for each modification. The Corps' method requires the definition of the status of the project at the time of each delay. Monthly schedule updates normally provide sufficient information for "base" schedule use. If performed carefully, the Corps of Engineers' method can produce results that are less subject to variation by the scheduler. As with all schedule analysis methods, however, reasonable judgment strongly influences the results of a schedule analysis under the Corps' method.

As-Planned Method

Another method of impacting a network diagram schedule to show the effect of changes and delays uses an as-planned schedule.[31] Under the as-planned method, as shown in Figure 12-5, the scope of changed work is first reviewed to determine where and how the revisions (or delays) should be incorporated into the original schedule. The activity revisions and additions are then sketched on the network. Revisions to subsequent activities caused by the change or delay are defined and made. The effect of the change on the schedule is determined by comparison of the approved schedules before and after all changes have been incorporated into the schedule. Under the as-planned method, the contractor is entitled to a time extension only if the scheduled completion is delayed beyond the extended contract completion date. The as-planned method uses the planned schedule to measure the delay, regardless of whether the actual construction in the field differs from the planned.

In addition to use of as-built information, the as-planned and the Corps of Engineers method differ in the timing of the analysis. The Corps requires a step-by-step analysis each time a delay occurs. The as-planned method may wait to include all information concerning all delays at one time rather than separate calculations for each delay. The analysis must wait until all contractor requests for changes have been submitted.

Some believe that, without substantiation, as-planned schedules do not provide sufficient basis to measure contractor delay claims. An initial schedule may be some evidence, but by no means conclusive evidence, of the length and effect of the delay. This view is most often expressed in England and

---

[31] Veterans Administration, *VACPM Handbook*, H-08-11, Washington, D.C.: Veterans Administration, January 1985.

FIGURE 12-5　Veterans Administration method of schedule analysis.

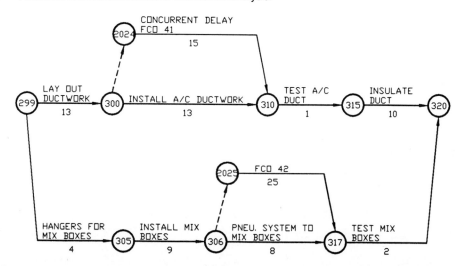

Canada, less frequently in the U.S. If this view is taken, the contractor must produce proof that performance according to the as-planned schedule was possible absent the delay. For example, in *W. A. Stevenson Construction (Western) Limited v. Metro Canada Limited*,[32] a Canadian decision, an as-planned schedule was demonstrated to be reliable by showing that the portion of the progress that was *not* affected by the delay was completed in the manner anticipated in the planned schedule. To measure the delay, the revised sequence was compared to the initial as-planned schedule.

Other Methods

Although both the Corps of Engineers and the as-planned methods state that only delays that extend the critical path are entitled to time extensions, contractors sometimes argue that delay in a noncritical activity should also justify a delay claim. This argument presumes that management of float is as important to project completion as management of critical activities. Thus, another way in which a schedule can be analyzed to determine the effect of a delay is to analyze the amount of float that a change requires.

One commentator has suggested that this can be done by allocating float to each activity and then determining time extensions by preserving the allocated float of unchanged activities.[33] Allocation of float to all activities in essence produces a CPM schedule in which every activity is "critical," thus producing an analysis basis with critical paths everywhere.

However, not all commentators agree on how float should be allocated. One position proposes that an activity "owns" only the free-float portion of its total float. In other words, an activity "owns" only the number of days it can be delayed without affecting any immediate successor activities. Yet, since free float exists only in the last activity in a chain (and other limited circumstances) it often occurs as a result only of action of the scheduler. Free float lends itself to some manipulation by arranging, say, for a particular activity to end a chain of activities.

Other commentators suggest that an activity should be assigned its float regardless of whether that activity has any free float. Another position is that float belongs to those activities that use it first, although the successor activities may lose their flexibility. This latter position is the assumption that underlies most court decisions which require a delay to extend the critical path before it is considered compensable. This approach is by far the simplest way to allocate float.

A compromise position is to allocate activities' float in a shared way.[34] An activity would be allocated a percentage of the float available to the chain based on the individual activity's duration. If an activity is delayed beyond its allo-

---

[32] (1987) 27 C.L.R. 113 (B.C.S.C.).

[33] G. Ponce de Leon, "Float Ownership—Some Recommendations," *Strategem*, Vol. 1, No. 1 (October 1982), p. 1.

[34] Ibid.

cated float, then a time extension may be justified to preserve the allocated float of other activities in the approved schedule. This suggested method of float allocation has been used by the U.S. Board of Contract Appeals.

Concurrent delays also play an important role in the use of a schedule to measure a delay. Generally, when two parties (i.e., owner and contractor) contribute to project delay, only the "critical" delay is entitled to additional performance costs or a time extension. If a delay is not critical, no liability for the delay may be imposed.

Some commentators believe that excusing one of several delays because it is not on the critical path is unfair. These commentators would assign some responsibility for the delay to the noncritical delay. To allocate responsibility to the noncritical activities, a compensable time extension is granted for a critical delay one day at a time, beginning at the last day of the claimed delay period. Additional days are granted until the noncritical concurrent delay caused by another party is forced onto a critical path. This is done by deducting time attributable to the delay from the original critical path until the alternate path becomes critical. At that time, the extension calculation on the initial critical path is interrupted and the second path is examined. If the delay on the concurrent second path is caused by the same party, the time extension continues. If the concurrent delay is caused by the other party, any compensation related to the time extension stops. The identified concurrent delay prevents a longer compensable delay claim.

This approach to concurrent delay analysis makes the assumption that delays on the project site are unrecognizable by the persons executing the construction contract. In reality, construction personnel are acutely aware of most delays occurring on the project. They adjust their work priorities in response to those delays. If they know the project is being delayed in some areas, they will not make required efforts in order to complete other work within the original time frame. This may appear as a delay when in fact it is an adjustment or change to the contractor's planned procedure. These adjustments should not affect the analysis of the delay or the project completion date. For example, a power plant project has a boiler to be installed that is expected to arrive 3 months late. There may be a 3-month delay. However, if the structural steel that supports that boiler is actually 5 months late, the scheduled boiler delivery will in fact be 2 months early. Neither the contractor's 2-month adjustment of the delivery of the boiler to coincide with the completion of the steel nor the "original" 3-month late delivery of the boiler is a concurrent delay. It will be impossible to know or to conclude retroactively that the boiler would have been 3 months late had the steel been on time. Had the steel been on time, the contractor might have put more pressure on the boiler manufacturer and in fact gotten the boiler delivered on its original delivery date.

There are other ways to prove a delay. Some are based on the economics of trial preparation rather than on a particular method of schedule analysis. Because of the expense involved, some contractors have attempted to prove a delay based on the ratio of the value of changes and change-order amounts to time. Because of the expense involved, these contractors do not use a critical path analysis to prove a delay but rather argue that the ratio of the cost of the changes or additional work when compared to the original contract amount correlates with their requested time extension of their performance. This approach has not been widely accepted, although if a contractor can prove the labor-intensive nature of the work, there is no reason why it could not form the basis for defining delay.

However, equating dollars to time on a construction project is not always reasonable. For example, assume that a contract for $10 million to build an office building is modified by a change order to increase the quality of the carpeting dramatically. The change order is executed considerably ahead of the time the carpet is needed at the project site. The increased carpet value increases the contract amount by $500,000. This is 5 percent of the original contract amount. The more expensive carpet will be installed in precisely the same time frame as the original carpet specified. Should there be a 5 percent increase in time to execute the project?

Assume that a change in the method of formwork for concrete is desired by the engineer. The revised formwork procedure will be equal in cost to the original procedure but will take considerably longer to execute. The net change for direct costs of this change order will be $0.00. However, the project completion date will be extended. Should no time extension be granted?

The reader should note that all methods of schedule analysis share common concepts, the most important of which is the premise that delay is measured from the project's end date rather than from any interim activity's completion date. Attempts to measure delay by comparing planned to actual completion of individual activities rather than the project completion date have been rejected regularly. For example, the U.S. Claims Court[35] rejected a contractor's claim for additional overhead caused by delay to completion of new storm sewer lines because the final date for completion of the project under the contract was not affected by the delay in completion of the new storm sewers. Similarly, the U.S. Armed Services Board of Contract Appeals[36] denied a contractor a greater recovery for delay than allowed by the contracting officer because no effect of the delay was shown on the overall construction schedule or operations.

Choosing a Delay Analysis Method

Each of the schedule analysis methods discussed can be used to analyze a delay. Which method is best suited to a particular situation depends on the type of delay claim presented, the circumstances of the delay, contract language, local laws and precedence cases, the rules of the particular scheduling clause, and availability of accurate information. Which method should be used

[35] *Tectronics, Inc. of Florida v. United States*, 10 Cl. Ct. 296 (1986).
[36] *Appeal of Hoyer Construction Co., Inc.*, ASBCA No. 31242, 86-1 BCA (CCH) 18731 (1986).

Accelerating activities so that projects can be finished early results in more resources and higher costs.

may often be determined only by a scheduling expert familiar with the presentation of delay claims. Nevertheless, general rules can be formulated to assist in the selection of schedule analysis methods.

Float adjustments may be used to analyze a claim if the scheduling clause permits float to be distributed among all activities. A float analysis may not be appropriate or accepted, however, if the contract grants the ownership of float to the owner.

The Corps of Engineers' method can be used to impact a schedule if the delays are recognized and can be analyzed in reasonable proximity to the date of the delay. This is not always possible, however, because some architects and engineers recommend that time extensions be postponed until the end of the project. In these circumstances, the Corps' modified as-built method may not be used, because the status of the job at the time of each delay may be too difficult to reconstruct. The Veterans Administration method of inserting all delays at once and analyzing the effect on the schedule as planned may be better in this situation.

Any method of impacting a schedule on a step-by-step basis as the Corps of Engineers suggests has other difficulties as well. Although the method may work well for a limited number of impacts, when 300 or more combinations of excusable, nonexcusable, compensable, and noncompensable delays exist, the mechanics of insertion, analysis, revised computations, and evaluations on a

10,000-activity schedule with resulting possibility of challenge to the judgment used in each and every step make the methodology unrealistic if not impossible. If the procedure is followed precisely, as many as 600 different computer runs and resulting printouts may be required. Even if impacts that occurred relatively closely to each other can be combined, too many computer runs may be necessary to analyze the delay under the Corps' method. In these situations, inserting all delays at once and producing one impacted schedule may be a better choice.

Use of the as-built method is difficult because of the limitations of the critical path method. Network schedules generally cannot represent the realities of construction in the field. Regardless of how the diagram shows activities beginning and ending, most subsequent activities in the field can begin before the preceding activity is complete. Also, field work is not always performed continuously. Work may move to other locations or activities at other parts of the project before a given activity is completed.

Further, because contractors do not regularly create a historical record of how a project was completed, reconstructing projects from job records is often impossible. The records either do not exist or, if they do exist, only infer rather than state when work was started, continued, or stopped. A considerable amount of judgment may be required to extrapolate key dates and actual sequences. This judgment can be distorted to change the delay implications. For these reasons, measuring delay from an as-planned schedule rather than an as-built one continues to be popular.

Regardless of whether the as-built method or the as-planned analysis is used, the underlying schedule must be reliable. To ensure reliability, the schedule should be checked and questionable areas corrected. It is important, however, for any subsequent use of the original schedule as the foundation of a delay analysis that all changes or corrections made to the schedule during this check be carefully recorded and documented, along with the reasons for the changes and corrections. Differences between the initial and a corrected schedule used to measure delay must be presented with the schedule analysis.

If challenged, the reliability of computer hardware and software may also have to be demonstrated to meet evidentiary standards of a court.

Which among the many different delay analysis methods available should be used is an important consideration because even with the same facts, different methods result in different delay lengths. As an experiment, the author created a 6-month project schedule and "impacted" the schedule using the as-built, modified as-built, and as-planned methods. Using identical delays and the same schedule, the as-built method resulted in a 4-week delay, the Corps of Engineers' method in a 3-week delay, and the as-planned method in a 2-week delay.

Whatever method is selected, using a network diagram to prove a delay depends as much on the skill of the scheduler as on any other factor. The scheduler must not only understand network schedules and be familiar with the details of the particular construction but must also understand the limitations of the scheduling process. Using a network schedule to prove delay is not as much a scientific effort as it is an exercise in judgment. But even after these intracacies have been mastered, the complex interrelationships must be made sufficiently simple so that the designer, judge, jury, arbitrator, or whoever else reviews the delay claims can understand the results of the analysis. This is why a bar chart is important for demonstrating the conclusions of the network analysis. The bar chart is an easily understood, instantly recognized visual demonstration of the delay.

APPORTIONING CONCURRENT DELAYS

What Are Concurrent Delays?

In network scheduling, "concurrency" is defined as a relationship which is neither precedent nor subsequent. A concurrent relationship between a pair of activities does not necessarily mean that the two activities will be performed simultaneously. It merely means that commencement of work on one of the activities is not contingent upon completion of some work on the other activity.

In delay claims, concurrent delays occur when there are two or more independent delays during the project completion period. As the term is now used, concurrent delays may occur during any part of the project performance period. A "concurrent" delay means more than one delay contributed to the *project* delay, not that the delays necessarily occurred at the same time. A concurrent delay may occur during the same period as another delay, but concurrent delay also includes any delays that have contributed to the overall project delay, whether the delay overlaps with another or not. The period of concurrency is the period of project performance, not just the period during which any individual delay may have occurred.

As a result of the expanded definition of concurrency in delay claims, one can see courts considering delays to consecutive activities "concurrent" delays. In *Raymond Constructors of Africa, Ltd. v. United States*,[37] for example, the court determined that three consecutive delays were "concurrent." The three consecutive delays occurred in the procurement, delivery, and operation of heavy equipment with little, if any, overlap. The court apportioned each delay among the three different parties contributing to the overall project delay. Defining concurrent delay in this manner expands the definition of concurrency used in network scheduling.

When concurrent delays involve any combination of excusable and nonexcusable events, time but not money for the delays may be awarded. When concurrent delays involve two nonexcusable delays, one by each party, then time but not money is awarded, as both parties are penalized for contributing to the overall delay.[38] In other words, if both the contractor and the owner

[37] 411 F.2d 1227 (Ct. Cl. 1969).
[38] *Appeal of Fischbach & Moore International Corporation*, ASBCA No. 18146, 77-1 BCA (CCH) 12,300 (1977).

contribute to the delay, the contractor receives a time extension, but not delay damages, and the owner does not receive liquidated damages.

Apportioning Concurrent Delays

How concurrent delays are apportioned is important not only in situations in which excusable and nonexcusable delays have been experienced, such as occurrences of owner-caused delay and unforeseen inclement weather, but also when two or more nonexcusable delays have been experienced such as concurrent owner-caused and contractor-caused delays. Apportioning concurrent, nonexcusable delays is also important to contractors distributing responsibility for delay among subcontractors or owners among multiple prime contractors. The ASBCA dealt with the apportionment of concurrent delays in *Fischbach & Moore International Corp.*, as extracted at the end of this chapter. Apportioning concurrent, excusable delays may also be important when the cost of the excusable delay may be recoverable from several insurers.

Methods of Apportioning Concurrent Delays

How concurrent delay should be apportioned results from the application of two rules. The first rule states that time for delay is awarded only when the delay has extended the project's completion date. The second states that when both parties contribute to a delay, neither can recover monetary damages unless clear proof of apportionment of delay is presented. If delay can be apportioned, the effect of the second rule is that one party's contribution to the overall delay may be used to reduce the other party's responsibility for the delay. If, after one party's delay has been offset by another's, some period of delay remains, that remaining period of delay is awarded to the aggrieved party. The second rule is meant to excuse the obligation to reimburse the delayed party to the extent the delayed party contributed to the delay.

It is uncertain in applying these rules, however, in what manner a concurrent delay must also have contributed to the project delay in order to be used as an offset. To be used as an offset, must the concurrent delay be on the critical path? Or should any concurrent delay, whether critical or not, be used as an offset for a claimed delay? Some cases require a concurrent delay to be on the critical path[39] or to "affect" the project completion[40] before being permitted to offset claimed delays. Other case law[41] permitted *any* offset to reduce a claimed delay.

In *Appeal of Fischbach & Moore International Corporation*,[42] the Armed Services Board of Contract Appeals required concurrent delays to be on the

---

[39] Ibid.

[40] *E. C. Ernst, Inc. v. Manhattan Construction Company of Texas*, 387 F.Supp. 1001 (S.D. Alabama 1974).

[41] *United States v. William F. Klingensmith, Inc.*, 670 F.2d 1227 (1982).

[42] ASBCA No. 18146, 77-1 BCA (CCH) 12,300 (1977).

critical path before they could reduce the claimed delay. Fischbach & Moore contracted to erect 104 steel radio towers in the Philippines. During fabrication of the towers, surface seams were noticed in the steel. The government stopped fabrication until an investigation of the cause of the seams had been completed and ultimately refused to permit Fischbach to continue until other steel without seam problems was substituted. Later, the Board of Contract Appeals determined that the government had wrongfully stopped fabrication and rejected the original steel. Fischbach claimed some 160 days delay as a result of the wrongful suspension and rejection.

The U.S. government defended by claiming that concurrent delays during the delay period should reduce the claimed 160-day delay. Among the concurrent delays presented by the government were weather, strikes, contractor inefficiencies, and subcontractor failures. The ASBCA examined all concurrent delays but permitted only those concurrent delays that were also critical, the strike and subcontractor failure, to be used to reduce Fischbach's claimed delay.

Although Fischbach received no credit for the strike, an excused delay normally entitling the contractor to an extension of time, overall project duration and the imposition of liquidated damages was not an issue in the dispute, as Fischbach had subsequently accelerated its work and completed within the contract duration. Presumably, if time of performance and the imposition of liquidated damages had been important, Fischbach would have been awarded time but no money for the excusable delay caused by the strike, and any liquidated damages would have been appropriately reduced.

The board reduced the claimed 160-day delay to 114 days because of concurrent, critical delays. Concurrent weather delays and contractor inefficiencies were found not to have affected the critical path and were not used to offset the claimed delay. The board specifically noted that Fischbach, when scheduling its activities, intentionally provided an additional 4 months' duration to allow for inefficiencies. However, a strike and subcontractor failures were determined to be critical and used to reduce the claimed delay period.

It is reasonable to require concurrent delays that do not occur during the same time as the claimed delay to be on the critical path before the duration of a claimed delay may be reduced. The claimed delay must affect the project completion date to be compensable. Thus, any claimed delay must be on the critical path before any additional time is recoverable. To require both the claimed and any earlier (or later) concurrent delay to be critical before a concurrent delay may be used to offset the delay period is equitable.

A delay that does not affect the overall project completion is incidental and does not qualify as a concurrent delay. For example, postponing scheduled delivery of material does not qualify as a concurrent delay if another, critical delay has occurred which makes the presence of the material unnecessary. Postponing delivery of the material does not affect project completion, while the critical delay does. Postponed delivery of material could not be used as a concurrent delay to offset the critical delay.

But what of the situation when concurrent delays contemporaneously over-

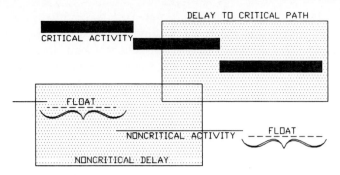

FIGURE 12-6   Critical and noncritical delays.

lap the claimed delay but are not both on a critical path? See Figure 12-6. Must only one of two simultaneous delays count because only one is critical? Because the second rule is meant to excuse the obligation to reimburse the delayed party to the extent the delayed party (or another event) has also contributed to the delay, it is not equitable in all situations to limit reducing delay damages unless the concurrent delay is on the critical path.

For any project, there are a number of different critical path method schedules which may be equally correct. Each may have a different critical path, because any project may be completed in a variety of ways. Any CPM schedule reflects only one way in which the project may be completed. Why should a simultaneous, concurrent delay be denied as an offset merely because one particular schedule indicates that the concurrent delay is not on the critical path? Might another schedule show the concurrent rather than the claimed delay on the critical path? The critical path could possibly be different on a different schedule. The critical path may also be changed by adding to, or subtracting from, an activity's duration, or by slight differences in the completion percentages. It does not seem equitable to deny any offset merely because of the manner an unknown scheduler has used to prepare the CPM schedule or analysis.

For two overlapping, concurrent delays to both be critical, a CPM schedule must have two critical paths. Although it is possible to have two critical paths and that both the concurrent and claimed delay are critical, this situation is infrequently encountered. Experience shows that as a schedule is updated to reflect the project's progress, the critical path tends to concentrate on a single rather than multiple chains of activities.

Reasonable Allocations for Concurrent Delays

With the assumptions that the CPM schedule used is reasonable and accurate, the allocation of concurrent delays can be reasonably apportioned using the following rule:

A concurrent delay by two differing parties on the project can be used to offset delays when both delays occur on the critical path or in reasonable proximity to it.

That is, if the most critical path is − 36, another delay concurrent with that delay of − 32 or − 31 could reasonably be considered as a concurrent critical delay. The accuracy of a CPM schedule can normally be anticipated to be within a 5-workday calculation. Delays beyond 5 days normally can be assumed not to be on the most critical path.

For example, a project schedule analysis indicates a 134-workday delay through a variety of different change-order items that have been imposed by the owner. These change-order items delay interior work on a substantially watertight building. A different delay due to excessive rains of 23 workdays to parking lot construction also occurs. The owner's delay caused by excessive change orders far exceeds the delays caused by the unanticipated rain days. The contractor might have absorbed the 23 days' delay due to weather at the parking lot had the owner delays not occurred. Had the other change-order delays not occurred, the contractor might have accelerated the parking lot work and completed the project on time. However, because the contractor was aware of the owner delays, it did not perform the extra effort to finish the parking lot within the original contract time frame. The 134 days should be granted as a compensable time extension without reduction for the 23 noncompensable rain days.

Those concurrent delays that are either on the critical path or very near the most critical path should be considered as an offset to delay claims. To consider every minor delay of 1 day or more that could have no effect on the critical path as a concurrent delay is neither reasonable nor fair to any party involved. However, all delays must be considered together. Viewing delays as independent events and occurrences on a construction project is not an equitable way to determine the length of construction delay.

Application of these suggested rules is equitable both in situations in which the parties have caused the delays and in situations in which not parties but unforeseen events such as hurricanes or earthquakes comprised one part of the concurrent delay. In situtations in which concurrent delays do not overlap, the unforeseeable excusable delay should be critical before it is allowed to offset any of the claimed delay unless a time extension for the unforeseen event has been granted earlier. In situations in which the concurrent delays overlap, the party causing the delay should also be allowed to take advantage of the fortuitousness of the unseen event to reduce the claimed delay as long as the unforeseen event is on a path that does not have any float.

Concurrent delay unrelated to the claimed delay under these rules would continue to be denied as an offset. In *Appeal of William M. Wilson's Sons, Inc.*,[43] the Armed Services Board of Contract Appeals properly found alleged government concurrent delay installing dedicated telephone lines unrelated to contractor delay installing automated refueling system and denied offset.

Thus, the suggested rule for apportionment of concurrent delay has two parts. Concurrent delays that do not overlap must be on the critical path in

[43] ASBCA No. 30553, 85-2 BCA (CCH) 18,028 (1985).

order for the duration of the claimed delay to be reduced. When concurrent delays overlap, an offset should be permitted when the concurrent delay is on any path which also does not have float.

The Armed Services Board of Contract Appeals' decision in *Appeal of Fischbach & Moore International Corporation* can be used to illustrate the application of the suggested rule to apportion concurrent delay. In Fischbach, concurrent critical delays for a strike and subcontractor failure were permitted to offset the claimed delay because they were also on the critical path. Other concurrent delays caused by weather and contractor inefficiencies were determined not to be critical and were not permitted to offset the claimed delay. All concurrent delays overlapped the claimed delay.

The board applied the suggested rule to the concurrent delays for the contractor inefficiencies, strike, and subcontractor failure. The strike and subcontractor failure were on a path that did not have any float. The board found both the strike and subcontractor failure to be critical. The concurrent delay for contractor inefficiency was found to be on a path which had 4 months float and was not permitted to offset. It is uncertain, however, whether the suggested rule was applied for the concurrent delay for weather. If the weather delay was also on a path without float, even though not critical, the suggested rule would permit the concurrent delay to offset. To apply the suggested rule correctly, the path on which the weather delay occurred would be reexamined. If no float was available, then the weather delay would also be permitted to offset the claimed delay.

The relevant portions of the board's decision in *Appeal of Fischbach & Moore International Corporation* [ASBCA No. 18146, 77-1 BCA (CCH) 12,3000 (1977)] follow.

December 13, 1976. Contract No. IA-11426.

INTRODUCTION

On 3 May 1966, appellant was awarded a contract to erect a complex of 104 steel radio towers, together with antennas, buildings, auxiliary structures, and site work, in the Philippines, for a fixed price of $16,580,639. (R4, Tab 1) The undertaking was given the code name, "Project Bamboo." During the course of appellant's performance, the Government, represented by the United States Information Agency (USIA), issued certain stop work orders and other directives relating to surface seams that appeared in some of the steel bars and which the Government considered injurious defects within the meaning of the contract specifications.

In *Fischbach & Moore International Corporation,* ASBCA No. 14216, 71-1 BCA ¶ 8775, affirmed on reconsideration, 71-2 BCA ¶ 9081, we held that the steel in question conformed to the applicable specifications, and that as a result the stop work orders and corrective work directed by the Government constituted a constructive change for which appellant was entitled to an equitable adjustment of the contract price. Having so decided, we remanded the matter to the parties to negotiate the amount of the equitable adjustment. In the negotiations that followed, appellant asserted its claim for additional compensation in the amount of $1,862,692, which the contracting officer disallowed in its entirety in his final decision of 5 January 1973. . . .

In view of the total failure of the negotiations both before and after the contracting officer's final decision, and the widely differing positions of the parties in respect to the allowability of the claim, a new hearing was held beginning in late May 1975 to enable us to decide the amount of appellant's recovery. At that hearing, and in the briefs that followed, it appeared that a number of entitlement issues, both legal and factual, remained for our consideration before we could proceed to quantum. The legal issues concern the effect of certain release agreements which appellant entered into with two of its subcontractors, and the kind of causal connection that must be established between a Government-caused delay and the increased costs alleged to result therefrom in order to charge the Government with those costs. The factual issues include certain affirmative defenses raised by the Government, such as the existence of concurrent delays for which appellant was responsible or which were coextensive with time extensions granted by the Government for causes unrelated to the steel seam problem. . . .

PART ONE
I. FINDINGS OF FACT ON THE GOVERNMENT'S LIABILITY

A. The Delay Claim—12 May thru 20 October 1967

1.   The seam problem which we considered in detail in our prior decision was first manifested to Government inspectors in early May 1967, at which time there appeared to be cracks on the tower leg bars adjacent to the guy gusset plate welds. Since these welds were critical to the stability of the towers after their erection, the Government issued its first stop work order on 12 May in relation to galvanizing the upper tower legs of a certain antenna then in fabrication. . . . On 16 May, the stop work order was broadened to include "all operations relating to the fabrication, galvanizing or erection of all tower sections.". . . The Government concedes that these stop work orders caused a complete cessation of work on the tower legs. . . .

2. Once the stop work orders had been issued, both parties took prompt action to ascertain the nature of the problem. In our previous decision, we found that samples of the steel bars were cut and air-freighted to the United States for review and analysis, and that on 19 May 1967, the initial inspection and analysis indicated that the surface imperfections were not cracks as had been feared, but were, rather, seams which are normally present in bars of the quality that had been specified. . . . A representative of U.S. Steel was also detailed to the jobsite in the Philippines to report on the matter. His report, which was presented to a meeting of the parties in Pittsburgh, Pa. on 6 June

1967, confirmed the initial analysis and advised that "the material inspected met the specification requirements and was as would have been expected of a shipment this size and specified quality.". . . Accordingly, on 9 June 1967, the Government authorized appellant to resume the work under certain specified conditions, none of which were called for by the contract specifications.

3. The Government contends that under the circumstances described in Findings 1 and 2 its issuance of the stop work orders of 12 and 16 May 1967 was a reasonable exercise of the contracting officer's authority under the terms of the contract and that the resulting suspension of work from 12 May thru 8 June 1967 was not for an unreasonable period of time so as to entitle appellant to additional compensation.

a. On this point, Article VIII of the contract's General Conditions, entitled "Price Adjustment for Suspension, Delay, or Interruption of the Work" provides in pertinent part as follows:

"A. The Contracting Officer may order the Contractor in writing to suspend all or any part of the work for such period of time as he may determine to be appropriate for the convenience of the Government.

"B. If, without the fault or negligence of the Contractor, the performance of all or any part of the work is, for an unreasonable period of time, suspended, delayed or interrupted by an act of the Contracting Officer in the administration of the contract, . . . an adjustment shall be made by the Contracting Officer for any increase in the cost of performance of the contract (excluding profit) necessarily caused by the unreasonable period of such suspension, delay, or interruption, and the contract shall be modified in writing accordingly. No adjustment shall be made to the extent that performance by the Contractor would have been prevented by other causes even if the work had not been so suspended, delayed, or interrupted. . . ."

b. In addition, the Government relies on the following provisions of Article II, entitled "Direction, Control, and Acceptance":

"B. . . . The AR/CO shall have the authority:

"(1) to order the Contractor to stop any work which in his opinion is not being performed in accordance with the requirements of the contract;

"(2) to require the Contractor to furnish samples of materials before and during the construction;

"(3) to delay work until receipt of results of any testing of samples when in his opinion such testing is necessary to ensure compliance with drawings and specifications;. . ."

4. From 23 May thru 6 June 1967, appellant's fabrication subcontractor, Pacific Engineering Company, Inc. (PECI), experienced a strike which caused the closing of its plant. . . The Government argues that the strike was a concurrent cause of appellant's delay. On or about 6 June 1967, PECI reduced its jobsite fabrication facilities somewhat by moving some equipment back to its main plant near Manila, leaving only a repair capability at the jobsite thereafter. . . .

5. In answer to the Government's argument that the contracting officer's suspension of the work from 12 May thru 8 June 1967 was a reasonable exercise

of his authority under the terms of the contract, appellant contends that since the suspension was caused by defective specifications, it is unreasonable *per se.* . . . However, there is no evidence in the present record that the specifications were defective. To the contrary, our previous decision found that the Government had in effect misinterpreted them and in doing so had imposed on appellant a suspension of work and extra work that was in the nature of a constructive change. Accordingly, we find that the specifications were not defective.

6. It will be recalled that on 9 June 1967 the Government authorized appellant to resume the fabrication of tower legs under certain conditions specified in the letter of authorization. . . . From that date until 5 September 1967, the Government issued a series of directives in an attempt to clarify and define the remedial work necessary to overcome the seam problem. . . . The ultimate effect of these directives, as discussed in our prior decision. . . , was to establish a seam-free requirement in all of the critical areas of the tower legs prior to welding. . . . This requirement was in the nature of a constructive change.

7. After having been authorized to resume work, appellant attempted to meet the seam-free requirement during the balance of June and July without success. Indeed, by the end of July appellant had not succeeded in having any critical tower legs repaired to the satisfaction of the Government. . . .

8. Faced by a complete disruption of the work of fabricating critical tower legs, appellant took a variety of actions.

a. On 28 June 1967, appellant ordered approximately 100 tons of steel for use in critical areas. . . . In early August, appellant submitted 1,207 bars then on hand for testing and inspection by the Government, but only 12 of this total were found acceptable for use in critical sections. . . . Accordingly, on 8 August 1967, appellant placed a second order for replacement steel in an amount sufficient for the fabrication of all critical tower sections. . . . All of the steel was required by contract to be obtained from the United States. . . .

b. Having been finally advised, by the Government's letter of 2 August 1967. . . , that a seam-free condition would be required on all critical sections, and realizing that the steel on hand could not be used, appellant made a management decision to proceed with the fabrication of noncritical tower legs, out of planned sequence, in the hope that the finished noncritical legs would be interchangeable when used in the erection of the towers. Appellant duly notified the Government of this decision on 2 August 1967, and advised that tower erection would be delayed until at least 30 days after the arrival of the replacement steel for the critical tower sections. . . . The decision also contemplated that use of PECI's shop facilities in this manner would mitigate damages and enable them to make only critical legs when the replacement steel arrived from the United States. . . . December 1968. . .illustrates the effect of appellant's management decision. While the production of critical legs remained virtually at a standstill until October 1967, when the replacement steel arrived, production of noncritical legs accelerated very sharply for three months, from August until November 1967. . . .

c. In addition to initiating productive work on the noncritical tower legs in August 1967, appellant further mitigated damages by performing work on transmission line supports and switch bay structures throughout the suspension period. . . .

9. The success of appellant's attempts to mitigate damages was charted by the Government in terms of appellant's monthly earnings under the contract. . . . This chart demonstrated a continuous flow of productive work throughout the period of the contract, without any significant peaks or valleys at least in terms of appellant's earnings. . . . To this thrust, appellant responds that all of the productive work performed during the period was not on the critical path toward timely completion of the project. That is, none of this work was pacing in relation to overall completion of Project Bamboo, although it undoubtedly facilitated appellant's subsequent acceleration efforts. In other words, appellant contends that the Government's stop work orders prevented the fabrication of critical tower sections, which in turn prevented erection of any towers other than four small rhombic towers which had no critical sections. . . . The unrebutted evidence on the effect of the stop work orders on the fabrication of critical tower legs is that as of 12 May 1967 there were 73 such legs, out of a total of 699 in the entire project, in various stages of fabrication and that no new critical legs were fabricated from that date until late October or early November 1967, after the replacement steel arrived. . . . On this record, we find that the productive work performed by appellant during the period from 9 June to 20 October 1967 was not on the critical path of the work as defined by appellant's expert, and hence it did not affect the delay in tower erection during the period. . . .

11. During the delay period asserted by appellant in its claim, the Government argues that there were a number of concurrent causes of delay for which the Government was not responsible.

a. In addition to the strike at PECI's plant early in the delay period, PECI experienced several other problems that seriously affected its work. Thus, the Government's inspector at the plant made numerous observations concerning the alleged inexperience and resulting inefficiencies of PECI's personnel. . . .

b. By 10 May 1967, PECI was experiencing severe financial difficulties, as evidenced by the fact that appellant agreed to provide the funds necessary for PECI to meet its Project Bamboo payrolls in exchange for the active management of the job by appellant's representatives. . . . In addition to financial troubles, PECI suffered low morale on the part of its personnel and an inability to purchase needed supplies. . . . Appellant's project manager agreed that PECI's management seemed then unable to cope with the problems it was facing. . . .

c. On 21 June 1967, appellant advised PECI that its performance was "completely unsatisfactory," and that it had been unable to comply with its production schedule. . . . On 28 June 1967, appellant consummated a "takeover" of PECI pursuant to the terms of a memorandum agreement between the two organizations. . . . As a result of the takeover, PECI was enabled to continue with the construction of Project Bamboo, and because appellant funded its operations from that time on, PECI was no longer handicapped by lack of funds. . . . The takeover did not end the technical problems which PECI encountered, as evidenced in the field notebooks of the Government's inspector throughout the period ending 12 September 1967. . . .

d. During the hearing, the Government questioned appellant's project manager regarding the latter's alleged failure to fabricate 17 towers which had legs of a diameter in excess of 3¾ inches and which could have been fabricated from the original "Duquesne" steel which did not suffer from the seam problem. . . . Although critical legs of Duquesne steel could indeed have been fabricated during the suspension period, appellant had investigated the possibility of changing its erection sequence in this respect, but found that all of the towers required the use of some "Youngstown" bars that were affected by the seam problem. . . . Accordingly, since appellant was at the time making noncritical legs and performing other structural work, we find that its failure to fabricate critical sections of Duquesne steel does not represent a material failure to mitigate damages. Furthermore, since all of the towers, excepting only the four small rhombic towers, contained critical legs made of the Youngstown steel, we find that fabrication of critical sections made of Duquesne steel would not significantly have advanced the work of tower erection.

e. During the 9 June to 5 September 1967 delay period, the Government has pointed out that another of appellant's subcontractors, United Construction Company, Inc. (UCCI), experienced numerous delays in performing its work, for which delays the Government was not responsible. The delays pointed out by the Government affected the erection of the transmitter and warehouse buildings, the installation of the 13.8 KVA line and the concrete for the RF transmission lines. . . . Since none of the work in question was on the critical path as charted by appellant's expert . . . , we find that the delays in UCCI's work had no effect on the timely completion of Project Bamboo.

f. Finally, the Government points to very heavy rainfall during the delay period which impeded the progress of the work. The effects of the adverse weather conditions were detailed in appellant's monthly narrative progress reports. . . , wherein it appears that it was the site work that was delayed. Since the site work was not then pacing, we find that the adverse weather did not delay the timely completion of Project Bamboo.

12. The facts relating to the Government's charge that PECI's inefficiencies and poor performance were a concurrent cause of appellant's delay are as follows.

a. The Government's stop work orders and related directives caused a complete disruption of PECI's work of fabricating critical tower legs, without which erection of the towers could not proceed. . . .

b. The productive work performed by PECI during the delay period was not on the critical path toward the timely completion of Project Bamboo, in that it was not related to tower erection. At the time, tower erection was delayed by the unavailability of critical tower legs. . . .

c. Appellant's project manager anticipated that PECI would encounter delaying problems in the course of its performance, and accordingly testified in its subcontract. . . . A comparison of the completion dates specified in the PECI subcontract . . . appellant's approved progress schedule . . . shows a contingency period of five months for both the fabrication of tower legs and the erection of the towers.

d. Appellant's concern with PECI's inefficiencies as of 10 May 1967 . . . was that the contingency factor that had been allowed for in PECI's subcontract would be soon used up. . . .

e. Since PECI's admitted inefficiencies and poor performance affected only work that was not in the critical path, and since the time lost on that account was within the five months' contingency period . . . no delay in the timely completion of the project during the period from 12 May to 20 October 1967 was caused by the inefficiencies and poor performance.

B.   The Impact Claim—1 November 1967 to 1 February 1968

13. The replacement steel which appellant had ordered on 28 June and 8 August 1967 arrived in the Philippines in two shipments, the first in early October and the second in November 1967. The first replacement steel arrived at the fabrication shop on 26 October, whereupon processing and fabrication began immediately. . . . The events that took place thereafter are summarized as follows.

a. It will be recalled that appellant had been authorized by the Government's letter of 5 September 1967 to repair the 73 critical legs on which the work had been stopped as of 12 May. . . . It took appellant most of September and October to accomplish the repair procedures. . . .

b. With 67 of the 73 critical legs repaired and available for tower erection, Tower 040 was erected and tensioned in late October. . . . When finally tensioned, inspection revealed that it had "some twist" as a result of numerous deficiencies in fabrication. . . .

c. As appellant continued to erect the towers in November, utilizing the critical legs fabricated of the original, conditioned tower steel, fabrication errors caused increasing concern. For example, a memorandum by one of appellant's supervisors on 11 November 1967 stated:

"I am aware of some of the causes of poor quality fabrication, but I feel that we have reached the end of the road on excuses.

"All sections received at the jobsite from this date should be within all contract tolerances.". . .

d. On 27 November 1967, appellant's project manager shut down all tower erection because of problems of fabrication and assembly that were then unsolved. Both the USIA in Washington and the project manager's superior in Dallas were alerted, and help was requested from the architect and engineering firm that had designed the towers. . . .

e. By early December, about 10 towers had been erected, and all were found to have serious twists and distortions. Accordingly, appellant sent for a consulting engineer, who was an expert in the fabrication and erection of radio antenna towers. . . . He arrived on 7 December to ascertain the cause of and correct the twisting and distortion of the towers that had been erected. . . . Upon his arrival, the expert found that the twisting and distortion was the consequence of fabrication errors; he recommended that all except one tower should be dismantled and taken down so as to permit necessary repairs to be made. Of the repairs that were ultimately ordered, minor ones were carried out at the jobsite, while major repairs in some cases had to be made in PECI's Manila plant.

14. On this state of the facts, the parties have developed competing theories with respect to the three months' delay and extra work of reconditioning tower legs which appellant experienced in the impact period from 1 November 1967 to 1 February 1968. For its part, appellant contends that the Government's stop work orders of 12 and 16 May 1967 denied appellant the opportunity to pursue the work of fabricating and erecting towers in an orderly sequence, as it had planned to do. . . . The resulting disruption of the work, it argues, is chargeable in substantial part to the Government, the balance being chargeable to the fabrication errors for which it concedes responsibility. The Government, on the other hand, argues that the delay, disruption and rework which appellant suffered in the impact period was due to fabrication errors for which appellant is solely responsible. . . .

d. In view of the complexity of Project Bamboo and the repetitive nature of the job, involving as it did similar towers, appellant originally planned to fabricate and erect prototype towers in order to prove out its techniques. For this purpose it selected towers 040 and 032. Tower 040 was the simplest tower in the project as far as the number of critical sections was concerned (excepting the four non-guyed rhombic towers); Tower 032 was selected as the second prototype because it had multiple guy levels and hence multiple critical sections in each tower. It was appellant's intention to assemble these two towers including the antennas so as to pinpoint any problems that might arise and to apply this knowledge in developing a learning curve for the erection of the remaining towers. . . .Appellant's project manager explained how this learning curve was built into its approved progress schedule, which depicted anticipated rates of production for both steel fabrication and tower erection that were significantly lower in the first two or three months than they were after learning had been accomplished and a steady production rate achieved. . . .

16. On the basis of the facts relating to appellant's impact claim as we have summarized them thus far, we make the following findings of fact.

a. It is not only good practice but also customary in the industry for towers of the sort called for in Project Bamboo to be erected in an orderly sequence.

b. Appellant had initially planned to fabricate and erect prototype towers in order to prove out its fabrication and erection procedures, and then to follow an orderly sequence in the erection of the remaining towers. This orderly sequence was necessarily abandoned because of the unavailability of critical tower legs resulting from the Government's suspension of work.

c. Appellant's unilateral decision of 2 August 1967 to fabricate noncritical tower legs out of sequence was reasonably intended to mitigate damages during the pendency of the Government's stop work orders.

d. From the time the first tower was erected and tensioned in late October, and was found to be defective, it required approximately three months, until late January 1968, for appellant to diagnose its problems and to prove out the necessary remedial measures.

C.   Acceleration Claim—1 March 1968 to 31 March 1969

17. It will be recalled that in late January 1968 Tower 090 was successfully erected and tensioned by way of demonstrating the effectiveness of the remedial measures which had been recommended by appellant's tower expert. . . . The fabrication and erection problems having been thus solved, appellant's progress in tower erection turned sharply upward after 1 March 1968. . . . Indeed, after the first tower was successfully erected, all of the remaining towers were erected in the nine months following January 1968, whereas 14 months had been allowed in appellant's original project schedule. . . . The substantial completion of tower erection appears to have been accomplished without further incident by the end of October 1968. . . . Thereafter, all of the remaining work covered by the contract as amended was certified by appellant's project manager as complete on 31 March 1969, with the exception of certain enumerated "punch list" items. The local representative of the contracting officer accepted the certification on the same day. . . .

18. Two of the contract articles have a special bearing on the issue of acceleration.

a. Article V of the contract, entitled "Time of Performance," provides in pertinent part as follows:

"A. Performance of work under this contract shall be completed within nine hundred (900) calendar days after the date of receipt of Notice to Proceed.

"B. It is specifically understood and agreed that the completion date or dates as established in accordance with this Article may be extended only as follows:

"(1) By formal written extension of time granted by the Contracting Officer for the convenience of the Government; and

"(2) By formal written extension of time granted pursuant to Standard Form 23-A clauses. . .

"The new completion date so fixed shall not be extended except in the same manner as provided in this paragraph. . . ."

b. Article VI of the contract, entitled "Liquidated Damages," provides in pertinent part as follows:

"A. Since time is of the essence of this agreement, and since it is difficult to determine exactly the damage the Government will sustain by a delay in performance, it is hereby stipulated that for each calendar day in excess of 900 the Contractor has failed to complete the performance of work under this Contract, the sum of $900.00 is hereby fixed, determined, and agreed upon as a reasonable assessment of liquidated damages to be charged against the Contractor. . . . Liquidated damages will be assessed from the completion date indicated in the contract or extension thereof to the date of actual completion. Actual completion as determined by the Contracting Officer shall be final and conclusive.''. . .

19. During the course of its performance of the contract, appellant experienced a number of delays that were unrelated to the steel seam problem, and for which the Government ultimately granted time extensions in a total amount of 134 days, thus extending the contract completion date from 17 November 1968 to 31 March 1969. Appellant contends that the Government failed to grant the time extensions within a reasonable time. . . . The Government responds that the requests for time extensions were at first not adequately substantiated, but when granted were ''unquestionably known to appellant in due course.''. . . The facts relating to these time extensions are as follows.

a. In November and December 1966, appellant experienced excusable delays because of a wharfage problem for which it was not responsible. Appellant's letter of 21 June 1967 requested a time extension and submitted the necessary justification. The contracting officer first indicated his approval of the request by letter of 5 February 1968 granting a time extension of 25 days . . . without requiring any supplementary justification. . . . This extension was included in Modification No. 10 to the contract executed on 26 September 1968. . . .

b. Two separate strikes of 14 days each occurred at PECI's plant in May and June 1967 and later in March and April of 1968. . . . Although the contracting officer's letter of 5 February 1968 rejected the time extension requested by appellant for the first strike for lack of adequate justification . . . , he ultimately approved a 28-day extension in Modification No. 10, dated 26 September 1968. . . . We have found nothing in the record to suggest that appellant supplemented its original justification.

c. On 18 March 1967, appellant addressed a letter to the Government regarding a field change order issued by the latter sometime prior to that date. . . . For this and some 34 other field change orders issued by the Government, appellant requested time extensions in a total amount of 516 days. In his letter of 11 July 1968, the contracting officer rejected the time extensions requested by appellant for these change orders because of an alleged lack of justification. . . . In reply, appellant noted the lack of requisite authority in the resident contracting officer . . . to negotiate time extensions for field change orders at the jobsite, and urged more timely action in order to avoid the unnecessary disruption that resulted from the long delays then being experienced in acting on its requests for time extensions. . . . Conceding that the total time extensions it had requested were more than required by the project as a whole because they ran concurrently, appellant indicated that the concurrency factor could easily have been eliminated in timely fashion by negotiations at the jobsite. By letter of 9 August 1968, following negotiations at the jobsite, the contracting officer offered a time extension of 25 days for the 35 field change orders . . . and formal action was completed in the matter by the inclusion of the time extension in Modification No. 10, dated 26 September 1968. . . .

d. In July and August 1968, correspondence between the parties referred to Change Order Proposal 97, embracing a number of change orders which presumably had been performed before that date. . . . The time extension of 23 days which the parties negotiated for the affected change orders was ultimately formalized in Modification 11, which was executed by the parties on 19 March 1969, less than two weeks before substantial completion of the contract work. . . .

e. For Field Change Order No. 57 and Domestic Change Order No. 5, the parties agreed to a time extension of 33 days, and the matter was formalized in Modification 13, dated 14 July 1969, three and one-half months after substantial completion of the contract. . . . Although the modification does not show when the delay occurred, the relevant change order proposal which was the subject of correspondence between the parties establishes the applicable dates as occurring in May and June 1968. . . .

f. The foregoing summary shows, and we find, that substantial periods of time elapsed between the time appellant suffered the excusable delays in question and the time when appellant was formally granted time extensions therefor. The summary also shows, and we find, that the contracting officer gave no indication, formal or otherwise, that he approved any of the time extensions before 5 February 1968. Finally, we find that of the time extensions totalling 134 days only 14 days allowed for the strike that occurred at PECI's plant in May and June 1969 ran concurrently with delays claimed by appellant in the delay and impact claim periods; the remainder of the events for which the time extensions were granted happened either before the delay claim period . . . or after the impact claim period. . . . Based on the record as a whole, we find that the approvals which were ultimately given would have been more timely had the Government been prepared to conduct a prompt review of appellant's requests and supporting justification and to negotiate the matter within a reasonable time.

20. Wholly apart from the delays discussed in the preceding paragraph, for which a time extension of 134 days was granted, appellant requested a time extension for the delay which it attributed to the steel seam problem. Since the Government did not concede liability for the delay experienced by appellant on this account before our entitlement decision was handed down, and because it considered that the 134 days of extensions otherwise granted were adequate for completion of the project. . ., no time extension on account of this delay was ever granted. The relevant facts are as follows.

a. The first request for a time extension for the delay caused by the steel seam problem was made in the fall of 1967. . ., when appellant proposed a revised schedule showing completion of tower erection on 1 October 1968, four months later than originally scheduled. . . . This request was summarily denied by the contracting officer on 5 December 1967. . . .

b. On 31 January 1968, appellant formally requested a time extension of 142 days and an increase in the contract price of $955,302 by way of compensation for the delays, disruption, and extra work which it attributed to the steel seam problem. . . . This letter detailed the events relating to the steel seam problem, but did not explicitly justify either the extension of 142 days or the dollar amount of the contract adjustment requested.

c. On 5 February 1968, the contracting officer requested appellant to submit a revised progress schedule for the completion of Project Bamboo, in such detail as to enable the Government to determine that the work would be prosecuted with such diligence as to insure its completion within the time specified in the contract. . . . Appellant complied with this request on 4 March 1968, submitting the necessary revised schedules which showed completion of all work except proof of performance by mid-November 1968. . . . In the letter transmitting these schedules, appellant noted that it did not agree with the Government's unilateral decision denying time extensions for contract changes to date, and added that in order to meet the existing contract completion date it would be forced to accelerate the construction effort. Accordingly, appellant stated that it would continue to accelerate the work, as it had been doing prior to receipt of the contracting officer's letter of 5 February 1968. For this acceleration appellant indicated that it would claim additional compensation. . . .

d. On 25 March 1968, the contracting officer specifically rejected appellant's claim of actual or constructive acceleration, and alleged that none of the time extensions that had then been requested, with the sole exception of the 25-day extension for wharfage delay granted on 5 February 1968, had been accompanied by an explanation of how the delays had affected the overall job scheduling. Such unsupported requests, the contracting officer concluded, were not sufficient either to justify extending the contract period or granting an acceleration claim. . . .

e. Because of the pending impasse on the subject of time extensions and acceleration, the parties met on 11 April and 17 May 1968 to discuss the matters that were then in disagreement. The contracting officer indicated that he felt the Government did not have adequate justification for all the time extensions that had been requested, which then totaled 912 days on a 900-day contract. . . . The contracting officer assured appellant that it would consider any reasonable request for time extensions, but that the Government had not ordered acceleration, did not want acceleration, and would not pay for acceleration. . . . On 22 May 1968, after the second meeting with the contracting officer, appellant withdrew its request of 31 January 1968 for a 142-day time extension relating to the steel seam problem, stating that it would resubmit the claim for both costs and time after detailed review based upon the parties' recent discussions. . . . These events were summarized by the contracting officer in a letter of 31 May 1968, which began with a recitation of the contract provisions relating to time of performance and liquidated damages for failing to meet the contract completion date. After taking note of appellant's failure to provide additional justification linking the time extensions to overall job completion, the contracting officer concluded:

"It would seem from the above chronology that F&M cannot factually support its requests for additional time extensions and, in the absence of a contract extension, cannot meet its revised contract performance date without acceleration. Any acceleration required to meet such date is solely F&M's responsibility. Similarly, a failure to complete the job in timely fashion is F&M's responsibility. Any extensions of time that can reasonably be attributed to contract changes or other appropriate causes on the overall job will be granted by the Agency when justified by F&M.''. . .

Appellant's response of 19 June 1968 . . . was to the effect that under the terms of subparagraph (d) of the Default clause the contracting officer had an affirmative duty to ascertain the facts and extend the contract time after receipt of an appropriate notice. Subparagraph (d) reads, in pertinent part, as follows:

''(d) . . . The Contracting Officer shall ascertain the facts and the extent of the delay and extend the time for completing the work when, in his judgment, the findings of fact justify such an extension.'' . . .

The record shows that when the contracting officer did investigate the facts relating to certain change orders in July 1968, he offered to grant a time extension within one month . . . and the matter was mutually agreed to on 26 September 1968. . . . Appellant also observed that the contracting officer's continual generalized requests for additional backup data did not waive his duty to investigate. . . .

f. On 21 September 1968, at a time when Modification No. 10 to the contract providing for time extensions totaling 78 days was in dispute, appellant wrote the contracting officer to reinstate and reconfirm its original request for a 142-day time extension due to the steel seam problem. . . . The letter stated that since Modification No. 10 had not been executed, appellant was still being held to the ''original completion date of November 20, 1968,'' under which circumstances it was forced to accelerate in order to minimize liquidated damages and to come as near as possible to meeting the original schedule. . . .

g. In a further letter of 11 October 1968, appellant recited the factual background and provided the legal reasoning supporting its request for an extension of 142 days. . . . The letter concluded that appellant had been delayed from 12 May 1967 until the arrival of the first shipment of replacement steel on 27 October 1967, plus a period thereafter to approximately 15 December 1967, when the second shipment of replacement steel arrived and it was finally able to reach the same point in the production process which existed as of 12 May 1967. Although it was entitled to a longer extension of time, appellant offered to limit its request to 142 days, consistent with its original request, in order to obtain a prompt settlement of the matter. . . .

h. The final event in this aspect of appellant's acceleration claim was the formal steel claim which appellant submitted on 13 December 1968. . . . This submission, which was voluminous, requested a time extension of six months and damages in the amount of $1,862,692, and supported the request with accounting data and a variety of charts which purported to show the impact of the steel seam problem on its performance of the contract.

21. On the basis of the record as a whole, we make the following findings of fact relating to appellant's acceleration.

a. In March, April and May 1968, after he had first learned of appellant's claim for actual and constructive acceleration, the contracting officer acted promptly to reject the claim as not supported by adequate justification, and to make it emphatically clear that the Government did not want appellant to accelerate and would not pay for any acceleration. . . .

b. The contracting officer never formally assessed liquidated damages against appellant. . ., although his letter of 31 May 1968 . . . considered in context implied that the Government might do so if appellant failed to meet the contract completion date.

c. The contracting officer consistently denied any time extension for the delay relating to the steel seam problem, starting with his denial of 5 December 1967. . . .

d. The contracting officer repeatedly failed to grant time extensions for excusable delays arising out of events other than the steel seam problem within a reasonable time after receiving appellant's requests. . . .

e. The contracting officer's denials and tardy approvals of time extensions, coupled with his insistence that appellant demonstrate, to his unspecified satisfaction, a causal connection between the events giving rise to the delays and the overall job scheduling, led appellant reasonably to believe that the requested time extensions would not be granted and that it would be held to the original contract completion date.

22. Appellant's project manager testified at the hearing that from the end of January 1968, when the first set of towers had been erected in full compliance with the contract specifications, it took nine months to complete the erection of all the remaining towers. He also testified that 14 months were allowed in the original project schedule for tower erection. . . . The record confirms this assessment, but with some qualifications. Thus, by the first of February 1968 appellant had been paid for 9 percent of Pay Item XIIB, that is, for the erection of towers and associated structures including guy lines, and on 1 November 1969 the tower erection was 99 percent complete, as shown on a chart based on the Government's pay records. . . . Turning to appellant's approved progress schedule, we find that appellant had scheduled 10 percent of tower erection (Pay Item XIIB) by 1 April 1967 and completion of tower erection by 1 June 1968, a period of 14 months. . . . On the basis of these facts, we find that appellant performed 90 percent of the tower erection in nine months, as compared with the originally scheduled period of 14 months for accomplishment of the same amount of work. . . .

E. The UCCI Claim

28. The United Construction Company, Inc., (UCCI) had a subcontract with appellant for performance of the site work, among other things, for Project Bamboo. The gist of the UCCI claim is that the site work was delayed and extended as a result of the constructive changes by the Government in relation to the steel seam problem. For its part, the Government defends on three

grounds: . . . second, that appellant failed to establish the necessary causal connection between the delay costs claimed by UCCI and the steel seam problem. . . .

30. The second ground of the Government's opposition to the UCCI claim is the lack of any causal connection between the delays experienced by UCCI and the constructive change relating to the steel seam problem. The relevant facts are as follows.

a. The UCCI subcontract called for the performance of site work, excavation for the towers and back fill, erection of a transmitter building and a number of other buildings that were necessary and incident to the project as a whole, and installation of certain underground cable and power equipment. . . . Appellant's project manager testified, and the record otherwise shows, that most of the UCCI work was unaffected by the suspension of work directed by the Government in relation to the steel seam problem. . . . Indeed, the president of UCCI testified that there were only two items of UCCI work that were delayed by the steel seam problem, namely, Item IIB—Road Paving, and Item IID—Finish Grade Site. . . . With regard to these items, he testified without contradiction that the road paving and finish grading had to await completion of tower erection because otherwise the paving and earthwork would have been damaged by the heavy equipment used in tower erection. . . . Since the last tower was not erected until the end of October 1968. . ., which also marked the end of the rainy season. . ., UCCI did not begin the paving and finish grading until November, according to a chart based upon official pay estimates submitted to the Government during the performance of the contract. . . .

b. As finally presented in its 1975 claim. . ., the UCCI claim was comprised of overhead and plant costs, additional road repair costs, and move-in/move-out costs. The first element was explained by the president of UCCI as covering the expenses of staff and equipment rentals for about six months longer than had been anticipated. . . . The record shows that UCCI had planned to complete its work by 1 October 1968. . ., and was in fact able to complete substantially all of its work, except for the road paving and finish grading, by 1 December 1968. . . . Because of the two excepted items, which were delayed by late tower erection, UCCI continued on the job "up to April of 1968," whereas it had not planned to work "after November 1968.". . . The extra time on the job, we find, was four months. The second element covered additional road repair costs which UCCI incurred during the rainy season from May through October 1968, while tower erection was being delayed and the work of paving the road could not begin. . . . Appellant's approved progress schedule shows that both the road paving and finish grading were scheduled for completion by 1 June 1968. . . . The final element of the claim, covering move-in/move-out costs incurred by UCCI, relates to the withdrawal of UCCI's roadwork subcontractor following the delay in tower erection that would have tied up its equipment for a lengthy period of time beyond its subcontractual commitment. UCCI permitted its subcontractor to move out its equipment rather than remain on the job at an exorbitant rental. . . . That equipment was replaced by equipment moved in by UCCI during the rainy season of 1968 until the project was completed. . . .

c. In contesting the lack of causal connection, the Government first contends that UCCI was able, during the delay period from 12 May 1967 through 20 October 1967, to perform productive work on most of the items comprising the UCCI subcontract. . . . Our study of the record shows, and we find, that UCCI performed a substantial amount of productive work during the delay period. . . .

d. During the hearing, the president of UCCI was asked to comment on a number of complaints made by appellant's project manager during the progress of the work. There were "probably about 50 complaints to me against the company," he said, because the project manager was always complaining about UCCI's performance. . . . Many of the complaints dealt with the fact that UCCI fell behind the schedule set forth in its subcontract. . . . However, UCCI ultimately accelerated of its own accord and completed all work, except the road paving and finish grading, on schedule. . . . Accordingly, we find that UCCI delays in question have no bearing on the UCCI claim for extra work after October 1968 and during the rainy season of 1968.

e. The final argument by the Government for the lack of any causal connection, is that only two items of the UCCI's subcontract were affected by the steel seam problem, and their contract value was, in the case of road paving and finish grading, $76,730 and $12,000 respectively. . . . Since 40% of the road paving and 60% of the finish grading had been completed by the end of November 1968, the Government's argument continues, the price for the remainder of the paving would have been $46,038 and for the remainder of the finish grading $4,800, or a total of only $50,838, whereas the UCCI claim is in the total amount of $217,163. The Government also noted that the $50,838 represents less than 2% of UCCI's amended subcontract price of approximately $2.8 million. . . . We fail to see the relevance of this argument. Whether or not the amount claimed by appellant on behalf of UCCI is proved, it is possible that a Government-caused delay affecting only a minor portion of appellant's work could, as in the present case, result in costs that far exceed the value of the work so delayed.

f. On the record thus summarized, we find that, through no fault of its own, UCCI's performance was extended for a period of four months, from 1 December 1968 to 31 March 1969. We also find that UCCI incurred additional costs in maintaining roads during the rainy season from 1 June to 31 October 1968, which it would have avoided by the scheduled completion of both road paving and finish grading on 1 June 1968, but for the delay in tower erection. Finally, we find that the move-in/move-out costs, which were incurred by UCCI and which related to the withdrawal of UCCI's subcontractor and UCCI's consequent need to move in new equipment for the balance of the work, were caused by the delay in tower erection. Notwithstanding these findings, which tie the UCCI claim to appellant's delay in completing tower erection, the liability of the Government on the UCCI claim must depend on the extent to which we find it responsible for that delay. We shall take that question up in our decision on the Government's liability. . . .

20F. The FMS Claim

32. Certain work of Project Bamboo, including antenna tuning among other things, was subcontracted by appellant to Fischbach and Moore Systems Company (FMS). The gist of the FMS claim is that the antenna tuning program was delayed as a result of constructive changes by the Government related to the steel seam problem. Specifically, appellant alleges that the towers on which the antennas were to be installed were erected later than scheduled, and as a result, FMS incurred extra costs in accelerating the work from and after September 1968, and its performance was extended from the scheduled completion date of November 1968 through January 1969. The present record contains the following relevant facts.

a. Appellant's approved project schedule called for antenna tuning to begin on 1 January 1968 and to be completed by 1 November 1968. . . . For this work, FMS planned to have two crews, supported by three tuning trailers, two of which were outfitted for antenna tuning and the third for work on the transmission lines. . . .

b. The critical path analysis of Project Bamboo performed by appellant's expert, to say nothing of ordinary common sense, demonstrates that tower erection, antenna erection, and antenna tuning and proof of performance must follow each other in orderly sequence. . . . In view of the sequential nature of the work, the delay in erecting the towers necessarily affected the erection of the antenna curtains and reflectors, as well as the tuning of the antennas after they were erected. . . .

c. Erection of the towers began in earnest in March 1968, after appellant had developed and proved out its new procedures for fabricating the tower legs. . . . Erection of the curtain antennas and reflectors, which had been scheduled to begin on 1 March 1967. . ., showed sporadic activity, as indicated by Government-approved pay estimates, from August 1967 to 1 April 1968, after which uninterrupted progress was made until the work was completed and paid for in February 1969. . . .

d. By 1 September 1968, the tower erection process was approximately 70% complete. . ., and the erection of the curtain antennas and reflectors had proceeded to a point at which the Government had paid for 64% of the work. . . . At this juncture, FMS added two additional crews, plus additional automatic test equipment in order to accelerate the tuning program. These two crews began operations in September 1968 and operated with the other two crews through January 1969.

e. In appellant's claim letter of 13 December 1968, the FMS claim was included in two pages as Cost Estimate No. 8, covering the estimated cost of two additional antenna test crews and related test equipment. . . . Although the same material was included as Section VII of the Steel Claim Repricing, dated 14 June 1973. . ., the FMS claim was subsequently revised and expanded to reflect actual costs purportedly incurred as a result of delays and acceleration related to the steel seam problem. The revised material supporting the FMS claim was forwarded to the contracting officer on 1 October 1973. . . .

33. In defending against the FMS claim, the Government argued that the record contained no competent evidence of what it termed compensable acceleration or of additional costs attributable thereto. . . . The relevant facts are as follows. . . .

d. On 11 June 1968, appellant's project manager addressed a letter to FMS in which he complained of a lack of manpower and the absence of a fully authorized supervisor as causes for delays by FMS in performing the work of transmitter testing. . . . The lack of progress on transmitter testing was also the subject to a TWX of 14 June 1968. . . . The context of these communications was explained by appellant's project manager, who testified that there was a problem in June 1968 involving certain transmitters furnished by the Government for Project Bamboo. Appellant had substantial difficulties with the testing, and the line item for such testing was eventually terminated for the convenience of the Government on 10 November 1968. . ., before FMS completed the testing. . . . There is nothing in the record to suggest that antenna tuning was delayed by, or otherwise related to, transmitter testing.

e. Finally, the Government argues that FMS originally underestimated the labor and materials necessary to perform the antenna testing, and that it was such underestimation rather than the steel seam problem which placed it in a delay posture. . . . FMS originally planned to have two antenna test crews and related equipment to perform the antenna tuning from 1 January to 1 November 1968, a period of 334 days. . . . On the other hand, the antenna system design engineers, who designed the antenna system and developed the antenna system test program, planned for three crews. . ., and in Proposal A, which the design engineers supplied to bidders on Project Bamboo, it was recommended that three tuning crews and one supervisory engineer could complete the tuning and proof of performance in an estimated 225 working days. . . .

34. On the basis of the foregoing summary of the record relating to the FMS claim, we make the following findings of fact.

a. The FMS claim was indeed supported by substantial, competent evidence, as conceded by the Government's auditor. . . .

b. The lack of manpower, coupled with the absence of a fully authorized supervisor for FMS work, which was the subject of appellant's project manager's communications in June 1968, related only to transmitter testing, which was terminated for the convenience of the Government on 10 November 1968. . ., and which bore no relation to, and hence did not delay, the antenna tuning which is the basis of the FMS claim.

c. Appellant's plan to perform the antenna tuning program with two crews and related test equipment for 334 days is approximately equivalent in crew days to the estimate of the antenna system design engineers that three crews and related test equipment would be required for 225 days. Accordingly, the

Government's contention that FMS originally underestimated the labor and materials needed for the job is without merit.

35. There remains the question raised by the Government whether appellant has established a causal connection between the steel seam problem and the additional costs incurred by FMS and included in its claim. We shall deal with this question in our decision on the Government's liability. . . .

II. DECISION ON THE GOVERNMENT'S LIABILITY

A. Introduction

The principal theme of the Government's defense against appellant's claim is that appellant has failed to prove a causal connection between the Government's suspension of work relating to the steel seam problem and the delays, disruption of work, and acceleration for which it incurred the costs it claims here. This issue, in turn, is affected by concurrent delays which the Government alleges were an intervening cause of the delay, disruption and acceleration of appellant's work, and for which either appellant or its first tier subcontractors are responsible or for which the Government granted adequate time extensions in due course.

On this point, the Government contends that any delays caused by it in relation to the steel seam problem were inextricably intertwined with other delays, such as those caused by the inefficiencies and poor performance of appellant's subcontractor, PECI, which appellant would have suffered even in the absence of the steel seam problem. When Government-caused delays are concurrent or intertwined with other delays for which the Government is not responsible, the argument continues, a contractor cannot recover delay damages. *Commerce International Co. v. United States*; *Hardeman-Monier-Hutcherson, A Joint Venture*, ASBCA NO. 11869, 67-2 BCA ¶ 6522.

We take no issue with the application of the *Commerce* rule to the facts of this case insofar as the concurrent delays for which appellant is responsible affected work in the critical path to timely completion of the contract. If the concurrent delays affected only work that was not in the critical path, however, they are not delays within the meaning of the rule since timely completion of the contract was not thereby prevented.

With regard to the alleged intertwining of Government-caused and concurrent delays in this case, we have found, in the critical path analysis offered by appellant, a ready and reasonable basis for segregating the delays. If the delays can be segregated, responsibility therefor may be allocated to the parties. *Chaney & James Construction Co., Inc. v. United States* [14 CCF ¶ 83,407], 190 Ct.Cl. 699; 421 F.2d 728 (1970). And if there is no basis in the record on which to make a precise allocation of responsibility, an estimated allocation may be made in the nature of a jury verdict. *Raymond Constructors of Africa, Ltd. v. United States* [13 CCF ¶ 82,155], 188 Ct.Cl. 147; 411 F.2d 1227 (1969). The seemingly contrary result in *Commerce* is explained by the fact that the Court was unable, on the record in that case, to separate delays for which the Government was not responsible from those for which it was. *Commerce International Co., supra*, 167 Ct.Cl. at 543, 338 F.2d at 89, 90. As will be seen in the discussion that follows, we have no such difficulty in the present case.

Although it is true, as the Government observes, that the critical path analysis made by appellant's expert is abstract, in the sense that a critical path method of scheduling was not called out in the contract, the validity of the analysis was not challenged in any way. We accept it as credible, based as it is on application of the construction experience of an expert to depict the orderly sequence of events that must be followed to accomplish a complex project such as Project Bamboo.

It is axiomatic that a contractor asserting a claim against the Government must prove not only that it incurred the additional costs making up its claim but also that such costs would not have been incurred but for Government action. The Government contends that appellant's proof is lacking in both respects. We shall consider, in our decision on the measure of the Government's liability, whether appellant's proof of costs is adequate. The existence of the necessary causal connection is the central subject of our decision on the Government's liability.

B. The Delay Claim

The Government's first stop work order was issued on 12 May 1967 and was shortly thereafter broadened to suspend all operations relating to the fabrication, galvanizing or erection of all tower sections. . . . From that time until 20 October 1967, a time span of 161 days, appellant claims that the work of tower erection, which was on the critical path to the timely completion of Project Bamboo, was effectively brought to a halt by the Government action. In our discussion of the record bearing on this point, we have so found. . . . We have also found that the 67 critical legs (that is, legs to which guy gusset plates are welded and which afford lateral support to the towers after their erection) which were in the shop of appellant's fabrication subcontractor, PECI, on 12 May 1967 were not repaired and ready for use until 20 October 1967, the end of the delay period claimed by appellant. . . . Thus, it appears that at the end of the delay period appellant had recovered to about where it had been at the beginning of the delay period so far as its readiness to begin tower erection was concerned.

On the other hand, PECI was able to perform a substantial amount of productive work during the delay claim period. However, we have found that this work was not on the critical path toward the timely completion of Project Bamboo. . . . Some of this productive work, such as the fabrication of non-critical legs after 2 August 1967, was undertaken by appellant in what we consider to be a reasonable attempt to mitigate damages while the work was otherwise delayed. PECI's admitted inefficiencies and poor performance during this period had no bearing on appellant's delay claim resulting from the Government's suspension of work, since only work that was not on the critical path was affected. . . .

The Government argues that the first part of the delay period, from 12 May to 9 June 1967, constitutes a reasonable period of suspension for which no compensation should be allowed. In support of its position, the Government cites Articles II and VIII of the contract, by which it reserved the right to suspend the work for a reasonable period of time for its convenience. . . . In opposing this argument, appellant contends that because the suspension was a result of defective specifications, it is unreasonable *per se*. Since we have found that the specifications were not defective. . ., we consider that there is no merit in appellant's argument. Accordingly, since the record shows that the Government took prompt and vigorous action to ascertain the nature of the steel seam problem and authorized a conditional resumption of the work on 9 June 1967. . ., we conclude that the 28 days in question constitute a reasonable period of suspension for which no compensation is allowable under the terms of the contract referred to above. *Cf. Merritt-Chapman & Scott Corp. v. United States* [16 CCF ¶ 80,187], 194 Ct.Cl. 461, 475; 439 F.2d 185, 193 (1971).

In addition to the foregoing, by 10 May 1967, PECI was experiencing severe financial difficulties which threatened its capability to continue its performance of the contract work and which led to the takeover agreement of 28 June 1967, by which appellant agreed to fund its operations and to provide management and other assistance. . . . Although the action taken by appellant was swift and effective in eliminating PECI's financial problems, we consider that before the takeover, PECI would have been unable to perform any effective work, whether on the critical path or not, even if the Government's stop work order had not been issued. Accordingly, the 28 days from 12 May to 9 June, as well as the 19 days between 9 June and 28 June 1967, represent a concurrent delay for which no compensation is allowable.

Several other concurrent delays are asserted by the Government. From 23 May through 6 June 1967, PECI closed its plant because of a strike. . . . On 6 June 1967, PECI moved some of its job site fabrication facilities to its main plant near Manila. . . . Since both of these delays fall within the period from 12 May to 28 June 1967, for which we have already denied liability, they require no further consideration.

We conclude that of the 161 days making up the delay period asserted by the Government, that is from 12 May to 20 October 1967, work on appellant's critical path was delayed by Government action from 28 June 1967 to the end of the period, a span of 114 days. The remaining 47 days within the delay claim period are neither allowable nor compensable for the reasons previously discussed.

C. The Impact Claim

This aspect of appellant's claim is grounded on the theory that the Government's suspension of work affecting the fabrication of critical tower legs disrupted the orderly sequence in which appellant had planned to fabricate and erect the towers. This disruption in turn created problems which led to appellant's unilateral work stoppage of 27 November 1967, which was not lifted until an expert could be summoned from the United States and appropriate remedial procedures devised. Appellant claims that the orderly progress of the work was delayed for three months, from 1 November 1967 to 1 February 1968, by the disruption of work so caused.

On the other hand, the Government argues that fabrication errors committed by appellant's fabrication subcontractor, PECI, were the cause of the disruption, and that since these errors were attributable solely to appellant and PECI the disruption is not chargeable to the Government.

The commission of numerous and serious fabrication errors by PECI is not in issue. Nor is it disputed that these errors caused the twisting of the towers when they were finally erected in late October and early November 1967, and which led to appellant's stop work order of 27 November 1967. Appellant does not allege that the specifications were defective, and there is no indication in the record to suggest that they were.

We have found that it is the practice in industry to erect towers of the sort called for in Project Bamboo in an orderly sequence. . . . We have also found that appellant's plan to fabricate and erect prototype towers, so as to prove out its fabrication and erection procedures, was frustrated by the unavailability of critical tower legs resulting from the Government's suspension of work. . . . The work sequence appellant had adopted was of its own devising, and not called for by the contract. However, we consider that, in the light of our findings, the adoption by appellant of such an orderly sequence was not only reasonable but also prudent. It is also plain that the Government's suspension of work was solely responsible for the frustration of appellant's plan of attack on tower erection.

Had appellant's plan not been so frustrated, it is reasonable to presume, as we do, that appellant would have detected the cause of the tower twisting during erection of the prototype towers, and would have been equally prompt and aggressive as it was in the present case in devising remedial measures that would have avoided the need for rework of tower legs built out of sequence and eliminated much of the disruption it suffered here. We make this presumption because of the nature of appellant's project manager, who, though often abrasive and exceedingly demanding, was always on top of the job. . . . Finally, if the remedial measures had been developed within a three month period after erection of the prototype tower, the delay would have been well within the five months' float time which appellant had allowed for PECI. . . .

The testimony at the hearing was conflicting on the question whether fabrication errors could be detected before the towers were erected and tensioned. The towers in Project Bamboo were unusually complex because of close tolerances and the tension applied by the guys. . . . Appellant's tower expert, who had considerable experience in the field, expressed the opinion that the fabrication problems experienced by PECI were beyond the range of what might be expected in an ordinary radio tower job, but only because of the unusual complexity of the towers in Project Bamboo. . . . On the other hand, the fabrication errors committed by PECI were of such nature and frequency that they could not have been solely the result of the complexity of the towers. With adequate quality control, we conclude that a significant portion of these errors could have been detected and avoided before the fabricated sections were taken to the site to be erected.

The allocation of responsibility for the disruption of work during the impact claim period depends on the extent to which we adopt the sequential theory advanced by appellant or the contrary theory that the disruption was occasioned solely by preventable fabrication errors on the part of PECI. Lacking the facts on which to make a more precise judgment, we decide, by jury verdict, on the basis of the record considered as a whole, that responsibility for the disruption of work during the impact claim period should be assigned to the two parties in a ratio of 40% to the Government and 60% to appellant.

Accordingly, of the total impact claim period from 1 November 1967 to 1 February 1968, a span of 92 days, we charge the Government with responsibility for disrupting appellant's work for 37 days. The balance of the period is the responsibility of appellant and hence not compensable.

D. The Acceleration Claim

During the delay and impact periods appellant suffered delays totalling 253 days that jeopardized the timely completion of Project Bamboo in accordance with its approved progress schedule. Some of these delays, as we have seen, were the result of inefficiencies and poor performance of its fabrication subcontractor, PECI, for which it alone was responsible. The remainder of the delays, amounting to 151 days, or 60% of the total, were caused by the steel seam problem and were such as to entitle appellant to a time extension which the Government refused to approve. Under these circumstances, appellant claims that it was compelled to accelerate so as to complete the work on time. The accelerated effort, it contends, lasted from 1 March 1968 until substantial completion of the work on 31 March 1969.

To this claim, the Government responds that it never approved appellant's requests for acceleration and took prompt steps after learning of appellant's intention to accelerate to deny any acceleration without adequate justification. It also argues that time extensions totalling 134 days for causes unrelated to the steel seam problem were sufficient to enable appellant to complete the work on time, as it eventually did.

The issue thus presented is whether appellant was compelled by action of the Government to accelerate the work, or whether, in the absence of an acceleration order from the Government, it voluntarily accelerated.

Thus far, we have decided that appellant was delayed in a total amount of 151 days by action of the Government relating to the steel seam problem in the delay and impact claim periods, a delay for which the Government consistently denied any time extension, beginning with its denial of 5 December 1967, and for which we held it responsible in our previous decision. . . . Since it was apparent that this delay affected steel fabrication and tower erection, work that was on the critical path to timely contract completion, the denial was, in the light of our previous decision, unreasonable. The productive work that appellant was otherwise able to complete during these periods no doubt had the effect of minimizing the acceleration that was required after the work stoppage ceased and appellant was able to proceed uninterruptedly with the erection of the towers after 1 March 1968.

The record shows that appellant successfully accelerated the work after 1 March 1968 in such a way as to complete the contract by 31 March 1969, the contract completion date as extended. . . . The degree of acceleration is indicated by our finding that appellant performed 90% of the work of tower erection in nine months, as compared with the 14 months' period that had originally been scheduled for accomplishment of the same amount of work. . . . There is nothing in the record to suggest that this acceleration was expressly ordered by the Government, and we have found that the contracting officer took prompt steps to make it clear that the Government did not want appellant to accelerate and would not pay for any acceleration. . . . However, at that very same time, the contracting officer was requiring appellant to submit a revised progress schedule showing how it proposed to complete the project within the time specified in the contract. . ., and he assumed that appellant was unable to meet the then revised contract completion date without acceleration. . . . Appellant continued to feel that it was being held to the original contract completion date until as late as 21 September 1968. . ., and we conclude that its position was reasonable under all of the circumstances.

As of 1 March 1968, the date on which appellant claims to have begun its acceleration, appellant had twice requested a time extension for delays growing out of the steel seam problem. . . . In addition, it had submitted a number of requests for time extensions based on other unrelated delays. Of the total of 134 days ultimately granted for these unrelated delays, we have found that 120 days did not run concurrently with the delay and impact claim periods. . . . Furthermore, the contracting officer summarily rejected most of the requests on 5 February 1968 and failed to grant any of appellant's requests within a reasonable time. . . .

In the months that followed, the parties discussed time extensions and acceleration further. The Government's attitude was to deny any liability for the steel seam problem and to refuse to grant any requests for the unrelated time extensions without additional justification. On the latter point, we agree that the contracting officer has an affirmative duty, under subparagraph (d) of the Default clause, to conduct a prompt review of requests for time extensions, together with the supporting justification, and to negotiate the matter within a reasonable time. . . . If the Government had discharged this duty, we are satisfied that the time extensions totalling 134 days would have been granted in a more timely fashion so as to permit appellant to make the necessary schedule adjustments.

We have found that the contracting officer consistently denied any time extension for delay caused by the steel seam problem and repeatedly failed to grant time extensions for other excusable delays within a reasonable time. . . . These denials and tardy approvals on the part of the contracting officer, we found, coupled with his insistence that appellant demonstrate, to his unspecified satisfaction, a causal connection between the events giving rise to the delays and the overall job scheduling, led appellant reasonably to believe that the requested time extensions would not be granted and that it would be held to the original contract completion date. . . . We have also found that although the contracting officer never formally assessed liquidated damages against appellant, his letter of 31 May 1968 implied that the Government might do so if appellant failed to meet the contract completion date. . . . In the face of such circumstances, we hold that appellant was compelled to accelerate in an attempt to meet the original contract completion date, and did not do so voluntarily.

Moreover, as a practical matter, if appellant had not accelerated, it would have suffered recoverable delay costs, including unabsorbed overhead, on account of the delay in project completion that would otherwise have resulted from the steel seam problem. Comparison of appellant's delay claim ($735,174) and acceleration claim ($154,516) shows that delay costs, including unabsorbed overhead less an allowance for productive work, would have considerably exceeded the cost of acceleration. Accordingly, under the circumstances of this case, we conclude that the effect of appellant's decision to accelerate was substantially to mitigate the damages resulting from the Government-caused steel seam problem.

For the foregoing reasons, we hold that this is a case of constructive acceleration for which appellant is entitled to be compensated to the extent of the delays chargeable to the Government during the delay and impact periods, that is, to 60% of whatever reasonable acceleration costs it can prove. See authorities discussed in Fermont Division, Dynamics Corporation of America, ASBCA No. 15806, 75-1 BCA ¶ 11,139, at pp. 52,966-53,000. . . .

F. The UCCI Claim

We have found that appellant's site work subcontractor, UCCI, was delayed in completion of its work for four months, that it incurred extra costs in maintaining roads during the rainy season because of that delay, and that it incurred certain move-in/move-out costs when its paving and grading subcontractor left the job because of the delay, all through no fault of UCCI. . . . The operative cause of the delay of UCCI's work was the delay in tower erection, since the UCCI work in question, namely, road paving and finish grading, necessarily followed tower erection in the schedule of events, and also because the rainy season which lasted through October prevented an earlier start of the work. . . . So far as the necessary causal connection between the steel seam problem and the UCCI claim is concerned, the latter is chargeable to the Government only to the extent that the Government is responsible for the delay in tower erection. To the extent that such delay is the fault of appellant, the claim is not compensable, since appellant's fault constitutes an intervening cause of the delay.

In our discussion of appellant's delay and impact claims, we have concluded that of delays of 161 and 92 days, respectively, or 253 days in all, a delay of 151 days resulted from the suspension of work ordered by the Government in relation to the steel seam problem and is compensable. This represents an allowance of approximately 60% of the total. Accordingly, the UCCI claim is compensable in the same proportion unless it is barred by the *Severin* doctrine, as the Government contends. . . .

G. The FMS Claim

In this claim, appellant seeks to recover for delays suffered by its subcontractor, FMS, in performing the task of antenna tuning and for the extra costs it incurred after accelerating the work from and after Steptember 1968. Notwithstanding the claimed acceleration, the performance of the FMS subcontract was extended from the scheduled completion date of November 1968 through January 1969. Appellant contends that FMS was delayed as a result of the delay in completing the erection of the towers and curtain antennas.

We have found that FMS was delayed for three months, from 1 November 1968 to 31 January 1969, because its antenna tuning program had to await completion of the erection of the towers and curtain antennas. . . . We have also found that there is no merit in the Government's argument that the delay experienced by FMS was the result of its underestimation of the work rather than the steel seam problem. . . .

In addition to the three-month delay suffered by FMS, we have found that FMS employed two additional crews and related automatic test equipment from September 1968 through January 1969 in order to accelerate the completion of the tuning program. . . . On the basis of the record considered as a whole, we decide that the acceleration effort of FMS was required by the Government's constructive acceleration order. . . .

The only remaining question is whether the extra costs claimed by FMS for delay and acceleration are the direct and necessary result of the suspension of work ordered by the Government in relation to the steel seam problem. On this point, we have decided that of the 253 days' delay and disruption of tower erection claimed by appellant, 151 days are the responsibility of the Government. Since the FMS claim is directly tied to appellant's delay in completing tower erection, and since the Government was responsible for 60% of that delay, we hold that appellant is entitled to recover on the FMS claim in the same proportion.

H. Conclusion

On the basis of the present record, we hold that the Government is liable to appellant for delays and disruption of appellant's work to the extent of 151 of the 253 days claimed as a result of the suspension of work ordered by the Government in relation to the steel seam problem. We also hold that the UCCI and FMS claims are allowable in the same proportion, since any delays to these subcontractors beyond 151 days are the responsibility of appellant rather than the Government.

In its claim of 13 December 1968, appellant sought a time extension of six months in addition to damages in the amount of $1,862,692. . . . Since appellant accelerated its work and that of its subcontractors from and after 1 March 1968, and achieved substantial completion of the work by the extended contract completion date of 31 March 1969, the case for a six months' time extension is now moot. However, we hold that appellant is entitled to recover whatever acceleration costs it may have incurred and can prove.

Finally, we hold that appellant is entitled to recover the extra-contractual costs of inspecting and repairing the Youngstown steel bars during the summer of 1967, in accordance with instructions received from the Government.

Insofar as the Government's liability is concerned, the appeal is sustained to the extent indicated. . . .

EXERCISES

1 Explain the special circumstances which would define and support a claim for an excusable delay.
2 What requirements are typically necessary for a delay to be compensable?
3 Why are network schedules used for delay analysis?
4 What requirements do courts typically impose on schedules if they are to be used for delay analysis?
5 Explain the primary differences between an as built and an as-planned delay analysis approach.
6 What is meant by the term "concurrent delay"? List several examples.
7 What methods are available to apportion concurrent delays? Explain.
8 List the excusable and unexcused delays discussed in the *Fischbach* case.
9 In *Fischbach*, could the board have calculated the length of delay differently? How?

CHAPTER
13

CPM
SCHEDULING
ON
MICRO
COMPUTERS

INTRODUCTION

Within the last few years the micro computer has become a commonplace tool to assist the scheduler with the complex and time-consuming calculations involved in determining schedule dates and other information related to the CPM scheduling process. A number of commercially available software programs have been developed to assist with this process. These software programs work on a variety of hardware configurations. This chapter explores the hardware and the software used in scheduling and provides an overview for the novice concerning the computer needs to support an ongoing scheduling function.

MICRO-COMPUTER HARDWARE

The physical parts that make up a computer are called the "hardware." A computer's hardware can be divided into three categories:

The processor and its associated memory
Data storage devices
Input and output devices

Central Processing Unit

The "central processing unit" is the main working component of the computer. The central processing unit (CPU) receives instructions from the operating program in the computer. The information is transmitted electronically in groups or "bits" of 8-, 16-, or 32-bit patterns. The CPU stores the bit patterns in various memory locations. The maximum number of memory locations available for use with a particular type of micro computer is determined by the design of the computer. The early computers were 8-bit machines. Each location consisted of 8 bits, which equals 1 byte. One kilobyte (1 KB) is 1024 bytes. The standard for the early micro computers was 64 KB of memory. Today most micro computers are either 16-bit or 32-bit computers. These central processing units are capable of processing more information faster than the earlier computers. It can be expected that the micro computers of the future will continue to show dramatic improvements in their ability to process larger amounts of information at faster rates.

The "operating system" is the set of instructions held in the computer which runs the application programs and provides commands to drive the various input and output devices. Before purchasing a computer, one should make sure that the operating system supports the types of software which will be used on that computer. There are several types of operating systems used by micro computers, and these are also being improved continually. The two early operating systems used by micro computers were the PC disk operating system (DOS) and MS DOS operating system. Today, new operating systems are being developed to speed the processing of information and provide more flexibility for control of application programs.

The central processing unit usually includes two types of storage. The processor has a "read-only memory" (ROM), which places certain instructions permanently in the machine, and a "random-access memory" (RAM), which holds instructions that are only temporarily stored. Memory storage in the central processing unit is seldom sufficient to hold all of the data and programs necessary for the various scheduling applications. The central processing unit memory is intended to provide an operating inventory for program and data. The central processing unit is analogous to a small factory with a number of machines that perform specialized operations. The various machines are similar to the processors. The memory in a micro-computer central processing unit is analogous to the storage stock piles that are placed near the machine in the factory and stock piles or space that is provided for stock piles around the machines for in-process goods and finished goods.

Memory

The central processing unit does not typically provide sufficient space for all the data with which the scheduler might typically need to work over the life of the computer. Additional storage for data and programs, known as "auxiliary storage," is required. Auxiliary storage holds data or programs that are not required for immediate use or that may be used only intermittently. Data from auxiliary storage can be fed to the central processing unit when required, and processed data or information and programs can be fed directly back to auxiliary storage when required.

There are several common types of auxiliary storage that can be used for micro computers. Two common types of auxiliary storage used with micro computers are "floppy disks" and "hard disks." The floppy disk looks like a small phonograph record. Floppy disks are typically $5\frac{1}{4}$ inches in diameter. These floppy disks are relatively inexpensive, costing less than $3. Another form of the floppy disk is a smaller, $3\frac{1}{2}$-inch plastic disk module used on the newer versions of micro computers. These disks have more storage space compressed into a smaller area and are easier to transport without fear of damaging the disk. The $3\frac{1}{2}$-inch floppy disks are slightly more expensive but still relatively cheap in comparison with other data storage mechanisms.

The "hard disk" is another type of storage device. It is a fixed part of the computer and allows more storage than floppy disks. Hard disks are more expensive because they are a hardware item. Hard disks are also faster for the transfer of data into and out of the central processing unit.

Other forms of auxiliary storage include magnetic tape and laser disks. Magnetic tape, similar to that used in tape recorders, can be used to store large quantities of data from files in the computer for later retrieval. Laser disks are a new technology that is being applied to the computer arena to store large amounts of data for copying and retrieval into the central processing unit. The hard disk, floppy disks, and tape data storage can be equated with the small factory's adjacent warehouse, its local distribution warehouse, and its faraway, long-storage warehouse.

Input and Output Devices

Data or instructions are input to the central processing unit by some type of input device. There are several types of commonly used input devices, including:

Keyboard
Mouse
Digitizer
Microphone/voice-recognition system

The keyboard is the most common device used to input data and instructions. The keyboard is similar to that found on a typewriter but also contains a numeric pad to one side and function keys across the top or down the other side. The keyboard is extremely versatile and easy to use to input all types of numeric and alphabetic information.

A mouse is a device that has been developed to input data by coordinating a reference point on a horizontal surface with a reference point on a display screen to permit use of a menu on the display screen to select various choices for input. The digitizer serves a purpose similar to the mouse but is usually a self-contained horizontal surface. The digitizer has a built-in wired coordinate system that can convey information of relative location to the central processing unit. When this digitizer board is referenced to a screen menu or a screen template, it can provide a means of inputting data based on the relative position at the point in time when a point is selected.

Increasingly used to input data are microphone devices in conjunction with voice recognition to permit audio input from the user. The microphone devices convert audio input into a keyboard type of command to the central processing unit. As these devices become more sophisticated they will eventually replace the keyboard, mouse, and digitizer as the simplest form of input device. Currently in scheduling programs, however, the most predominant input device remains the keyboard.

There are also several types of output devices used to display information that has been created by the central processing unit. The output devices include cathode-ray tubes, printers, plotters, and voice synthesizers. The cathode-ray

FIGURE 13-1   Dell System 310.

tube is a screen much like a typical television screen that displays information for the user to see. These cathode tubes provide varying qualities of display. The two most common types of display units are the EGA enhanced graphics and the VGA screens. The screens provide a temporary means of output but do not provide a permanent record of the output. Video screens have limited utility for the scheduler, because they cannot usually display all the data needed to study interrelationships among activities. Video monitors are used primarily to define individual activities. Figure 13-1 shows a central processing unit, CRT screen, keyboard, and mouse for a micro computer.

The use of a printer is essential, both when permanent records of the computer output are required and when interrelationships among activities must be studied. Printers come in a variety of types and quality, ranging from a dot matrix printer to a laser printer. These printers also come in a variety of widths to accommodate various types of paper and media for permanent output. Printers are extremely good at producing output of tabular data and very simple graphics. The quality and speed of the printer determine the cost of the printing device. Schedulers usually require faster printers to facilitate transfer of the calculated data to the field managers. Figure 13-2 shows a wide-carriage printer used to produce CPM printouts.

Plotters are another kind of output device. Plotters are similar to printers but can work with a variety of sizes of paper and operate at very different speeds. Plotters are typically used for graphic output rather than alphabetic or numeric output. Plotters are most useful to draw logic diagrams and bar charts for the network. Figure 13-3 shows a plotter used to produce logic diagrams.

FIGURE 13-2   A Hewlett Packard printer.

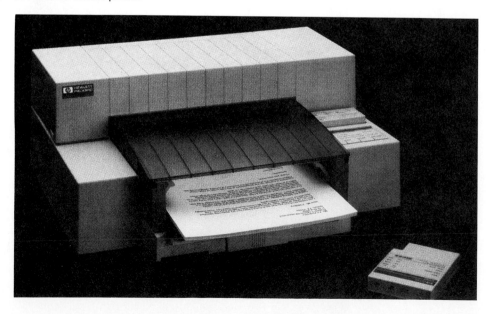

FIGURE 13-3 Hewlett Packard drafting plotter models 7580B, 7585B, and 7586B.

A fourth type of output device is the voice synthesizer. The voice synthesizer can produce audio output that can enhance or augment other forms of output such as that shown on a screen or provided with a plotter. The voice synthesizer is seldom seen with scheduling programs, but may be used in conjunction with programs that use a voice-recognition system for input.

Hardware for Scheduling Systems

Hardware for scheduling should not be chosen until a determination has been made which software will be used for scheduling and any other associated software that may be used by the scheduler has also been selected. Other software might include spreadsheet software and word processing software that could be used in conjunction with the scheduling system software to provide high-quality narrative reports. All scheduling system software will specify the minimum hardware configuration required to utilize the features of that software. The scheduling system software will specify this hardware in its user manual.

Computer hardware is associated commonly only with the CPU, the video monitor, and the keyboard. However, some type of a printing device is required for schedules, and in most cases a plotter will also be necessary to produce the desired quality of logic diagrams.

In addition to minimum hardware requirements, the scheduler needs to assess the amount of use that will be required of the scheduling system and determine the acceptable limits of processing speed that can be accommodated. If the central processing unit is too slow, a large amount of time will be tied up waiting for the computer to calculate the desired results. Similarly, if large reports are required it will be necessary to have a fast printer available. Also, the desired quality of printout and types of printout that might be required will influence the selection of the hardware for the system.

The size of projects that may be anticipated will determine the amount of computer memory required. Because programs differ in the way they store data, there is not a simple solution to determining the amount of memory storage that is required for a network of a particular size. For each software, certain guides will be provided in the literature as to the amount of storage space that is required for a particular size of network.

EVALUATING AVAILABLE SOFTWARE

This section will not attempt to actually perform an evaluation of software, but rather will focus on a methodology for evaluating various types of software.

Software is changing so rapidly and being improved at such a pace that it is imperative that software evaluations be made with the most recent version of software packages. Additionally, the evaluation of scheduling software should include the scheduler and staff.

The evaluation of the software should include analysis of a number of factors:

Input characteristics
Output characteristics
Interfacing capability
Format characteristics
Cost

In evaluating scheduling software, the types of devices that can be used for inputting data should be identified and the skill level required to input data and to validate the inputs should also be analyzed. The novice scheduler and data processor will require simple instructions with fewer choices than an experienced scheduler. Too many variables or too much flexibility will confuse the novice and delay schedule preparation.

Scheduling software literature will refer to interaction between files, records, and data sets. A "record" is a logical grouping of related fields or areas of information. A logical grouping of records may be called either a "file" or a "data set." A collection of data sets or files with specified relationships between them is referred to as a "database."

According to a recent survey there are more than 200 micro-computer-based project management or scheduling systems from which one may choose when purchasing network scheduling software.[1] When selecting software, one must understand how the software implements a network schedule and potential scheduling problems which may occur during preparation of the schedule. This section describes specific technical aspects of micro-computer-based scheduling software to provide an understanding of the operation of these programs.

All software is accompanied by documentation which explains how to install and activate the program. The documentation usually includes reproduced "menus," which will also appear on the monitor's screen. The menus provide the features of the software system and list the choices for use of the software's features. A well-documented scheduling software system will provide the user with descriptions of each menu and the outcome of each selection. The first option which must be selected to create a new network is usually called "add a new project" on the menu. There are generally three types of menus which may be used in scheduling software: standard menus, top- or bottom-line menus, and pull-down menus.

Standard menus are recognized as a list, usually in the middle of the computer screen. The user is prompted at the bottom of the screen to make a selection. To select one of the options available, a number or letter corresponding to one of the selections is pressed. As with every style of menu, this selection may either prepare the system to accept information from the user or access other menus.

Top- or bottom-line menus list their options on either the top or bottom two lines of the computer screen. The pull-down menu is a form of top or bottom menu, but more graphically interesting and may be used with a mouse. The user selects the action to be performed in two steps. The first is to move a highlighted bar or mouse over the top line, where the general menu is typically located. Once in place, another menu "pulls down" and appears. As the highlighted bar or mouse is moved, the menu disappears.

Once the user has executed the "add a new project" command, the user must usually identify a project number. Most systems require a four-character project number. Careful consideration of project numbers is necessary if the project may span several years. The user must also consider the way the schedule will be updated when assigning a project number. Some systems require a new project number for each update of the schedule.

All activity data are entered into the database through data entry screens. Screens are presented in a wide variety of formats. More complex programs allow users to customize screens to their own requirements. For the infrequent user the more powerful programs may not be suitable if customization is required for easy access to activity data.

The selection of a single activity for modification on the screen is usually accomplished by entering the activity's identification number and pressing a key or a series of keys which instruct the program to find and display the

[1] Kenneth Stepman, *A Buyer's Guide to Project Management Software*, Milwaukee: New Issues, Inc., 1986.

activity's information. When a large number of related activities are modified, this process is tedious and requires the user to constantly refer to a printed report to find each related activity identification number. To assist the user in these types of operations, software vendors have incorporated a number of time-saving features in data-retrieved screens.

Several programs allow the user to select a group of activities to be accessed. For example, to update the schedule the user may select all activities scheduled to begin on a certain date. The essential element is the ability to "select" and "sort" information according to conditions supplied by the user. The ability to search for a subset of activities selected is extremely useful and should be a feature included in the scheduling software selected. Some programs allow the user to scroll through and access all activities within a table instead of performing computer searches for related or particular activities. These "table editors," while useful, are not as helpful as the ability to search for particular activities, since the user must look instead of the computer.

Scheduling software provides several tools which allow the user to define the hours of work each day, workdays in a week, and specialized calendars. Some programs allow shift work to be scheduled over a 24-hour period. Although this type of customization may not be needed on all projects, some may need this flexibility and should be part of the user's selection criteria. A variable work week is another feature to be considered. Some programs allow the user to provide different calendars for different types of work. While useful on a complex job, simple projects may not need this complexity.

To increase the computer's ability to search for a group of related activities, the user may code activities in the activity description to permit more field searches. Assignment of codes determines the degree of analysis possible. There is a wide variation in possible coding systems among available scheduling software. A few allow the user to create any codes necessary, provided the total number of characters does not exceed a fixed ceiling. This type of coding is the most useful because it permits communication in words.

All programs provide a minimum of responsibility and work area codes. These two codes generally appear on the activity information data entry form. These fields may be used for any code desired, despite the labels "responsibility" or "location." Other systems provide fields labeled "code 1" or "code 2," etc. The user may define the code, but the label code 1 is used. A code dictionary is needed by all involved in the schedule to make this system work. No codes are defined by the software, but the user must anticipate the important factors in the schedule and code them appropriately.

Many software systems provide the ability to allocate resources to an activity and algorithms to assist in planning the most efficient use of the resources. The most versatile use of resource codes allows the contractor to create a "library." Libraries are a user-defined set of information. The library may contain production rates, materials, or equipment required. Once crews are assigned to specific activities, then the crew schedule, material lists, equipment lists, and cost-related reports may be generated. The ability to study the utilization of resources on a project is important.

The sequence in which data are input can have a significant impact on the time required for inputting data. Some software may use a sequential menu system for inputting each element of data. This can extend significantly the time involved in inputting data. Software with sequential menu systems may have very rigid requirements with respect to fields for inputting data, while others may have more flexibility in the field in which data can be input and the formats with which data can be input. Figures 13-4a and 13-4b show sequential menu screens.

Flexibility may permit the data processor to skip steps or develop new, quicker data entry fields, while rigidity will require each step to be completed

FIGURE 13-4   Example of sequential screens: AlderGraf Version 4.2, Level 4G.

```
04 MAY 1990              ENTER OR CHANGE ACTIVITIES/RESOURCES    A,4,A-CPU100-4.3
CCL CONSTRUCTION CONSULTANTS
    JOB NUMBER SAMPLE CPM   SAMPLE PRECEDENCE CPM SCHEDULE
                                                    ┌───────GRAPHICS───────┐
        ACTIVITY - [        13]                     │ Pen [ ]   Fill [ ]   │   LINE
                                                    └──────────────────────┘
                    ....    ....v    ....    ...v         ....    ....
    1=DESCRIPTION -[SUBMIT PAVEMENT MIX DESIGN                              ]
    2=DURATION     -[  10] DAYS
    3=ORGNZL CODE -[GC  ] [PROC]           20=ACTUAL START  -[MMDDYY  ]  ACTL DUR
    4=IMPSD START -[MMDDYY  ]              21=ACTUAL FINISH -[MMDDYY  ]  [      ]
    5=IMPSD FIN   -[MMDDYY  ]              22=% COMPLETE    -[    ]
                                          23=DAYS REMAINING-[  10]
    6=           -[         ]             24=            -[    ]
    7=           -[         ]             25=            -[    ]
    8=           -[         ]             26=            -[    ]
    9=           -[         ]             27=            -[    ]
   10=           -[         ]             28=            -[    ]

ALL OK? (Y/N/C/D):[ ]      ITEM#:[ ]
      ENTER TIES? [ ] (Y/N)


Clear date-<F1>  Activ notes--<F3>              Prev-<F7>  Main Menu--<F9>
Set Mlstn--<F2>  Chg Pen/Fill-<F4>              Next-<F8>  Calculate--<F10>
                                (a)

04 MAY 1990                                          A,4,A,T-CPU100-4.3
CCL CONSTRUCTION CONSULTANTS
    JOB NUMBER SAMPLE CPM   SAMPLE PRECEDENCE CPM SCHEDULE

  ACTIVITY - [        13] DUR-[ 10]     CODE-[GC  PROC] LINE-[    ]
   DESCRIPTION - [SUBMIT PAVEMENT MIX DESIGN                       ]

P:F-S        10   0          START PROCUREMENT
S:F-S        23   0          APPROVE PAVEMENT MIX DESIGN




A/1/2/3/D/M/Esc:  Pred,Succ St,Fin  St,Fin     Activity#     Lead/Lag
    [A]             [.]  :  [.]  -  [.]     [..........]  [....]
1=ADD P:F-S
2=ADD P:S-S
3=ADD P:F-F

PREVIOUS ACTIVITY - <F7>            NEXT ACTIVITY - <F8>
                                (b)
```

before the next step can begin. However, the greater the flexibility, the greater the complexity of the software and the increased potential for errors and confusion. Figure 13-5 shows a flexible menu screen.

These types of characteristics need to be fully understood and evaluated when comparing software packages. Because the input process is time-consuming, yet is only an intermediate step for the use of scheduling software, the amount of time required for input is a critical decision element in selecting a software system.

There are several output characteristics that also need to be evaluated. These include the flexibility in output that is possible with the software package; the combination of output that is possible; the graphic and tabular options that are possible; and the format of the output that the software system produces. With some software packages only a limited number of output reports are possible. Other software packages provide flexible reports for output.

The format in which the results are presented is also of interest and should be evaluated. Software packages vary in the type and quality of graphics they can produce from the input data. Along the same line, the output devices that the software supports vary substantially from one software system to another. Where it is desired to produce multicolor or high-quality output, the software must be evaluated to ensure that the system is designed to support printers and plotters capable of producing the quality and the size of desired reports and plots.

Similarly, the speed with which the software can produce outputs can also be a significant cost factor, as will be discussed later.

The interfacing capability should also be evaluated for a scheduling software system. In some cases it is necessary to select a scheduling system that is

FIGURE 13-5   Example of a flexible screen: AlderGraf Version 4.2, Level 4G, and Dbase III Plus Version 1.0.

```
ACTIV----- DESC_1--------------------------------- O1_4 O5_8 MDUR
        2 NOTICE TO PROCEED 04/02/90               OWN           0
        4 MOBILIZATION                             GC            2
       10 START PROCUREMENT                        GC            0
       11 SUBMIT DRAINAGE STRUCTURE REINF. STEEL   GC    PROC    3
       12 SUBMIT BASE GRADATION                    GC    PROC    5
       13 SUBMIT PAVEMENT MIX DESIGN               GC    PROC   10
       21 APPROVE DRAINAGE STRUCTURE REINF. STEEL  OWN   PROC   10
       22 APPROVE BASE GRADATION                   OWN   PROC   10
       23 APPROVE PAVEMENT MIX DESIGN              OWN   PROC   10
       31 FAB/DEL DRAINAGE STRUCTURE REINF. STEEL  GC    PROC    4
       32 FAB/DEL BASE AGGREGATE                   GC    PROC    2
       33 FAB/DEL PAVEMENT/ASPHALT                 GC    PROC    5
       50 BEGIN CLEAR & GRUB (0+00 TO 20+00)       GC            2
       52 COMP CLEAR & GRUB (20+00 TO 120+00)      GC           10
       54 POLE RELOCATION                          UTIL          2
       56 BEGIN EXCAVATION (0+00 TO 96+00)         GC           12
       58 COMP. EXCAVATION (96+00 TO 120+00)       GC            3


BROWSE           ║<C:>║ACTFLE              ║Rec: 1/26              ║      ║NumCaps
                            View and edit fields.
```

compatible with other users or with other parties involved. In such a case, it is usually necessary to either select the same software or to select a system that can create a data structure that can be read by other software packages. Compatibility among scheduling software systems is important when the scheduler is different from the schedule user or when interaction among multiple locations is required, such as between contractor field and management offices. Some systems allow data to be input from a floppy disk which may be copied from another file. Currently, however, there is no standard file format for data exchange among different systems.

Another aspect of the interfacing capability that is becoming more and more pronounced as the construction industry moves toward more integrated project management is the ability of the systems to interface with other software programs. Quite typical today is the need for software scheduling systems to interface with database or spreadsheet software such as accounting, estimating, word processing, or simulation software. A recent development has been the interfacing of scheduling software with expert system shell software and with computerized design software. While attempts at complete integration are in their infancy, it is expected that in the future, complete system integration will become a key criterion in the selection of scheduling software.

There are several formatting characteristics that must be evaluated to ensure that the software will match the particular application or range of applications for which it will be utilized. These formatting characteristics include such things as the scheduling method to be employed, the size limits for the schedule, the necessity and extent of resource loading that may be required, the project control characteristics, and the number of calendars that will be utilized. If it is anticipated that both arrow diagramming and precedence diagramming methods will be employed, then the scheduling system should be capable of dealing with both methods. In some cases it may be necessary to establish a PERT system as a criterion. Not all I-J and PDM systems will also permit PERT.

Most software system literature will define the size limits. These are usually expressed in terms of the maximum number of activities or number of activities and number of relationships that the software is capable of accepting. The user needs to have an estimate of the number of activities that are expected on the largest of projects and to this should add a sizable contingency factor to ensure that the selected software will have enough capability available. In some software the size limits can be further expanded through the use of fragnets or subnets, which break a large project into several smaller projects.

If resource loading and resource leveling are needed, these features should be evaluated for the scheduling software system. The method of inputting resources can influence the amount of time that is required for input and may also affect the ability to use this feature economically. Similarly, with resource leveling the algorithm that is used can be quite time-consuming and complex for very large projects. Extended calculation time may delay completion of the schedule and prevent the use of the resource-leveling feature. This aspect of resource leveling should also be carefully evaluated.

When the scheduling software will be used as a control tool, it is important to evaluate the way in which the software is updated and the methods of calculation used to determine progress and percent complete. Additionally, if the software will be used for cost control, the way in which the costs are input and used to define status is of concern. This format characteristic should be evaluated to determine that it is consistent with the way in which the firm wishes to calculate these items.

For companies that work worldwide, it is necessary to have a variety of calendars available, as work weeks differ throughout the world. Even for a simple project it is possible to require a separate calendar for various crafts working on the project. Thus the number of available calendars can affect the accuracy of the scheduled dates for a particular craft. It is usually desirable to have as large a number of calendars as possible so as to be able to create realistic assessments of actual calendar dates from workday calculations in the scheduling software.

Another important aspect of scheduling software to be considered is the way the software uses the date of the update, remaining duration, and actual start and finish dates to calculate progress. Although progress-measuring approaches may be valid, the different approaches provide different answers because they evaluate different items.

An activity may be considered "in progress" if the remaining duration is between the original duration and zero. Although remaining duration is used frequently as the indication of progress, another way is to indicate expended duration. Time-based percent complete is an important concept and is often used interchangeably with remaining or expended duration, although different totals result from the different calculation methods (also significant if activities are dollar-loaded).

Although actual dates may be entered into the system, the exact date is not always known by the scheduler. Software systems differ widely in their method of calculating progress if actual dates are not provided. There are two situations which may arise if actual dates are not provided. The first occurs when the date of the update is the same date as an activity's early start date. In this case the program assigns the update date as the actual start date and calculates the early finish date as the actual start date plus the remaining duration indicated by the scheduler. Previously calculated early start dates and original durations are disregarded. The default calculation used in the first situation masks actual progress, since the update date is usually not indicated as a substitute for the actual date.

In the second situation, the actual start date of an activity precedes the date of the update. Two different scheduling algorithms are possible, depending on the software. In one algorithm the actual start date is considered to be the update date and the early finish is determined to be the sum of the update date and the remaining duration. The second algorithm calculates a totally different early finish date. This algorithm presumes the actual start date to be the previously calculated early start date. The early finish is the sum of the early start date and the remaining duration, earlier than the first algorithm by the difference between the early start date and the date of the update. If the difference is larger than the remaining duration, some software will indicate that the activity has been completed. The user must understand how the software uses default dates if actual dates are not known. The understanding is particularly important if the schedule is to be used to justify time extensions or allocate delay.

It is common for actual construction sequence to differ from initial network schedules as contractors add, delete, or discover better methodology. The realities of field construction may also permit two activities scheduled sequentially to be performed simultaneously. This is termed "out-of-sequence" work. Software systems generally allow one of two ways to show out-of-sequence work. The first method calculates the early start of the first activity by adding the remaining duration to the date of the update, sets a temporary start date for the second activity to be the previous activity's early start, and calculates the second activity's early finish as the sum of the temporary start date and the second activity's remaining duration. Figure 13-6 shows this calculation method.

The second method calculates the early finish of the second activity by adding the remaining duration of the out-of-sequence activity to the data date. This method ignores the logical restraints imposed by the first activity. Figure 13-7 shows this calculation method.

Users need to understand how the software system displays out-of-sequence work in order to determine what actions must be taken to modify the logic diagram which has out of sequence work.

Different scheduling software systems also utilize "plug" dates differently. There are three different levels of restraint which plug dates may impose on a schedule. The minimum control is the "target date." Target dates do not affect the schedule but only provide a reference for comparison. The intermediate

FIGURE 13-6   First method of displaying out-of-sequence work.

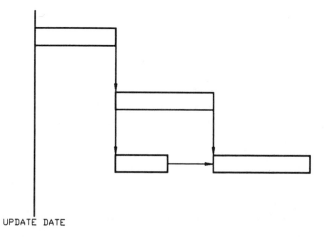

UPDATE DATE

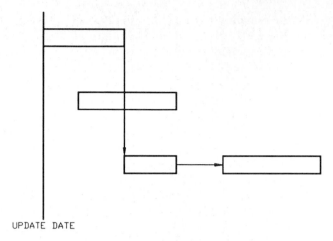

UPDATE DATE

FIGURE 13-7   Second method of displaying out-of-sequence work.

level of control forces use of a plug date only if the calculated date goes beyond the necessary plug date. The third level of restraint forces the schedule to show a plug date regardless of the calculated date. When a conflict arises, the software will ignore the network logic and utilize the plug date. In some circumstances this may force an activity to show complete on the schedule prior to its calculated start date. Users need to be concerned with the potential disruption to the schedule when using this third level of plug date.

The ability to provide time-scaled network diagrams is also an important consideration when evaluating scheduling software. Some software systems require additional programs and expensive equipment to drive a plotter. Other programs limit the number of activities which may be plotted. Even the manner in which the logic diagram is drawn varies widely. The user needs to review all aspects if ability to plot is an important consideration.

The final yet ultimate component that needs to be evaluated is the overall cost of the scheduling system. This cost has three major components: the software cost, the hardware cost, and the cost of operating the software. The software cost includes the purchase price of the software along with the annual support cost that may be necessary to secure updated versions of the software and to provide technical support for questions related to the use of the software. In addition, training at a remote location may be necessary for users of the software to be able to fully utilize the features related to the scheduling system. It must be determined whether the training cost is included in the price of the software, and if not, an estimate of the travel and attendance fee should be considered.

With each software system there is a minimum configuration of hardware that is required. This configuration is unique for each software package and results in a unique set of costs. To the extent that the hardware will be used solely for the scheduling system, this hardware cost should be assigned to the cost of the scheduling system. If the hardware is a part of the configuration that can be used and will be used for other functions, a pro-rated portion of the hardware cost should be assigned to the cost of the scheduling system. Where various hardware configurations might work, an evaluation of the time savings associated with higher-cost, faster-processing units should be made and the optimal solution chosen. Further, hardware that is flexible for several uses may require a higher investment to gain all of the features needed for the various software packages on which it will be used. Additional costs that do not reflect needs of the scheduling system should not be allocated to the cost of the hardware system for schedule.

Finally, an assessment of the operations cost is necessary for use in evaluating which scheduling system is the most cost-effective.

It is suggested that a typical network be utilized for a trial run among several software systems. This affords an opportunity for comparison of input time, output characteristics and time, and the technical features of the software among the various software packages being evaluated. Table 13-1 is a sample network that can be used to evaluate the capabilities of various software packages and determine the amount of time required to input, process, and provide output for the scheduling system.

TABLE 13-1 SAMPLE NETWORK FOR EVALUATING SOFTWARE

Sample project activity list and relevant resources*

I	J	Duration in days	Description	Activity code	Predecessor code	Follower code	Rel. type	Lead time	Resources and cost categories
Phase I: Site preparation and utilities									
1	3	0	Notice to proceed (milestone)	1000		1010			
3	7	3	Clear site	1010	1000	1020	S-S	2	LA.5 E01 D01 $P987
7	9	2	Survey and lay out	1020	1010	1030	F-S		LA1 CP2 $P618
9	11	2	Rough grade	1030	1020	1040	F-S		E01 LA.5 D01 $P521
						1050	F-S		
						1070	F-S		
						1080	F-S		
11	13	1	Excavate electrical manholes	1040	1030	1090	F-S		LA2 E01 BL1 $P745
11	19	4	Water tank foundation	1050	1030	1100	F-S		LA4 SW.25 E0.25 EV.25 CN.25 RD2 CP2 PT1 $P1946
11	15	10	Excavate sewer	1070	1030	1110	F-S		LA1 E01 TC1 $P548
11	17	15	Drill well	1080	1030	1130	F-S		LA5 DR1 $P1108
13	21	5	Install electrical manholes	1090	1040	1140	F-S		LA1 E0.5 EL2 CE.5 $P892
15	21	5	Install sewer and backfill	1110	1070	1140	F-S		LA1 PL2 $P684
17	23	2	Install well pump	1130	1080	1150	F-S		LA1 PL2 $P684
19	29	10	Install water tank	1100	1050	1120	F-S		LA1 CE.5 SW2 $P902
21	25	3	Install electrical ductbank	1140	1090	1160	F-S		LA3 RD2 CP1 PT1 $P1281
					1110				
23	27	8	Underground water piping	1150	1130	1170	F-S		PL2 LA1 $P684
25	31	5	Pull in power feeder	1160	1140	1180	F-S		EL2 $P506
27	31	2	Connect water pipe	1170	1150	1180	F-S		PL2 LA1 $P684
29	31	10	Tank piping and valves	1120	1100	1180	F-S		PL2 LA1 $P684
31		0	Phase I completed (milestone)	1175	1170	1180			
Phase II: Plant and office foundation work									
31	33	1	Building layout	1180	1120	1190	F-S		CP2 PT1 LA1 $P634
					1160	1200	F-S		
					1170				
33	35	10	Drive and pour piles	1200	1180	1230	S-S	8	PD5 E02 001 CB1 LD1 HA1 AC1 AH1 $P2913
33	37	3	Excavate office building	1190	1180	1210	F-S		E02 001 EX1 LA.5 TD2 DT2 D01 $P2996
						1230	F-S		

i	j	Dur	Description	t1	t2	t3	Rel	Lag	Resource loading
35	39	5	Excavate plant building	1230	1190	1260	S-S	3	E02 001 EX1 LA.5 TD2 DT2 D01 $P2996
37	35	0	DUMMY		1200				
37	41	4	Spread office footings	1210	1190	1220	F-S		LA3 CP2 RD2 $P1480
39	43	5	Pour pile caps	1260	1230	1290	S-S	3	SW4 LA1 CC1 $P1323
41	45	6	Form and pour office grade beams	1220	1210	1240	F-S		LA4 E0.25 EV.5 CN.25 RD2 CP2 PT1 $P1886
43	47	10	Form and pour plant grade beams	1290	1220	1290	F-S		
						1300	S-S	9	LA4 E0.25 EV.5 CN.25 RD2 CP2 PT1 $P1886
45	43	0	DUMMY		1260				
45	49	1	Backfill and compact office	1240	1220	1310	F-S		LA3 $P527
47	51	3	Backfill and compact plant	1300	1290	1250	F-S		LA3 $P527
47	59	5	Form and pour truck loading dock	1310	1290	1320	F-S		LA3 RD2 CP1 PT1 $P1281
49	53	2	Underslab plumbing office	1250	1240	1340	F-S		PL2 LA1 $P684
51	55	5	Underground plumbing plant	1320	1320	1270	S-S		PL2 LA1 $P684
53	57	2	Underslab conduit office	1270	1300	1330	F-S		EL3 $P760
55	49	0	DUMMY		1250				
55	59	5	Underground conduit plant	1330	1330	1280	F-S		EL3 $P760
57	61	3	Form and pour slabs office	1280	1320	1270	S-S	3	LA4 SW.25 E0.25 EV.5 CN.25 RD2 CP2 PT1 $P1946
58	53	0	DUMMY		1270				
58	59	0	DUMMY		1310				
59	61	10	Form and pour slabs plant	1340	1330	1340	F-S		LA4 SW.25 E0.25 EV.5 CN.25 RD2 CP2 PT1 $P1946
61		0	*Phase II completed (milestone)*	1345	1330 1340	1350	F-S		

Phase III: Plant close-in

i	j	Dur	Description	t1	t2	t3	Rel	Lag	Resource loading
61	63	10	Install structure steel	1350	1280	1370	F-S		SB6 E01 WD1 001 CD1 WM1 T01 $P3238
63	67	3	Roof joists plant	1370	1350	1380	F-S		CP4 LA1 PT1 $P1077
						1390	F-S		
67	69	3	Roof decking plant	1380	1370	1400	F-S		CP4 LA3 PT2 $P1444

TABLE 13-1 Cont'd

I	J	Duration in days	Description	Activity code	Predecessor code	Follower code	Rel. type	Lead time	Resources and cost catagories
67	71	10	Install metal siding	1390	1370	1410 1430 1440 1470	F-S F-S F-S F-S		SW2 LA2 PT1 $P822
69	71	5	Built-up roofing	1400	1380	1410 1430 1440 1470	F-S F-S F-S F-S		CP4 LA3 AE1 $P1508
71		0	*Phase III completed (milestone)*	1405	1390	1470	F-S		

Phase IV: Plant building

I	J	Duration in days	Description	Activity code	Predecessor code	Follower code	Rel. type	Lead time	Resources and cost catagories
71	75	10	Interior masonry	1470	1390 1400	1500	S-S	9	BR2 LA2 $P802
71	77	5	Install exterior doors	1430	1390 1400	1550	F-S		CP2 SB2 $P969
71	85	15	Install heat and vent units	1440	1390 1400	1640	F-S		SM2 EL1 $P657
71	89	2	Install electrical load room	1410	1390 1400	1480	F-S		EL2 $P506
75	77	5	Install frame ceilings	1500	1470	1550 1640 1630 1620 1610	S-S S-S S-S F-S F-S	3	CP4 LA2 PT2 $P1268
77	87	8	Install dry wall	1550	1430 1500	1720	S-S	5	CP4 LA2 PT2 $P1268
85	97	15	Ductwork	1640	1440 1550	1690	S-S	12	SM2 $P404
87	85	8	Ceramic tile	1630	1550	1750	S-S	6	SW2 LA1 $P631
87	93	0	DUMMY						
87	111	5	Install interior doors	1620	1550	1560	F-S		CP2 LA1 $P619
89	91	2	Install power conduit	1480	1410	1610	F-S		EL3 $P760
91	93	0	DUMMY						
91	101	10	Pull wire	1560	1480	1660	F-S		EL2 $P506
93	103	5	Install outlets	1610	1550	1650	F-S		EL2 $P506
95	99	5	Paint rooms	1690	1630	1700 1710	S-S F-S	7	P02 LA1 $P590

97	111	10	Insulate heat and vent system	1720	1640	1750	F-S		LA2 $P351
99	105	10	Floor tile	1700	1690	1740	S-S	7	SW2 LA1 $P631
99	111	10	Install plumbing fixture	1710	1690	1750	F-S		PL2 $P508
101	107	3	Install panel internals	1660	1560	1680	F-S		EL1 $P253
103	109	10	Install electrical fixtures	1650	1610	1730	F-S		EL2 $P506
105	111	10	Install furnishings	1740	1700	1750	F-S		CP4 LA2 PT2 $P1268
107	109	10	Electrical trimout	1680	1660	1730	F-S		EL2 $P506
109	111	1	Connect power	1730	1650	1750	F-S		EL1 $P253
					1680				
111		0	*Phase IV completed (milestone)*	1745	1740	1750			

Phase V: Office building

111	113	5	Install precast columns and beams	1750	1620	1760	F-S		SW3 LA2 CE1 $P1130
					1710				
					1720				
					1730				
					1740				
113	115	5	Roof joists and decking	1760	1750	1770	F-S		CP4 LA2 PT2 $P1268
						1780	F-S		
						1790	F-S		
						1800	F-S		
115	117	10	Exterior masonry	1800	1760	1820	S-S	8	BR3 LA3 CP.25 $P1259
						1830	S-S	8	
						1840	F-S		
						1850	F-S		
115	119	5	Built-up roofing	1770	1760	1810	F-S		CP4 LA3 AE1 $P1508
						1820	F-S		
						1830	F-S		
						1840	F-S		
						1850	F-S		
115	121	5	Install exterior doors	1780	1760	1810	F-S		CP2 LA1 $P619
						1820	F-S		
						1830	F-S		
						1840	F-S		
						1850	F-S		
115	139	5	Install package AC units	1790	1760	1890	F-S		EL2 $P506
117	123	5	Glazing	1850	1770	1900	S-S		GL2 LA1 $P614
					1780				
					1800				
117	125	4	Install conduit	1820	1800	1860	F-S		EL2 $P506
					1770	1870	S-S	4	
					1780				

TABLE 13-1 Cont'd

I	J	Duration in days	Description	Activity code	Predecessor code	Follower code	Rel. type	Lead time	Resources and cost categories
117	127	5	Install piping	1830	1800 1770 1780	1870	S-S	4	PL2 $P508
117	133	10	Ductwork	1840	1800 1770 1780	1920	F-S		SM2 $P404
119	117	0	DUMMY						
119	145	5	Paint exterior	1810	1770 1780	1990	F-S		P02 LA1 $P590
121	119	0	DUMMY						
123	129	10	Plaster coat	1900	1850 1870	1920 1930 1940 1950	F-S S-S F-S F-S	8	PA2 GE1 LA1 $P706
125	127	0	DUMMY						
125	137	5	Pull wire	1860	1820	1880 1890	F-S F-S		EL2 $P506
127	123	5	Lath partitions	1870	1820 1830	1900	S-S S-S	4 4	LH2 $P424
129	131	4	Paint interior	1930	1900	1960 1970 1980	F-S F-S S-S	3	P02 LA1 $P590
129	133	0	DUMMY						
129	135	5	Ceramic tile	1940	1900	1970	F-S		SW2 LA1 $P631
129	145	5	Interior doors	1950	1900	1990	F-S		CP2 LA1 $P614
131	135	0	DUMMY						
131	143	0	DUMMY						
131	145	5	Floor tile	1960	1930	1990	F-S		SW2 LA1 $P631
133	143	5	Install ceiling grid	1920	1840 1900	1980	S-S	3	CP2 LA1 PT1 $P634
135	145	5	Install toilet fixtures	1970	1930 1940	1990	F-S		PL2 $P508
137	139	0	DUMMY						
137	141	3	Install panel internals	1880	1860	1910	F-S		EL1 $P253
139	141	3	Electrical connections	1890	1790 1860	1910	F-S		EL1 $P253
141	145	5	Electrical trimout	1910	1880 1890	1990	F-S		EL2 $P506
143	145	10	Install acoustic tiles	1980	1920 1930	1990	S-S S-S	3 3	CP2 LA1 PT1 $P634

i	j	Dur	Description	1985	1980	1990	Rel.	Resources
145	72	0	*Phase V completed* (milestone)					

Phase VI: Project outside work

i	j	Dur	Description	1985	1980	1990	Rel.	Resources
71	72	8	Access road	1520	1390	1490	F-S	LA6 E02 GR1 PM1 $P3134
72	111	3	Paving parking area	1490	1520 / 1400	1570	F-S	LA6 E02 GR1 PM1 $P3134
71	147	5	Sidewalks and curbs	1510	1390 / 1400	1570	F-S	SW2 LA2 CP2 $P1250
71	145	12	Area lighting	1540	1390 / 1400	990	F-S	EL2 $P506
71	153	10	Perimeter fencing	1530	1390 / 1400	1990	F-S	SW2 LA2 $P806
111	151	4	Fine grade	1570	1490 / 1510	1600	F-S	E02 GR1 TR1 $P1105
151	145	3	Finish landscape	1600	1570	1390	F-S	LA3 $P527
145		0	*Phase VI completed* (milestone)	1605	1600 / 1530 / 1540	1990		
145	155	3	Final project clean-up	1990	1530 / 1540 / 1500 / 1910 / 1970 / 1960 / 1950 / 1980 / 1810 / 1990	2000	F-S	LA4 TD.5 LT.5 $P822
155	157	0	*Project completion* (milestone)	2000	1990			

Y.C. Tang, The Validation of Scheduling Software Packages, class project, University of Kansas, 1985.

Sample Network Resource Data

Name	Full name	Type	Maximum available	Rate	Per	Accrual	Level
AC	Air compressor, 600 CFM	Variable cost		$ 249.70	1 day	Pro-rate	0
AE	Application equipment	Variable cost		95.00	1 day	Pro-rate	0
AH	Two 50-ft air hoses, 3' diam.	Variable cost		18.40	1 day	Pro-rate	0
BL	Backhoe loader, 80 HP	Resource	1.00	249.00	1 day	Pro-rate	1
BR	Brick layer	Resource	4.00	225.60	1 day	Pro-rate	3
CA	25-ton hyd. crane	Resource	1.00	419.80	1 day	Pro-rate	1
CB	40-ton crane	Resource	1.00	527.30	1 day	Pro-rate	1
CC	Crane, 80 ton, and tools	Resource	1.00	948.40	1 day	Pro-rate	1

Sample Network Resource Data

Name	Full name	Type	Maximum available	Rate	Per	Accrual	Level
CD	90-ton crane	Resource	1.00	860.20	1 day	Pro-rate	1
CE	S.P. crane, 5 ton	Resource	1.00	192.00	1 day	Pro-rate	1
CN	Conc. pump (small)	Variable cost		494.30	1 day	Pro-rate	0
CP	Carpenter	Resource	8.00	221.60	1 day	Pro-rate	3
DO	200-HP dozer	Resource	1.00	671.90	1 day	Pro-rate	1
DR	Truck and drill rig	Resource	1.00	229.50	1 day	Pro-rate	1
DT	Dumper truck	Resource	2.00	303.60	1 day	Pro-rate	1
EL	Electrician	Resource	5.00	253.20	1 day	Pro-rate	3
EO	Equipment operator	Resource	4.00	227.60	1 day	Pro-rate	4
EV	Two gas engine vibrators	Variable cost		54.50	1 day	Pro-rate	0
EX	1.5-CY hyd. excavator	Resource	1.00	617.20	1 day	Pro-rate	1
FE	F.E. Loader, T.M., 2.5 CY	Resource	1.00	634.70	1 day	Pro-rate	1
GE	Grouting equipment	Variable cost		249.00	1 day	Pro-rate	0
GL	Glazier	Resource	2.00	219.00	1 day	Pro-rate	3
GR	Grader, 30,000 lb	Resource	1.00	460.70	1 day	Pro-rate	1
HA	Hammer, 15K ft-lb	Variable cost		249.00	1 day	Pro-rate	0
HG	Vibratory hammer and generator	Variable cost		1,081.30	1 day	Pro-rate	0
LA	Construction laborers	Resource	9.00	175.60	1 day	Pro-rate	2
LD	60 LF loads 15K ft-lb	Variable cost		48.20	1 day	Pro-rate	0
LH	Lathers	Resource	2.00	212.00	1 day	Pro-rate	3
LT	Light truck	Variable cost		15.50	1 day	Pro-rate	1
MG	Mixing machine and grinder	Variable cost		124.80	1 day	Pro-rate	0
OO	Equipment operator, oiler	Resource	2.00	194.00	1 day	Pro-rate	4
PA	Plaster	Resource	3.00	212.00	1 day	Pro-rate	3
PD	Pile drivers	Resource	5.00	243.00	1 day	Pro-rate	4
PL	Plumber	Resource	6.00	254.00	1 day	Pro-rate	4
PM	Paving machine and equipment	Variable cost		1,164.90	1 day	Pro-rate	1
PO	Painter	Resource	2.00	207.20	1 day	Pro-rate	3
PT	Power tools	Variable cost		15.40	1 day	Pro-rate	0
RD	Rodmen (reinforcing)	Resource	6.00	254.80	1 day	Pro-rate	3
SB	Structure steel worker	Resource	6.00	262.80	1 day	Pro-rate	4
SM	Sheet metal worker	Resource	2.00	202.00	1 day	Pro-rate	4
SW	Skilled worker	Resource	6.00	227.60	1 day	Pro-rate	4
TC	Trencher chain, 12 HP	Variable cost		145.00	1 day	Pro-rate	1
TD	Truck driver	Resource	2.00	180.80	1 day	Pro-rate	4
TO	Torch, gas and air	Variable cost		50.60	1 day	Pro-rate	0
TR	Tandem roller, 10 ton	Variable cost		188.80	1 day	Pro-rate	1
WD	Welder	Resource	2.00	262.80	1 day	Pro-rate	4
WM	Gas weld machine	Variable cost		61.20	1 day	Pro-rate	0
$P	Cost in dollars						

CHAPTER 14

THE
SCHEDULING
SPECIFICATIONS

A SCHEDULING OR REPORTING REQUIREMENT?

Many owners and designers now recognize the potential benefits of a network schedule and include a network specification in their contracts. General contractors will also often include network scheduling requirements in their subcontracts. Whether the scheduling obligation is to be imposed on a subcontractor, general contractor, or designer, the scheduling specification should be as carefully considered and written as any other technical specification.

Unfortunately, the scheduling specification is not so carefully considered in many instances. Often, the designer's specification writer has never prepared or used a detailed construction schedule, but still prepares the scheduling clause (or copies one prepared earlier by another equally unfamiliar with schedules). As one can imagine, this is the major reason why scheduling clauses regularly make useless, inappropriate, or unnecessary demands.

A scheduling specification can attempt to fulfill one of two goals. A construction scheduling specification can either attempt to force a contractor to plan and manage a project or impose a detailed reporting requirement permitting the designer or owner to schedule the project. A scheduling specification used in the second way attempts to facilitate the designer or owner's administration function by requiring the contractor to use a method and program which not only exist in the designer's office but which also give the designer the information needed to prepare a schedule and record the contractor's actual progress.

Attempts to impose a planning and scheduling obligation on an unwilling contractor usually fail. If a contractor does not want to invest the time necessary to plan its work so that the contractor's home and field office can manage its performance or achieve any of the other advantages of scheduling, or if the contractor refuses to prepare its schedule in the form dictated by the specifications, the contractor can always find a reason to confound or avoid the obligation. The schedule can be submitted late, and even then, incomplete and with minimum detail. Required updates can be ignored. The schedule can be full of logic errors, violate good scheduling practice, and not reflect the contractor's actual intent. Even with obvious errors and omissions, the recalcitrant contractor can argue that its poor schedule should be accepted. Finally, the unwilling contractor can offer an attractive credit to release the obligation that an exasperated and frustrated designer may readily recommend the owner accept. When the designer decides that its efforts to make the contractor conform to the schedule specification are useless, even an owner committed to the scheduling process will usually agree to abandon the requirement.

Recognizing the difficulty of making an unwilling contractor schedule its work, many owners, including several U.S. government agencies, have attempted to make their scheduling specification more detailed and longer, and to include illustrations of how the contractor's completed schedule should appear. An "iterative" process has been adopted, beginning with contractor receipt of an "example" network from the agency or its scheduling consultant, which the contractor is expected to revise and return for review as many times as necessary to make the contractor accept its obligation to plan and schedule. The iterative process may require months before an approved schedule has been submitted. Quite often, the lengthy iterative process results only in increased fees for the scheduling consultant and the designer's recommendation that the network scheduling clause be changed to a bar chart. No matter how well prepared a scheduling specification may be, it is nearly impossible to force an unwilling contractor to schedule its work.

Recognizing the difficulty of forcing a contractor to schedule, some owners decide to schedule the project themselves without contractor input. An owner-prepared schedule has many advantages. An owner's organization may know the project better than a contractor, after participating in a development process that may have taken several years. An owner-prepared schedule may include milestone and interim completion dates that satisfy owner needs but that may not be included in a contractor-prepared schedule. An owner schedule may include contingencies for possible changes that the contractor may not recognize.

An owner-prepared schedule without contractor input also has many disadvantages. An owner-prepared schedule may not use the most economical method or trade sequence. The contractor may be better able to judge the most efficient duration for activities. An owner-prepared schedule may include an ill-advised or erroneous sequence or duration, which could be used to support claims for additional cost when the actual sequence or duration inevitably varies.

A better way is to impose a reporting requirement that defines what form the report is to take and what information is to be included to permit either the designer or owner to administer the contractor's progress. In this way, the contractor remains responsible for managing its own performance, the owner does not assume unnecessary responsibilities, but the information necessary to correct contractor scheduling errors and administer contractor progress is available. Once the specified information has been provided in the required manner, the designer gains the information necessary to prepare an accurate schedule and the information needed to follow the progress of the project or to anticipate any potential problems in a contractor's plan of action.

A reporting requirement can achieve these goals despite any contractor inclination to either ignore or frustrate the otherwise worthy effort to plan or schedule. Scheduling specifications prepared to define a reporting requirement have a better chance of success than clauses which attempt to impose an obligation to schedule. The difference between the two types of specifications is significant and require a philosophical acceptance of the impossibility of making anyone do many things against his or her wishes.

The most effective way to force a contractor to schedule is to encourage the contractor. A clause that elects to encourage a contractor to plan and schedule its work will be short and general. It will not attempt to impose a certain kind of schedule on the contractor. A clause that encourages contractor scheduling will permit the contractor to use the planning system with which the contractor feels most comfortable. A clause that attempts to impose a scheduling obligation will be sufficiently flexible to permit any schedule as long as the contractor indeed prepares and uses one. The disadvantage of a clause intended to encourage a contractor to schedule is in the brevity of the clause. A brief clause can be easily satisfied and risks that a contractor who does not want to schedule will ignore the scheduling obligation altogether by submitting a minimal schedule reflecting little or no effort.

On the other hand, a successful detailed clause will assume that the contractor will not schedule and concentrate on defining what information is necessary from the contractor to permit the designer to prepare the schedule or monitor the contractor's progress. A successful detailed clause will not concentrate on devising a procedure to ensure that the contractor cannot avoid the obligation to schedule.

The American Institute of Architects' Document A201 contains an example of a very brief scheduling specification:

3.10   Contractor's Construction Schedules

3.10.1   The Contractor, promptly after being awarded the Contract, shall prepare and submit for the Owner's and Architect's information a Contractor's construction schedule for the Work. The schedule shall not exceed time limits current under the Contract Documents, shall be revised at appropriate intervals as required by the conditions of the Work and Project, shall be related to the entire Project to the extent required by the Contract Documents, and shall provide for expeditious and practicable execution of the Work.

3.10.2   The Contractor shall prepare and keep current, for the Architect's approval, a schedule of submittals which is coordinated with the Contractor's construction schedule and allows the Architect reasonable time to review submittals.

3.10.3   The Contractor shall conform to the most recent schedules.[1]

The AIA clause does not impose a particular scheduling obligation on a contractor or define what schedule information is necessary for the designer to follow the contractor's progress. The contractor preparing a schedule under AIA A201 3.10 selects the kind of schedule it wishes to employ and uses its own system, if the contractor indeed elects to sincerely schedule its work.

In contrast, the U.S. Postal Service recently included the detailed scheduling specification shown in Example 14-1 at the end of this chapter. Other U.S. government organizations, such as the Corps of Engineers and the Veterans Administration, also regularly use detailed, lengthy scheduling specifications. The detailed specifications of the government, if enforced, can provide the designer with the information necessary to follow the contractor's progress, although they may not successfully force the contractor to schedule.

A scheduling clause intended to specify what information the contractor is required to provide in order for the designer or owner to prepare the schedule or track the project's progress may include the following items:

1   Whether a network or bar chart schedule is required
2   The use of a particular software system
3   The use of a particular graphical representation method (PDM, I-J, or LSM)
4   The boundaries for the number of activities to be included in the schedule
5   The minimum and maximum duration of activities
6   Whether the activities are to be cost-loaded
7   Whether the activities are to be man-loaded (or loaded with any other resource)
8   Whether the information must be delivered in a particular form (floppy disk, magnetic tape, or hard copy)
9   Who will determine that the submitted information or schedule is approved
10 Determination of the frequency with which the schedule or the information should be updated
11 Whether the update should be accompanied by a narrative report
12 What information should be included in the updated schedule.

The selection of which of these items will be required to be reported will depend on the software, hardware, and scheduling experience of the designer and owner, along with the size and complexity of the project. A detailed clause

[1] American Institute of Architects, *General Conditions of the Contract for Construction*, 14th ed, Document A201, Washington, D.C.: AIA, 1987.

will answer the questions, "In what form does the designer need the information?" "What information is needed?" and "When and how often is the information needed?"

The scheduling clause can be used to provide the owner with a particular software for following the contractor's progress. If the owner has not yet acquired a scheduling software program, the scheduling specification can be used to purchase one with construction project funds. For example:

> In addition, the contractor shall furnish to the engineer one complete functioning set of _____ software licensed to the owner. This program shall be used by the contractor in preparing and maintaining the CPM schedule. The software shall be operable with IBM or IBM-compatible personal computers.

Designers or construction managers often use schedules to determine resource requirements or for calculating progress payments. Detailed clauses intended to produce information to be used to measure either progress payments or resource requirements need special consideration. If the schedule is to be used to fulfill these functions, it should be clearly stated in the specifications. Not all activities in a schedule will necessarily be income-producing or require craftspeople to complete. In addition, for cost-loaded schedules, not all contractor costs can be associated with field activities. For man-loaded schedules, many activities require combinations of tradespeople that make resource scheduling difficult. If the schedule is to meet the intended purpose, the scheduler must select and organize the activities to permit calculation of progress payments or resource scheduling. The ability to match the schedule to the needs of the user requires that the use of the schedule be clearly established in the scheduling specification.

Detailed clauses are not without disadvantage, however. One major disadvantage is the potential for challenge to a specification that requires a particular software be used in preparing the system. A software vendor whose system is not specified may protest the requirement to use a particular software. A contractor may protest the required use of a particular software if it does not presently own it. A detailed specification that identifies precisely what information is required and how it is to be reported may be contrary to the procurement policy of a public owner or simply illegal if few software systems are capable of meeting the requirements. Publicly bid projects may not be able to directly or indirectly specify a particular software or delivery system in the scheduling clause.

HOW SOON SHOULD THE INFORMATION BE PRESENTED?

Another important question that should be considered when drafting a scheduling specification is how quickly the information required in the clause should be submitted. The purpose of the scheduling clause will also influence the speed with which the contractor is required to respond. If the clause is to encourage the contractor to schedule the work, it is probably wise to give the contractor sufficient time to plan the work and prepare a good schedule. Although some planning by the contractor would have taken place at the time the contractor's bid was prepared, the majority of the planning effort will not begin until the contractor has been awarded the job and usually will not begin until the contractor has received a notice to proceed. A contractor needs to be given sufficient time to formulate a good plan, develop a good schedule, and complete it according to good scheduling practice and technique.

Many scheduling specifications do not provide time for schedule preparation sufficient to encourage a contractor to plan and schedule its work. For example, review the U.S. Postal Service time limits under Section D in Example 14-1. The section permits only 14 days for submission of a "preliminary" or interim schedule describing the first 90 days of the project and requires the final schedule to be submitted within 70 days. Most specifications require contractor schedule submission within similar periods. If the scheduling specification is intended to encourage contractor planning and scheduling, more time should be allowed for schedule preparation. Ninety or 120 days is more reasonable for major projects.

Preparation of a good network schedule is a deliberate, time-consuming task. As described in Chapter 6, time is required to study the plans and contract documents; draft and input the data into the computer; distribute summary schedules among subcontractors for comment; revise the summary schedules to include subcontractor response; review the schedule with the contractor's project manager, estimator, and purchaser; and finally, after all this information has been gathered and integrated, time for everybody (contractor, subcontractor, owner, and designer) to review and agree that the distributed schedule will indeed represent the manner in which all the parties required to complete the project intend to complete the project.

On the other hand, if the scheduling specification is intended only to impose a reporting requirement on the contractor, a shorter period between notice to proceed and submission of the data necessary for the designer to prepare a schedule to be able to monitor progress is acceptable.

A deadline is needed, however, regardless of what period is permitted for contractor response. The scheduling specification should identify the maximum time permitted for preparation of the schedule. General contractors need the deadline in their subcontracts in order to request subcontractor information in time to be able to submit the schedule to the designer within the time required in the owner-contractor contract. Owners need the deadline in their contract with the general contractor in order to ensure that the contractor will finish its preparation and submit a schedule at some time.

Often establishing a pay estimate item related to schedule submittals will encourage parties to meet the contract deadlines. In any event, withholding approval of payment requests until the contractor's schedule submittal requirements have been met will ensure compliance. Although contractual authority is not necessary in order to withhold progress payments for failure to schedule, the schedule specification may make the importance of the schedule clear. An example clause is:

> Failure to provide the engineer with an updated schedule every month will be cause to withhold any partial payment due the contractor during the course of the contract until the deficiency has been remedied.

Even when not paid, a contractor sometimes may continue to ignore the scheduling requirement, often increasing the designer's and owner's administration costs. Contractors can be encouraged to satisfactorily complete the schedule as quickly as possible by including the following language in the schedule clause:

> If more than two submittals are required to approve the contractor's initial construction schedule, the owner shall have the right to impose liquidated damages for every day after the second submittal is rejected. In addition, the engineer shall have the right to impose liquidated damages for every day that the initial contractor's construction schedule or the updated construction schedule is submitted late.

INTERIM SCHEDULES

During the time given to the contractor to complete the project schedule, one may choose to require the contractor to submit an interim schedule. As described in Chapter 6, an interim schedule is a manifestation of the contractor's plan to complete the work involved in the first few months of the project. An interim schedule is intended to provide the owner and designer with an indication of how the contractor intends to proceed while the contractor works to complete the schedule for the entire project.

Typically, an interim schedule includes a limited number of activities and shows work for only the first 90 or 120 days of the project. It is usually due within 30 days of the contractor's receipt of notice to proceed.

The primary advantage of an interim schedule is to demonstrate to the owner or designer that the contractor has started its required planning and scheduling effort and to indicate whether the contractor is "on the right track" toward timely project completion. The interim schedule can also help the owner or designer develop their CPM for the early phases of the project, if they choose to prepare the schedule themselves. Omissions or errors discovered in the interim schedule should alert the designer, or whoever reviews the schedule, to be sure the errors or omissions are corrected in the project schedule.

An interim schedule can also permit the contractor time to properly evaluate the project schedule. At the beginning of a job, many factors necessary for a complete schedule are not available; even if they are available, time may not permit their proper evaluation. Little emphasis may be placed on important factors, while too much emphasis may be placed on others that turn out to be inconsequential. Each job has its own characteristics, which may not be known at the start of a project. An interim schedule permits the contractor opportunity to prepare a comprehensive construction schedule while giving the owner or designer an indication of the contractor's initial intent to proceed.

An interim schedule can help the contractor plan by starting the schedule preparation in a discrete, easily satisfied area. An interim schedule can also provide the owner or designer an early indication that efforts to encourage contractor scheduling are either a success or failure. Thus, an interim schedule requirement is appropriate in specifications intended to encourage contractor scheduling, although it is probably unnecessary in clauses intended only to impose a reporting requirement.

DISPUTE-AVOIDANCE CONSIDERATIONS

Directions or instructions in the scheduling specification can also assist in avoiding disputes. The schedule can play an important role in dispute resolution. It can be used to measure delays, determine time extensions, demonstrate impact or out-of-sequence work, or anticipate the effect of possible change orders on project completion. There are a number of factors that must be considered if the schedule is to be used to accomplish any of these things. Float can be consumed before time extensions are granted or maintained to permit continued contractor flexibility. Even whether float is positioned at the beginning or end of an activity can influence the length of a delay or time extension. Delay can be measured by early or late dates. But how the schedule can be used to accomplish any or all of these goals is often not considered by the specifier while preparing the scheduling specification. A scheduling specification can define these variables and thereby better contribute to the avoidance of disputes on the project.

A scheduling specification can define how time extensions or delay will be determined. Will completion dates of individual activities be used, or will an extension be granted only if the project's completion date is extended? Traditionally, time extensions are granted only when the project completion date has been extended. However, some courts, when considering delay claims, measure time extensions from individual activity's completion dates, apparently interpreting the dates in a schedule as a series of separate contracts. To avoid misinterpretation by courts or arbitrators unfamiliar with the scheduling process, the scheduling clause should define how delay or extensions should be measured, for example:

> No time extensions shall be granted or delay damages paid unless the delay can be clearly demonstrated by the monthly update current as of the month the change is issued or the delay occurred, and that the delay cannot be mitigated, offset, or eliminated through such actions as revising the intended sequence of work or other means.

> As a condition precedent to the release of retained funds, the contractor shall, after substantial completion of the work has been achieved, submit a final contractor's construction schedule that accurately reflects the manner in which the project was constructed and includes actual start and completion dates of all work activities on the construction schedule.

Will early completion or late completion dates be used to determine the length of a delay? Without any contrary indication in the scheduling specification, most courts will measure extensions from late completion dates, although the parties are free in the specification to choose either late or early dates.

Can actions of the engineer or owner that reduce the critical path be used to reduce contractors' time extension requests? The scheduling clause can so indicate:

> Owner-caused delays on the project may be offset by owner-caused time savings. Critical path submittals returned in less time than allowed by the contract shall be used to reduce contractor time extension requests. Contractor shall not be entitled to receive a time extension or delay damages until all owner time savings are exceeded and the contract completion date is also exceeded.

In some situations, scheduled completion dates for activities can be anticipated to change. The time scheduled for activity completion can be anticipated to change as the design is completed or prime contracts are awarded, particularly in fast-track construction or when multiple trade contracts are used. Without some arrangement in the contract, however, trade contractors may attempt to measure their performance and any resulting claims from the scheduled dates regardless of any intended flexibility in them. It may be appropriate to identify flexibility in the scheduling specification in order to allow the contract manager to deal with time-of-performance adjustments and avoid later disputes over delay. For example, this clause was used in a scheduling specification to allow the contract manager flexibility but maintain the prime contractor's ability to plan:

> 4.1.10 The Subcontractor recognizes the revisions in the planned schedule are inherent in the nature of construction. This schedule may result in revisions of Contractor's schedule of the work during the progress of construction. Subcontractor agrees that Contractor cannot guarantee Subcontractor can start work on any particular date or continue without interruption once started.
>
> The Subcontractor agrees that at the option of only the Contractor, the start date and/or finish date of each activity can be adjusted. The start date is subject to a one month advance. The finish date is subject to an extension of up to three months. The advance and extension can occur separately or at the same time. The actual time extension can be up to four months. Any such changes will not result in an extra cost to Contractor.

To avoid disputes, a scheduling clause should also consider the role of float in the measurement of delay. Who owns float? Will float be required to be completely consumed before an extension will be granted? Or can a contractor maintain any float in a delayed or affected activity in the calculation of the time extension? Float represents a contractor's "flexibility" and is important to its planning and execution. Float permits the contractor to manage its business by shifting necessary equipment and crews among projects to best complete all the contractor's obligations. Float also permits the contractor to make a reasonable number of mistakes without incurring a performance penalty. But float is also important to an owner because it reduces the owner's obligation for delay or to extend a contract completion date for a change or variation. Murray Hohns, in his book, *Preventing and Solving Construction Disputes*,[2] suggests that float belongs to the project and implies that it should be allocated among activities in a "shared" way. Courts that have considered the ownership of float without direction in the contract have sometimes awarded it to the contractor and sometimes to the owner, although a majority of decisions have given the float to the owner when the contract is silent. By defining who owns float in the contract, the parties can reduce the amount of disputes on the project.

One recent scheduling specification included this clause giving the contractor the "administration" of float on behalf of subcontractors:

> . . .Furthermore, float or slack in all schedules will be administered by the Contractor and may not be used by the Subcontractor without written approval by the Contractor. . . .

Ownership of float can also be identified by stating that time extensions will be granted only for delays occurring to critical activities.

The specification may also define how the schedule may be used to define time extensions. Often, the effect of a delay can be most clearly seen from a schedule revised to include a fragnet representing the additional logic of the change. The scheduling specification may require the submission of a fragnet of the revised or additional logic. One specification had this requirement for showing the effect of changes on the schedule:

> . . . The impact of change orders to this contract shall be included in the project schedule. . . . As a part of his proposal for each change order involving a request for a time extension or otherwise, the contractor shall submit a network diagram showing the detailed work involved in the change and the impact on other work of the proposed adjustment of the schedule.

This type of clause helps to prevent claims for schedule impact after the change order has been submitted.

Contractor disputes involving early completion claims can also be troublesome. The scheduling specification can be used to avoid this kind of dispute:

> If the contractor submits a schedule showing completion of the work more than thirty calendar days in advance of the contract completion date, the owner will at no cost to the owner decrease the contract duration by issuance of a change order that changes the contract completion date to match that shown on the contractor's schedule. Unless the contractor can demonstrate that the project was bid on an early completion schedule, the owner will also deduct the home office and field office overhead costs associated with the amount of time the contract duration was decreased.

[2] New York: Van Nostrand Reinhold, 1979.

The schedule update is often the best place for delays or anticipated delays to be identified. Once identified, the designer or owner can better evaluate the possible impact on the contractor's performance. The schedule specification may thus include:

> The narrative report accompanying the updated schedule shall also include a description of all current and anticipated delays not resolved by change order, the cause of the delay, corrective action and schedule adjustments to correct the delay, and known or potential impact of the delay on other activities, milestones, and project completion.

MUTUAL OBLIGATIONS RESULT FROM THE SCHEDULE

A requirement to prepare a schedule results not only in the imposition of a scheduling or reporting requirement on the contractor, but also imposes obligations on the owner and designer. In its schedule, the contractor commits itself to completion in a certain way and within a certain time. But it also commits the owner and designer to perform their work within the time alloted in the contractor's schedule. Some courts have even concluded that designers must conform to the time limits in the contractor's schedule for the designer's work, whether the designer agrees to the scheduled durations or not. For example, review time for contractor submittals identified in the schedule may be enforced whether the designer agreed to them or not. To avoid this possibility, the schedule specifications may identify the duration of submittal for review:

> Engineers shall be allowed fifteen calendar days to review contractor submittals unless a larger period of time is specified elsewhere in these contract documents.

Similarly, general contractors who require subcontractors to submit a schedule of subcontractor's performance obligate themselves to perform within the time limits for the general construction work stated in the subcontractor's schedule.

A requirement to approve a schedule can be an opportunity for the owner or designer to control the time for their performance in the contractor's schedule. General contractors may use a required approval of a subcontractor schedule to force conformance to the general's schedule and to control the time in which the general must perform its work. Although withholding approval may not affect the ability of a contractor or subcontractor to use its schedule to measure delay or impact, the review and approval process should be used as an opportunity to study the obligations imposed on a schedule.

These mutual obligations under the schedule cannot be avoided or minimized in the scheduling specification. They are a necessary part of the construction planning and scheduling process. All parties involved in the construction must cooperate with each other in order to complete the project. The schedule recognizes and defines the necessary cooperation. By recognizing these mutual obligations and cooperating, each party to a construction contract can help the other achieve timely completion of the project.

EXAMPLE 14-1: U.S. POSTAL SERVICE SCHEDULE SPECIFICATIONS

1.05 CPM NETWORK ANALYSIS SCHEDULING SYSTEM The progress chart to be utilized by the Contractor pursuant to the General Provisions Clause "Construction Progress Chart" shall consist of a precedence network CPM schedule and narrative report as described hereinafter. In preparing this system, the scheduling logic of construction activities and their durations is the responsibility of the Contractor. The requirement of a CPM schedule is included to assure adequate planning and execution of the work, and to assist the Contracting Officer in evaluating the reasonableness of the proposed schedule and progress of the work. The method of submitting the Contractor's CPM schedule shall consist of the following steps:

A The U.S. Postal Service's Construction Support Contractor (CSC) will prepare a guideline schedule to be used as a guide by the Contractor. The guideline schedule will provide the activity numbering code, a logic network, and all activities with an arbitrary duration of three days each. This is provided purely as a guide to the Contractor. Included in the information given to the Contractor, will be a logic diagram and a tabular report in LOTUS or PRIMAVERA format on a diskette which includes activity numbers, the activity descriptions, the successor activities (i.e., the activities that must follow this activity), the activity duration (arbitrarily set at three days), the activity cost allocation (left blank), activity manpower allocation (left blank), and subcontractor responsibility code (left blank).

1 The Contractor will take the guideline schedule and:

    a Revise the logic to reflect his means and method of doing the work as he had planned by entering the logic changes as changes in the activity successor column.

    b Revise the activity breakdown so that no activity is longer than two weeks or greater than $30,000.

    c Revise the durations to reflect his means and methods.

    d Allocate the cost by material and labor to each activity and planned mandays to each activity.

This same information is then compiled by the Contractor in the tabular format utilizing either LOTUS or PRIMAVERA software (IBM compatible) and returned to the CSC on a diskette as a first submission.

2 The CSC will process the schedule data provided by the Contractor and return the results in the form of a tabular report and time scaled logic diagram to the Contractor. If there is a logic error and/or the scheduled completion date does not satisfy the contract required completion date then the CSC will coordinate with the Contractor in revising the logic and/or activity durations. The CSC will debug any loops.

    3 This iterative process is continued until the Contractor's scheduled completion date (early finish) is on or before the contract completion date. The contract completion date will be the late finish date. The CSC will be responsible for all computer network analysis and schedule plots.

    4 When the Contractor has a contract conforming schedule, as determined by the CSC, he is required to formally submit this schedule to the Contracting Officer for approval as the baseline schedule with a signed statement that the schedule represents his planned approach to doing the contract work and will be used by the Contractor for planning, organizing and directing the work, reporting progress, and requesting payment of work accomplished. In accordance with the General Provisions, Clause, "Construction Progress Chart," the Contractor shall take necessary corrective action whenever the progress of the work falls behind that which is scheduled.

    5 Once the schedule is accepted and approved by the Contracting Officer, the CSC in cooperation with the Contractor will update the schedule monthly to show current progress. This progress update will be the basis for approving the Contractor's monthly invoice payment request.

B The detailed CPM network will include construction activities, activities for submittal and procurement of critical materials, and activities for off-site fabrication of specialty items. All activities assigned to the Postal Service or its representatives and the contract required milestone will be clearly identified. The Contractor should anticipate that the total number of activities will range between 2,000 and 3,500, depending on the construction options and methods selected by the Contractor. The selection and number of activities shall be subject to the Contracting Officer's approval.

C The CPM logic network will be computer generated by the CSC using activity on arrow methods. The following information will be shown either on the logic network or on a tabular schedule report, based on input provided by the Contractor:

    1 Activity description and activity number.

    2 Activity duration in work days.

    3 Activities shall be coded by location and CSI specification section.

    4 Activity cost as an allocation of the bid cost.

    5 Activity manpower loading as planned by the Contractor.

    6 Activities planned to be work [sic] on an overtime basis as defined in this specification shall be identified.

    7 Total float for each activity and critical path.

D Submission and approval of the schedule shall be as follows:

    1 A preliminary network or bar chart defining the Contractor's planned operations for the first 90-days shall be submitted within 14 calendar days after receipt of Notice to Proceed for construction. If construction is phased, the preliminary network shall be submitted to USPS Contracting Officer within 14 calendar days after Notice to Proceed for first phase of construction, and shall cover first 90 days of phased construction or partially completed before submission and approval of the whole schedule should be included.

2 Within 14 calendar days after receipt of Notice to Proceed, the CSC will provide a guideline schedule to the contractor.

3 Activity descriptions, schedule logic, and duration data for the complete CPM network analysis shall be submitted by the Contractor to the CSC within 21 calendar days after receipt of the guideline schedule from the CSC. This information shall be provided on an IBM compatible diskette in either LOTUS or PRIMAVERA.

4 Within 7 calendar days after receipt by the CSC of the schedule described in Item 3 above, the CSC will provide comments and/or concurrence in the data submitted.

5 Activity cost and manpower data for the complete CPM schedule shall be submitted by the contractor to the CSC within 14 days after approval of the activity logic and durations. This information shall be provided on an IBM compatible diskette in either LOTUS or PRIMAVERA format.

6 Within 7 calendar days after receipt of the cost and manpower date, the CSC will return a complete CPM to the contractor for his review and concurrence. When CSC concurrence is given, the Contractor will submit the schedule for approval as required by Paragraph 1.05 A.4 above.

7 If the Contractor fails to meet the required schedule submission dates, the Contractor may be required to make his superintendent available to the CSC for half days until the submittal has been completed.

8 The final schedule must be submitted by the contractor to the Contracting Officer for approval within 70 calendar days after receipt of Notice to Proceed. If the schedule is not submitted within that time frame, due to delays of the Contractor, subsequent payment requests after the 90-day preliminary schedule has expired will be approved based on work-in-place estimates determined by the Contracting Officer.

9 See Attachment 1—Simplified Schedule for Submittal and Approval of CPM.

E The CPM schedule shall be thoroughly reviewed and updated on a regular basis, as follows:

1 On a monthly basis, the Contractor and CSC shall meet to review progress. The most current schedule activity listing will be marked up to reflect current activity percent completes and remaining durations. This mark-up will be mutually agreed to by both the CSC and Contractor.

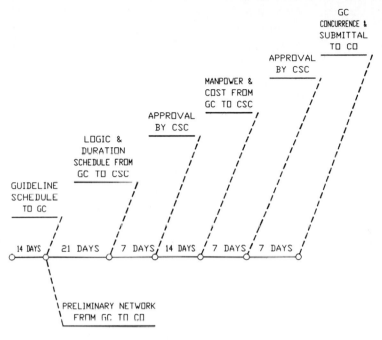

FIGURE 14-1 Simplified schedule for submittal and approval of CPM.

2. The CSC will input the update information into the computer and perform a complete network analysis. The following reports will be submitted to the Contractor by the CSC:

> Earnings report sorted by subcontractor.
> Earnings report sorted by major categories of work. Activities listing by activity number (lowest to highest).
> Activity listing by early start.
> Activity listing by late finish.
> Activity listing by float (lowest to highest).

Accompanying the schedule and earnings reports, a narrative report describing the CSC assessment of progress, a description of problem areas, and the current anticipated delaying factors will be provided by the CSC.

3 If the Contractor's scheduled completion date, as calculated by the CSC, goes beyond the contract required completion date and there have been no contract time extensions approved by the Contracting Officer, the Contractor will have to develop a recovery plan demonstrating how to get the project completed within contract limits. If the Contractor wishes to change his sequence of work or estimated durations of future activities, these changes must be requested in

writing, stating the reason for the change and must be made before or during the monthly schedule update. The CSC will input the schedule adjustments identified in the recovery plan and simplified schedule for submittal and approval [by the Contracting Officer]. If the Contracting Officer finds the proposed plan not acceptable, he may require the Contractor to submit a new plan. If a satisfactory plan is not agreed upon, the Contracting Officer may require the Contractor to increase the work force, the construction plan and equipment, or the number of work shifts, without additional cost to the Postal Service.

4 This updated schedule will be reviewed, approved and submitted by the Contractor as part of his monthly payment request. The submittal item is mandatory to process the invoice.

5 During the period between major monthly schedule updates, the CSC *may* perform weekly or bi-weekly schedule review, by assessing activity percent complete and remaining durations inputting the information and running a network analysis. The Contractor shall make himself available to assist the CSC in assessing the activity percent complete and remaining duration.

F Activity link to Submittals, RFI/inquiries, and Changes: The Contractor is required to provide a list of submittals required under the Contract specifications. He is also required to identify the earliest activity impacted by each of these submittals. The contractor shall also provide activities impacted when he submits an RFI/inquiry for all potential work scope changes. If any impacted activity has less than ten days float, it will be designated as critical and given priority treatment. It is the Contractor's responsibility to ensure that all submittals, RFI/inquiries, and potential work scope changes have been linked with the most critical activity impacted by the item. It is also the Contractor's responsibility to notify the Contracting Officer when the impacted activity has less than ten days of float.

G When Notice to Proceed with changes in the work must be issued prior to settlement of price and/or time to avoid delay, the contractor will revise the network logic and/or duration time estimates of all activities affected by the modification on the next succeeding update report after the date of Notice to Proceed. These revisions will be submitted for concurrence of the Contracting Officer prior to inclusion in the network.

1 If the Contractor fails or refuses to submit or include acceptable revisions within 30 days after the date of Notice to Proceed, the Contracting Officer may furnish to the Contractor the suggested logic and/or duration time revisions to be entered into the network and used in all subsequent updating reports until such time as the revisions have been settled or until actual dates supersede the portion of the schedule represented by the revisions.

2 In the event the Contracting Officer has furnished the suggested logic

and/or duration time revisions because of the Contractor's failure to furnish acceptable revisions on time and the contractor has any objections to the revisions furnished by the Contracting Officer, the Contractor shall notify the Contracting Officer in writing of such objections fully supported by his own counterplan. If the Contractor fails to submit, in writing, his objections to the revisions along with supporting data and counterplan within 20 days after the date such suggested revisions have been furnished by the Contracting Officer, it will be deemed that the Contractor has agreed with Contracting Officer suggested logic/duration time revisions, which revisions then will be the basis for equitable adjustment of the time of performance of the work.

3 Inclusion in the network and use of revised logic and/or duration time estimates for updating, whether furnished by the Contractor or by the Contracting Officer, will not be construed as extensions of time to the dates required in the contract. These revisions are for the purpose of keeping the schedule up-to-date to reflect the work to be accomplished and to include the best time estimates for work yet to be done.

H Float or slack is defined as the amount of time between the early finish date and the late finish date of any of the activities in the network analysis system schedule. Float or slack time is not time for the exclusive use of or benefit of either the Postal Service or the Contractor. Extensions of time for performance required under the GENERAL PROVISION clause entitled "Changes," "Differing Site Conditions," "Termination for Default—Damages for Contractor Delay—Time Extensions," or "Suspension of Work" will be granted only to the extent that equitable time adjustments for the activity or activities affected exceed the total float or slack time along the channels involved at the time Notice to Proceed was issued for change.

I The Contractor shall incorporate into the Network Analysis System any and all dates required by the U.S. Postal Service for early occupancy of certain areas of the building for the purpose of installation of U.S. Postal Service owned non-fixed mechanization equipment prior to the schedule completion date. Such occupancy will require the completion of items within the designated occupancy areas such as flooring, lighting, water coolers, power, restrooms, storage areas, sprinkler and alarm systems, fire extinguishers, dock areas and doors, maneuvering and personnel parking areas, painting, HVAC operation (65 minimum— 78 maximum), and compressed air system, etc. The installation of said equipment requires the presence of U.S. Postal Service employees, live mail and other contractors for testing and servicing prior to beneficial occupancy of the facility.

[1.06 not shown]

1.07 NOTICE REQUIREMENT: If the Contracting Officer or his representative is responsible for an activity which is delaying a critical path activity, the Contractor must provide the Contracting Officer with written notice and permit him 48 hours to resolve the problem causing the delay.

EXERCISES

1 Study the U.S. Postal Service scheduling specification shown in Example 14-1 and answer the follwing questions.
 (*a*) What is the intent of the specifications? A reporting requirement? A scheduling requirement?
 (*b*) How can the specification be improved?
 (*c*) Are early or late completion dates to be used to determine delay in the U.S. Postal Service Scheduling Specification in Example 14-1?
2 Obtain and study a scheduling specification from a local designer, F. W. Dodge plan room, or government agency used for a current project. What suggestions can you make to improve the specification?
3 Who should own the schedule's float? Why?

APPENDIX:
WORK DAY/CALENDAR DAY CALCULATION CALENDAR

This calendar can be used in a variety of ways to calculate calendar days, work days and work days with holidays. The calendar provides for four different calculation columns. The first column lists calendar days. Every single calendar day has a numerical value. The second column, seven-day work week with holidays, has a numerical value for every single day except for the standard six holidays, which are: New Year's Day, Memorial Day, July 4/Independence Day, Labor Day, Thanksgiving, and Christmas. The next column, six-day work week with holidays, is a calendar which allows for every calendar day except for Sundays and the six standard holidays previously listed. The final column, five-day work week with holidays, is probably the most commonly used calendar. It includes a numeric value for every work day, Monday through Friday, except for the six holidays listed above. When calculating work days for a schedule which has different or additional holidays, adjustments must be made.

To use the calendar:

Example #1—A project has been given 920 calendar days' duration. The anticipated notice to proceed is July 26, 1991. Reviewing the calendar day column, note that the 7-26-91 date is given a numeric value of 572. Adding the 920 calendar days' duration to 572 yields a numeric value of 1492. Turning in the calendar to find the value of 1492 in the calendar day column, we find the anticipated completion date to be 1-31-94.

Example #2—For the same project indicated in example #1, determine the number of work days available for this contract. The first step is to go to the completion date of 1-31-94 and determine which work week the contractor anticipates using. For this example we will assume a five-day work week with the standard six holidays. Looking to the far right-hand column, note that 1-31-94 has a numeric value of 1041. Then, referring to the start date of the project, which is 7-26-91, note that the numeric value of this date is 401. Subtracting 401 from 1041 gives a result of 640 work days for this project.

DATE	DAY OF THE WEEK	CALENDAR DAY	7-DAY WORK WEEK W/HOLIDAYS	6-DAY WORK WEEK W/HOLIDAYS	5-DAY WORK WEEK W/HOLIDAYS
01/01/90	MO	1	H	H	H
01/02/90	TU	2	1	1	1
01/03/90	WE	3	2	2	2
01/04/90	TH	4	3	3	3
01/05/90	FR	5	4	4	4
01/06/90	SA	6	5	5	NO WORK
01/07/90	SU	7	6	NO WORK	NO WORK
01/08/90	MO	8	7	6	5
01/09/90	TU	9	8	7	6
01/10/90	WE	10	9	8	7
01/11/90	TH	11	10	9	8
01/12/90	FR	12	11	10	9
01/13/90	SA	13	12	11	NO WORK
01/14/90	SU	14	13	NO WORK	NO WORK
01/15/90	MO	15	14	12	10
01/16/90	TU	16	15	13	11
01/17/90	WE	17	16	14	12
01/18/90	TH	18	17	15	13
01/19/90	FR	19	18	16	14
01/20/90	SA	20	19	17	NO WORK
01/21/90	SU	21	20	NO WORK	NO WORK
01/22/90	MO	22	21	18	15
01/23/90	TU	23	22	19	16
01/24/90	WE	24	23	20	17
01/25/90	TH	25	24	21	18
01/26/90	FR	26	25	22	19
01/27/90	SA	27	26	23	NO WORK
01/28/90	SU	28	27	NO WORK	NO WORK
01/29/90	MO	29	28	24	20
01/30/90	TU	30	29	25	21
01/31/90	WE	31	30	26	22
02/01/90	TH	32	31	27	23
02/02/90	FR	33	32	28	24
02/03/90	SA	34	33	29	NO WORK
02/04/90	SU	35	34	NO WORK	NO WORK
02/05/90	MO	36	35	30	25
02/06/90	TU	37	36	31	26
02/07/90	WE	38	37	32	27
02/08/90	TH	39	38	33	28
02/09/90	FR	40	39	34	29
02/10/90	SA	41	40	35	NO WORK
02/11/90	SU	42	41	NO WORK	NO WORK
02/12/90	MO	43	42	36	30
02/13/90	TU	44	43	37	31
02/14/90	WE	45	44	38	32
02/15/90	TH	46	45	39	33
02/16/90	FR	47	46	40	34
02/17/90	SA	48	47	41	NO WORK
02/18/90	SU	49	48	NO WORK	NO WORK
02/19/90	MO	50	49	42	35
02/20/90	TU	51	50	43	36
02/21/90	WE	52	51	44	37
02/22/90	TH	53	52	45	38
02/23/90	FR	54	53	46	39
02/24/90	SA	55	54	47	NO WORK
02/25/90	SU	56	55	NO WORK	NO WORK
02/26/90	MO	57	56	48	40
02/27/90	TU	58	57	49	41
02/28/90	WE	59	58	50	42
03/01/90	TH	60	59	51	43
03/02/90	FR	61	60	52	44
03/03/90	SA	62	61	53	NO WORK
03/04/90	SU	63	62	NO WORK	NO WORK
03/05/90	MO	64	63	54	45
03/06/90	TU	65	64	55	46
03/07/90	WE	66	65	56	47
03/08/90	TH	67	66	57	48
03/09/90	FR	68	67	58	49
03/10/90	SA	69	68	59	NO WORK
03/11/90	SU	70	69	NO WORK	NO WORK
03/12/90	MO	71	70	60	50
03/13/90	TU	72	71	61	51
03/14/90	WE	73	72	62	52
03/15/90	TH	74	73	63	53
03/16/90	FR	75	74	64	54
03/17/90	SA	76	75	65	NO WORK
03/18/90	SU	77	76	NO WORK	NO WORK
03/19/90	MO	78	77	66	55
03/20/90	TU	79	78	67	56
03/21/90	WE	80	79	68	57
03/22/90	TH	81	80	69	58
03/23/90	FR	82	81	70	59
03/24/90	SA	83	82	71	NO WORK
03/25/90	SU	84	83	NO WORK	NO WORK
03/26/90	MO	85	84	72	60
03/27/90	TU	86	85	73	61
03/28/90	WE	87	86	74	62
03/29/90	TH	88	87	75	63
03/30/90	FR	89	88	76	64
03/31/90	SA	90	89	77	NO WORK
04/01/90	SU	91	90	NO WORK	NO WORK
04/02/90	MO	92	91	78	65
04/03/90	TU	93	92	79	66
04/04/90	WE	94	93	80	67
04/05/90	TH	95	94	81	68
04/06/90	FR	96	95	82	69
04/07/90	SA	97	96	83	NO WORK
04/08/90	SU	98	97	NO WORK	NO WORK
04/09/90	MO	99	98	84	70
04/10/90	TU	100	99	85	71
04/11/90	WE	101	100	86	72
04/12/90	TH	102	101	87	73
04/13/90	FR	103	102	88	74
04/14/90	SA	104	103	89	NO WORK
04/15/90	SU	105	104	NO WORK	NO WORK
04/16/90	MO	106	105	90	75

DATE	DAY OF THE WEEK	CALENDAR DAY	7-DAY WORK WEEK W/HOLIDAYS	6-DAY WORK WEEK W/HOLIDAYS	5-DAY WORK WEEK W/HOLIDAYS
04/17/90	TU	107	106	91	76
04/18/90	WE	108	107	92	77
04/19/90	TH	109	108	93	78
04/20/90	FR	110	109	94	79
04/21/90	SA	111	110	95	NO WORK
04/22/90	SU	112	111	NO WORK	NO WORK
04/23/90	MO	113	112	96	80
04/24/90	TU	114	113	97	81
04/25/90	WE	115	114	98	82
04/26/90	TH	116	115	99	83
04/27/90	FR	117	116	100	84
04/28/90	SA	118	117	101	NO WORK
04/29/90	SU	119	118	NO WORK	NO WORK
04/30/90	MO	120	119	102	85
05/01/90	TU	121	120	103	86
05/02/90	WE	122	121	104	87
05/03/90	TH	123	122	105	88
05/04/90	FR	124	123	106	89
05/05/90	SA	125	124	107	NO WORK
05/06/90	SU	126	125	NO WORK	NO WORK
05/07/90	MO	127	126	108	90
05/08/90	TU	128	127	109	91
05/09/90	WE	129	128	110	92
05/10/90	TH	130	129	111	93
05/11/90	FR	131	130	112	94
05/12/90	SA	132	131	113	NO WORK
05/13/90	SU	133	132	NO WORK	NO WORK
05/14/90	MO	134	133	114	95
05/15/90	TU	135	134	115	96
05/16/90	WE	136	135	116	97
05/17/90	TH	137	136	117	98
05/18/90	FR	138	137	118	99
05/19/90	SA	139	138	119	NO WORK
05/20/90	SU	140	139	NO WORK	NO WORK
05/21/90	MO	141	140	120	100
05/22/90	TU	142	141	121	101
05/23/90	WE	143	142	122	102
05/24/90	TH	144	143	123	103
05/25/90	FR	145	144	124	104
05/26/90	SA	146	145	125	NO WORK
05/27/90	SU	147	146	NO WORK	NO WORK
05/28/90	MO	148	H	H	H
05/29/90	TU	149	147	126	105
05/30/90	WE	150	148	127	106
05/31/90	TH	151	149	128	107
06/01/90	FR	152	150	129	108
06/02/90	SA	153	151	130	NO WORK
06/03/90	SU	154	152	NO WORK	NO WORK
06/04/90	MO	155	153	131	109
06/05/90	TU	156	154	132	110
06/06/90	WE	157	155	133	111
06/07/90	TH	158	156	134	112
06/08/90	FR	159	157	135	113
06/09/90	SA	160	158	136	NO WORK
06/10/90	SU	161	159	NO WORK	NO WORK
06/11/90	MO	162	160	137	114
06/12/90	TU	163	161	138	115
06/13/90	WE	164	162	139	116
06/14/90	TH	165	163	140	117
06/15/90	FR	166	164	141	118
06/16/90	SA	167	165	142	NO WORK
06/17/90	SU	168	166	NO WORK	NO WORK
06/18/90	MO	169	167	143	119
06/19/90	TU	170	168	144	120
06/20/90	WE	171	169	145	121
06/21/90	TH	172	170	146	122
06/22/90	FR	173	171	147	123
06/23/90	SA	174	172	148	NO WORK
06/24/90	SU	175	173	NO WORK	NO WORK
06/25/90	MO	176	174	149	124
06/26/90	TU	177	175	150	125
06/27/90	WE	178	176	151	126
06/28/90	TH	179	177	152	127
06/29/90	FR	180	178	153	128
06/30/90	SA	181	179	154	NO WORK
07/01/90	SU	182	180	NO WORK	NO WORK
07/02/90	MO	183	181	155	129
07/03/90	TU	184	182	156	130
07/04/90	WE	185	H	H	H
07/05/90	TH	186	183	157	131
07/06/90	FR	187	184	158	132
07/07/90	SA	188	185	159	NO WORK
07/08/90	SU	189	186	NO WORK	NO WORK
07/09/90	MO	190	187	160	133
07/10/90	TU	191	188	161	134
07/11/90	WE	192	189	162	135
07/12/90	TH	193	190	163	136
07/13/90	FR	194	191	164	137
07/14/90	SA	195	192	165	NO WORK
07/15/90	SU	196	193	NO WORK	NO WORK
07/16/90	MO	197	194	166	138
07/17/90	TU	198	195	167	139
07/18/90	WE	199	196	168	140
07/19/90	TH	200	197	169	141
07/20/90	FR	201	198	170	142
07/21/90	SA	202	199	171	NO WORK
07/22/90	SU	203	200	NO WORK	NO WORK
07/23/90	MO	204	201	172	143
07/24/90	TU	205	202	173	144
07/25/90	WE	206	203	174	145
07/26/90	TH	207	204	175	146
07/27/90	FR	208	205	176	147
07/28/90	SA	209	206	177	NO WORK
07/29/90	SU	210	207	NO WORK	NO WORK
07/30/90	MO	211	208	178	148
07/31/90	TU	212	209	179	149

DATE	DAY OF THE WEEK	CALENDAR DAY	7-DAY WORK WEEK W/HOLIDAYS	6-DAY WORK WEEK W/HOLIDAYS	5-DAY WORK WEEK W/HOLIDAYS
08/01/90	WE	213	210	180	150
08/02/90	TH	214	211	181	151
08/03/90	FR	215	212	182	152
08/04/90	SA	216	213	183	NO WORK
08/05/90	SU	217	214	NO WORK	NO WORK
08/06/90	MO	218	215	184	153
08/07/90	TU	219	216	185	154
08/08/90	WE	220	217	186	155
08/09/90	TH	221	218	187	156
08/10/90	FR	222	219	188	157
08/11/90	SA	223	220	189	NO WORK
08/12/90	SU	224	221	NO WORK	NO WORK
08/13/90	MO	225	222	190	158
08/14/90	TU	226	223	191	159
08/15/90	WE	227	224	192	160
08/16/90	TH	228	225	193	161
08/17/90	FR	229	226	194	162
08/18/90	SA	230	227	195	NO WORK
08/19/90	SU	231	228	NO WORK	NO WORK
08/20/90	MO	232	229	196	163
08/21/90	TU	233	230	197	164
08/22/90	WE	234	231	198	165
08/23/90	TH	235	232	199	166
08/24/90	FR	236	233	200	167
08/25/90	SA	237	234	201	NO WORK
08/26/90	SU	238	235	NO WORK	NO WORK
08/27/90	MO	239	236	202	168
08/28/90	TU	240	237	203	169
08/29/90	WE	241	238	204	170
08/30/90	TH	242	239	205	171
08/31/90	FR	243	240	206	172
09/01/90	SA	244	241	207	NO WORK
09/02/90	SU	245	242	NO WORK	NO WORK
09/03/90	MO	246	H	H	H
09/04/90	TU	247	243	208	173
09/05/90	WE	248	244	209	174
09/06/90	TH	249	245	210	175
09/07/90	FR	250	246	211	176
09/08/90	SA	251	247	212	NO WORK
09/09/90	SU	252	248	NO WORK	NO WORK
09/10/90	MO	253	249	213	177
09/11/90	TU	254	250	214	178
09/12/90	WE	255	251	215	179
09/13/90	TH	256	252	216	180
09/14/90	FR	257	253	217	181
09/15/90	SA	258	254	218	NO WORK
09/16/90	SU	259	255	NO WORK	NO WORK
09/17/90	MO	260	256	219	182
09/18/90	TU	261	257	220	183
09/19/90	WE	262	258	221	184
09/20/90	TH	263	259	222	185
09/21/90	FR	264	260	223	186
09/22/90	SA	265	261	224	NO WORK
09/23/90	SU	266	262	NO WORK	NO WORK
09/24/90	MO	267	263	225	187
09/25/90	TU	268	264	226	188
09/26/90	WE	269	265	227	189
09/27/90	TH	270	266	228	190
09/28/90	FR	271	267	229	191
09/29/90	SA	272	268	230	NO WORK
09/30/90	SU	273	269	NO WORK	NO WORK
10/01/90	MO	274	270	231	192
10/02/90	TU	275	271	232	193
10/03/90	WE	276	272	233	194
10/04/90	TH	277	273	234	195
10/05/90	FR	278	274	235	196
10/06/90	SA	279	275	236	NO WORK
10/07/90	SU	280	276	NO WORK	NO WORK
10/08/90	MO	281	277	237	197
10/09/90	TU	282	278	238	198
10/10/90	WE	283	279	239	199
10/11/90	TH	284	280	240	200
10/12/90	FR	285	281	241	201
10/13/90	SA	286	282	242	NO WORK
10/14/90	SU	287	283	NO WORK	NO WORK
10/15/90	MO	288	284	243	202
10/16/90	TU	289	285	244	203
10/17/90	WE	290	286	245	204
10/18/90	TH	291	287	246	205
10/19/90	FR	292	288	247	206
10/20/90	SA	293	289	248	NO WORK
10/21/90	SU	294	290	NO WORK	NO WORK
10/22/90	MO	295	291	249	207
10/23/90	TU	296	292	250	208
10/24/90	WE	297	293	251	209
10/25/90	TH	298	294	252	210
10/26/90	FR	299	295	253	211
10/27/90	SA	300	296	254	NO WORK
10/28/90	SU	301	297	NO WORK	NO WORK
10/29/90	MO	302	298	255	212
10/30/90	TU	303	299	256	213
10/31/90	WE	304	300	257	214
11/01/90	TH	305	301	258	215
11/02/90	FR	306	302	259	216
11/03/90	SA	307	303	260	NO WORK
11/04/90	SU	308	304	NO WORK	NO WORK
11/05/90	MO	309	305	261	217
11/06/90	TU	310	306	262	218
11/07/90	WE	311	307	263	219
11/08/90	TH	312	308	264	220
11/09/90	FR	313	309	265	221
11/10/90	SA	314	310	266	NO WORK
11/11/90	SU	315	311	NO WORK	NO WORK
11/12/90	MO	316	312	267	222
11/13/90	TU	317	313	268	223
11/14/90	WE	318	314	269	224

DATE	DAY OF THE WEEK	CALENDAR DAY	7-DAY WORK WEEK W/HOLIDAYS	6-DAY WORK WEEK W/HOLIDAYS	5-DAY WORK WEEK W/HOLIDAYS
11/15/90	TH	319	315	270	225
11/16/90	FR	320	316	271	226
11/17/90	SA	321	317	272	NO WORK
11/18/90	SU	322	318	NO WORK	NO WORK
11/19/90	MO	323	319	273	227
11/20/90	TU	324	320	274	228
11/21/90	WE	325	321	275	229
11/22/90	TH	326	H	H	H
11/23/90	FR	327	322	276	230
11/24/90	SA	328	323	277	NO WORK
11/25/90	SU	329	324	NO WORK	NO WORK
11/26/90	MO	330	325	278	231
11/27/90	TU	331	326	279	232
11/28/90	WE	332	327	280	233
11/29/90	TH	333	328	281	234
11/30/90	FR	334	329	282	235
12/01/90	SA	335	330	283	NO WORK
12/02/90	SU	336	331	NO WORK	NO WORK
12/03/90	MO	337	332	284	236
12/04/90	TU	338	333	285	237
12/05/90	WE	339	334	286	238
12/06/90	TH	340	335	287	239
12/07/90	FR	341	336	288	240
12/08/90	SA	342	337	289	NO WORK
12/09/90	SU	343	338	NO WORK	NO WORK
12/10/90	MO	344	339	290	241
12/11/90	TU	345	340	291	242
12/12/90	WE	346	341	292	243
12/13/90	TH	347	342	293	244
12/14/90	FR	348	343	294	245
12/15/90	SA	349	344	295	NO WORK
12/16/90	SU	350	345	NO WORK	NO WORK
12/17/90	MO	351	346	296	246
12/18/90	TU	352	347	297	247
12/19/90	WE	353	348	298	248
12/20/90	TH	354	349	299	249
12/21/90	FR	355	350	300	250
12/22/90	SA	356	351	301	NO WORK
12/23/90	SU	357	352	NO WORK	NO WORK
12/24/90	MO	358	353	302	251
12/25/90	TU	359	H	H	H
12/26/90	WE	360	354	303	252
12/27/90	TH	361	355	304	253
12/28/90	FR	362	356	305	254
12/29/90	SA	363	357	306	NO WORK
12/30/90	SU	364	358	NO WORK	NO WORK
12/31/90	MO	365	359	307	255

DATE	DAY OF THE WEEK	CALENDAR DAY	7-DAY WORK WEEK W/HOLIDAYS	6-DAY WORK WEEK W/HOLIDAYS	5-DAY WORK WEEK W/HOLIDAYS
01/01/91	TU	366	H	H	H
01/02/91	WE	367	360	308	256
01/03/91	TH	368	361	309	257
01/04/91	FR	369	362	310	258
01/05/91	SA	370	363	311	NO WORK
01/06/91	SU	371	364	NO WORK	NO WORK
01/07/91	MO	372	365	312	259
01/08/91	TU	373	366	313	260
01/09/91	WE	374	367	314	261
01/10/91	TH	375	368	315	262
01/11/91	FR	376	369	316	263
01/12/91	SA	377	370	317	NO WORK
01/13/91	SU	378	371	NO WORK	NO WORK
01/14/91	MO	379	372	318	264
01/15/91	TU	380	373	319	265
01/16/91	WE	381	374	320	266
01/17/91	TH	382	375	321	267
01/18/91	FR	383	376	322	268
01/19/91	SA	384	377	323	NO WORK
01/20/91	SU	385	378	NO WORK	NO WORK
01/21/91	MO	386	379	324	269
01/22/91	TU	387	380	325	270
01/23/91	WE	388	381	326	271
01/24/91	TH	389	382	327	272
01/25/91	FR	390	383	328	273
01/26/91	SA	391	384	329	NO WORK
01/27/91	SU	392	385	NO WORK	NO WORK
01/28/91	MO	393	386	330	274
01/29/91	TU	394	387	331	275
01/30/91	WE	395	388	332	276
01/31/91	TH	396	389	333	277
02/01/91	FR	397	390	334	278
02/02/91	SA	398	391	335	NO WORK
02/03/91	SU	399	392	NO WORK	NO WORK
02/04/91	MO	400	393	336	279
02/05/91	TU	401	394	337	280
02/06/91	WE	402	395	338	281
02/07/91	TH	403	396	339	282
02/08/91	FR	404	397	340	283
02/09/91	SA	405	398	341	NO WORK
02/10/91	SU	406	399	NO WORK	NO WORK
02/11/91	MO	407	400	342	284
02/12/91	TU	408	401	343	285
02/13/91	WE	409	402	344	286
02/14/91	TH	410	403	345	287
02/15/91	FR	411	404	346	288
02/16/91	SA	412	405	347	NO WORK
02/17/91	SU	413	406	NO WORK	NO WORK
02/18/91	MO	414	407	348	289
02/19/91	TU	415	408	349	290
02/20/91	WE	416	409	350	291
02/21/91	TH	417	410	351	292
02/22/91	FR	418	411	352	293

DATE	DAY OF THE WEEK	CALENDAR DAY	7-DAY WORK WEEK W/HOLIDAYS	6-DAY WORK WEEK W/HOLIDAYS	5-DAY WORK WEEK W/HOLIDAYS
02/23/91	SA	419	412	353	NO WORK
02/24/91	SU	420	413	NO WORK	NO WORK
02/25/91	MO	421	414	354	294
02/26/91	TU	422	415	355	295
02/27/91	WE	423	416	356	296
02/28/91	TH	424	417	357	297
03/01/91	FR	425	418	358	298
03/02/91	SA	426	419	359	NO WORK
03/03/91	SU	427	420	NO WORK	NO WORK
03/04/91	MO	428	421	360	299
03/05/91	TU	429	422	361	300
03/06/91	WE	430	423	362	301
03/07/91	TH	431	424	363	302
03/08/91	FR	432	425	364	303
03/09/91	SA	433	426	365	NO WORK
03/10/91	SU	434	427	NO WORK	NO WORK
03/11/91	MO	435	428	366	304
03/12/91	TU	436	429	367	305
03/13/91	WE	437	430	368	306
03/14/91	TH	438	431	369	307
03/15/91	FR	439	432	370	308
03/16/91	SA	440	433	371	NO WORK
03/17/91	SU	441	434	NO WORK	NO WORK
03/18/91	MO	442	435	372	309
03/19/91	TU	443	436	373	310
03/20/91	WE	444	437	374	311
03/21/91	TH	445	438	375	312
03/22/91	FR	446	439	376	313
03/23/91	SA	447	440	377	NO WORK
03/24/91	SU	448	441	NO WORK	NO WORK
03/25/91	MO	449	442	378	314
03/26/91	TU	450	443	379	315
03/27/91	WE	451	444	380	316
03/28/91	TH	452	445	381	317
03/29/91	FR	453	446	382	318
03/30/91	SA	454	447	383	NO WORK
03/31/91	SU	455	448	NO WORK	NO WORK
04/01/91	MO	456	449	384	319
04/02/91	TU	457	450	385	320
04/03/91	WE	458	451	386	321
04/04/91	TH	459	452	387	322
04/05/91	FR	460	453	388	323
04/06/91	SA	461	454	389	NO WORK
04/07/91	SU	462	455	NO WORK	NO WORK
04/08/91	MO	463	456	390	324
04/09/91	TU	464	457	391	325
04/10/91	WE	465	458	392	326
04/11/91	TH	466	459	393	327
04/12/91	FR	467	460	394	328
04/13/91	SA	468	461	395	NO WORK
04/14/91	SU	469	462	NO WORK	NO WORK
04/15/91	MO	470	463	396	329
04/16/91	TU	471	464	397	330
04/17/91	WE	472	465	398	331
04/18/91	TH	473	466	399	332
04/19/91	FR	474	467	400	333
04/20/91	SA	475	468	401	NO WORK
04/21/91	SU	476	469	NO WORK	NO WORK
04/22/91	MO	477	470	402	334
04/23/91	TU	478	471	403	335
04/24/91	WE	479	472	404	336
04/25/91	TH	480	473	405	337
04/26/91	FR	481	474	406	338
04/27/91	SA	482	475	407	NO WORK
04/28/91	SU	483	476	NO WORK	NO WORK
04/29/91	MO	484	477	408	339
04/30/91	TU	485	478	409	340
05/01/91	WE	486	479	410	341
05/02/91	TH	487	480	411	342
05/03/91	FR	488	481	412	343
05/04/91	SA	489	482	413	NO WORK
05/05/91	SU	490	483	NO WORK	NO WORK
05/06/91	MO	491	484	414	344
05/07/91	TU	492	485	415	345
05/08/91	WE	493	486	416	346
05/09/91	TH	494	487	417	347
05/10/91	FR	495	488	418	348
05/11/91	SA	496	489	419	NO WORK
05/12/91	SU	497	490	NO WORK	NO WORK
05/13/91	MO	498	491	420	349
05/14/91	TU	499	492	421	350
05/15/91	WE	500	493	422	351
05/16/91	TH	501	494	423	352
05/17/91	FR	502	495	424	353
05/18/91	SA	503	496	425	NO WORK
05/19/91	SU	504	497	NO WORK	NO WORK
05/20/91	MO	505	498	426	354
05/21/91	TU	506	499	427	355
05/22/91	WE	507	500	428	356
05/23/91	TH	508	501	429	357
05/24/91	FR	509	502	430	358
05/25/91	SA	510	503	431	NO WORK
05/26/91	SU	511	504	NO WORK	NO WORK
05/27/91	MO	512	H	H	H
05/28/91	TU	513	505	432	359
05/29/91	WE	514	506	433	360
05/30/91	TH	515	507	434	361
05/31/91	FR	516	508	435	362
06/01/91	SA	517	509	436	NO WORK
06/02/91	SU	518	510	NO WORK	NO WORK
06/03/91	MO	519	511	437	363
06/04/91	TU	520	512	438	364
06/05/91	WE	521	513	439	365
06/06/91	TH	522	514	440	366
06/07/91	FR	523	515	441	367
06/08/91	SA	524	516	442	NO WORK

DATE	DAY OF THE WEEK	CALENDAR DAY	7-DAY WORK WEEK W/HOLIDAYS	6-DAY WORK WEEK W/HOLIDAYS	5-DAY WORK WEEK W/HOLIDAYS
06/09/91	SU	525	517	NO WORK	NO WORK
06/10/91	MO	526	518	443	368
06/11/91	TU	527	519	444	369
06/12/91	WE	528	520	445	370
06/13/91	TH	529	521	446	371
06/14/91	FR	530	522	447	372
06/15/91	SA	531	523	448	NO WORK
06/16/91	SU	532	524	NO WORK	NO WORK
06/17/91	MO	533	525	449	373
06/18/91	TU	534	526	450	374
06/19/91	WE	535	527	451	375
06/20/91	TH	536	528	452	376
06/21/91	FR	537	529	453	377
06/22/91	SA	538	530	454	NO WORK
06/23/91	SU	539	531	NO WORK	NO WORK
06/24/91	MO	540	532	455	378
06/25/91	TU	541	533	456	379
06/26/91	WE	542	534	457	380
06/27/91	TH	543	535	458	381
06/28/91	FR	544	536	459	382
06/29/91	SA	545	537	460	NO WORK
06/30/91	SU	546	538	NO WORK	NO WORK
07/01/91	MO	547	539	461	383
07/02/91	TU	548	540	462	384
07/03/91	WE	549	541	463	385
07/04/91	TH	550	H	H	H
07/05/91	FR	551	542	464	386
07/06/91	SA	552	543	465	NO WORK
07/07/91	SU	553	544	NO WORK	NO WORK
07/08/91	MO	554	545	466	387
07/09/91	TU	555	546	467	388
07/10/91	WE	556	547	468	389
07/11/91	TH	557	548	469	390
07/12/91	FR	558	549	470	391
07/13/91	SA	559	550	471	NO WORK
07/14/91	SU	560	551	NO WORK	NO WORK
07/15/91	MO	561	552	472	392
07/16/91	TU	562	553	473	393
07/17/91	WE	563	554	474	394
07/18/91	TH	564	555	475	395
07/19/91	FR	565	556	476	396
07/20/91	SA	566	557	477	NO WORK
07/21/91	SU	567	558	NO WORK	NO WORK
07/22/91	MO	568	559	478	397
07/23/91	TU	569	560	479	398
07/24/91	WE	570	561	480	399
07/25/91	TH	571	562	481	400
07/26/91	FR	572	563	482	401
07/27/91	SA	573	564	483	NO WORK
07/28/91	SU	574	565	NO WORK	NO WORK
07/29/91	MO	575	566	484	402
07/30/91	TU	576	567	485	403
07/31/91	WE	577	568	486	404

DATE	DAY OF THE WEEK	CALENDAR DAY	7-DAY WORK WEEK W/HOLIDAYS	6-DAY WORK WEEK W/HOLIDAYS	5-DAY WORK WEEK W/HOLIDAYS
08/01/91	TH	578	569	487	405
08/02/91	FR	579	570	488	406
08/03/91	SA	580	571	489	NO WORK
08/04/91	SU	581	572	NO WORK	NO WORK
08/05/91	MO	582	573	490	407
08/06/91	TU	583	574	491	408
08/07/91	WE	584	575	492	409
08/08/91	TH	585	576	493	410
08/09/91	FR	586	577	494	411
08/10/91	SA	587	578	495	NO WORK
08/11/91	SU	588	579	NO WORK	NO WORK
08/12/91	MO	589	580	496	412
08/13/91	TU	590	581	497	413
08/14/91	WE	591	582	498	414
08/15/91	TH	592	583	499	415
08/16/91	FR	593	584	500	416
08/17/91	SA	594	585	501	NO WORK
08/18/91	SU	595	586	NO WORK	NO WORK
08/19/91	MO	596	587	502	417
08/20/91	TU	597	588	503	418
08/21/91	WE	598	589	504	419
08/22/91	TH	599	590	505	420
08/23/91	FR	600	591	506	421
08/24/91	SA	601	592	507	NO WORK
08/25/91	SU	602	593	NO WORK	NO WORK
08/26/91	MO	603	594	508	422
08/27/91	TU	604	595	509	423
08/28/91	WE	605	596	510	424
08/29/91	TH	606	597	511	425
08/30/91	FR	607	598	512	426
08/31/91	SA	608	599	513	NO WORK
09/01/91	SU	609	600	NO WORK	NO WORK
09/02/91	MO	610	H	H	H
09/03/91	TU	611	601	514	427
09/04/91	WE	612	602	515	428
09/05/91	TH	613	603	516	429
09/06/91	FR	614	604	517	430
09/07/91	SA	615	605	518	NO WORK
09/08/91	SU	616	606	NO WORK	NO WORK
09/09/91	MO	617	607	519	431
09/10/91	TU	618	608	520	432
09/11/91	WE	619	609	521	433
09/12/91	TH	620	610	522	434
09/13/91	FR	621	611	523	435
09/14/91	SA	622	612	524	NO WORK
09/15/91	SU	623	613	NO WORK	NO WORK
09/16/91	MO	624	614	525	436
09/17/91	TU	625	615	526	437
09/18/91	WE	626	616	527	438
09/19/91	TH	627	617	528	439
09/20/91	FR	628	618	529	440
09/21/91	SA	629	619	530	NO WORK
09/22/91	SU	630	620	NO WORK	NO WORK

DATE	DAY OF THE WEEK	CALENDAR DAY	7-DAY WORK WEEK W/HOLIDAYS	6-DAY WORK WEEK W/HOLIDAYS	5-DAY WORK WEEK W/HOLIDAYS
09/23/91	MO	631	621	531	441
09/24/91	TU	632	622	532	442
09/25/91	WE	633	623	533	443
09/26/91	TH	634	624	534	444
09/27/91	FR	635	625	535	445
09/28/91	SA	636	626	536	NO WORK
09/29/91	SU	637	627	NO WORK	NO WORK
09/30/91	MO	638	628	537	446
10/01/91	TU	639	629	538	447
10/02/91	WE	640	630	539	448
10/03/91	TH	641	631	540	449
10/04/91	FR	642	632	541	450
10/05/91	SA	643	633	542	NO WORK
10/06/91	SU	644	634	NO WORK	NO WORK
10/07/91	MO	645	635	543	451
10/08/91	TU	646	636	544	452
10/09/91	WE	647	637	545	453
10/10/91	TH	648	638	546	454
10/11/91	FR	649	639	547	455
10/12/91	SA	650	640	548	NO WORK
10/13/91	SU	651	641	NO WORK	NO WORK
10/14/91	MO	652	642	549	456
10/15/91	TU	653	643	550	457
10/16/91	WE	654	644	551	458
10/17/91	TH	655	645	552	459
10/18/91	FR	656	646	553	460
10/19/91	SA	657	647	554	NO WORK
10/20/91	SU	658	648	NO WORK	NO WORK
10/21/91	MO	659	649	555	461
10/22/91	TU	660	650	556	462
10/23/91	WE	661	651	557	463
10/24/91	TH	662	652	558	464
10/25/91	FR	663	653	559	465
10/26/91	SA	664	654	560	NO WORK
10/27/91	SU	665	655	NO WORK	NO WORK
10/28/91	MO	666	656	561	466
10/29/91	TU	667	657	562	467
10/30/91	WE	668	658	563	468
10/31/91	TH	669	659	564	469
11/01/91	FR	670	660	565	470
11/02/91	SA	671	661	566	NO WORK
11/03/91	SU	672	662	NO WORK	NO WORK
11/04/91	MO	673	663	567	471
11/05/91	TU	674	664	568	472
11/06/91	WE	675	665	569	473
11/07/91	TH	676	666	570	474
11/08/91	FR	677	667	571	475
11/09/91	SA	678	668	572	NO WORK
11/10/91	SU	679	669	NO WORK	NO WORK
11/11/91	MO	680	670	573	476
11/12/91	TU	681	671	574	477
11/13/91	WE	682	672	575	478
11/14/91	TH	683	673	576	479
11/15/91	FR	684	674	577	480
11/16/91	SA	685	675	578	NO WORK
11/17/91	SU	686	676	NO WORK	NO WORK
11/18/91	MO	687	677	579	481
11/19/91	TU	688	678	580	482
11/20/91	WE	689	679	581	483
11/21/91	TH	690	680	582	484
11/22/91	FR	691	681	583	485
11/23/91	SA	692	682	584	NO WORK
11/24/91	SU	693	683	NO WORK	NO WORK
11/25/91	MO	694	684	585	486
11/26/91	TU	695	685	586	487
11/27/91	WE	696	686	587	488
11/28/91	TH	697	H	H	H
11/29/91	FR	698	687	588	489
11/30/91	SA	699	688	589	NO WORK
12/01/91	SU	700	689	NO WORK	NO WORK
12/02/91	MO	701	690	590	490
12/03/91	TU	702	691	591	491
12/04/91	WE	703	692	592	492
12/05/91	TH	704	693	593	493
12/06/91	FR	705	694	594	494
12/07/91	SA	706	695	595	NO WORK
12/08/91	SU	707	696	NO WORK	NO WORK
12/09/91	MO	708	697	596	495
12/10/91	TU	709	698	597	496
12/11/91	WE	710	699	598	497
12/12/91	TH	711	700	599	498
12/13/91	FR	712	701	600	499
12/14/91	SA	713	702	601	NO WORK
12/15/91	SU	714	703	NO WORK	NO WORK
12/16/91	MO	715	704	602	500
12/17/91	TU	716	705	603	501
12/18/91	WE	717	706	604	502
12/19/91	TH	718	707	605	503
12/20/91	FR	719	708	606	504
12/21/91	SA	720	709	607	NO WORK
12/22/91	SU	721	710	NO WORK	NO WORK
12/23/91	MO	722	711	608	505
12/24/91	TU	723	712	609	506
12/25/91	WE	724	H	H	H
12/26/91	TH	725	713	610	507
12/27/91	FR	726	714	611	508
12/28/91	SA	727	715	612	NO WORK
12/29/91	SU	728	716	NO WORK	NO WORK
12/30/91	MO	729	717	613	509
12/31/91	TU	730	718	614	510

DATE	DAY OF THE WEEK	CALENDAR DAY	7-DAY WORK WEEK W/HOLIDAYS	6-DAY WORK WEEK W/HOLIDAYS	5-DAY WORK WEEK W/HOLIDAYS
01/01/92	WE	731	H	H	H
01/02/92	TH	732	719	615	511
01/03/92	FR	733	720	616	512
01/04/92	SA	734	721	617	NO WORK
01/05/92	SU	735	722	NO WORK	NO WORK
01/06/92	MO	736	723	618	513
01/07/92	TU	737	724	619	514
01/08/92	WE	738	725	620	515
01/09/92	TH	739	726	621	516
01/10/92	FR	740	727	622	517
01/11/92	SA	741	728	623	NO WORK
01/12/92	SU	742	729	NO WORK	NO WORK
01/13/92	MO	743	730	624	518
01/14/92	TU	744	731	625	519
01/15/92	WE	745	732	626	520
01/16/92	TH	746	733	627	521
01/17/92	FR	747	734	628	522
01/18/92	SA	748	735	629	NO WORK
01/19/92	SU	749	736	NO WORK	NO WORK
01/20/92	MO	750	737	630	523
01/21/92	TU	751	738	631	524
01/22/92	WE	752	739	632	525
01/23/92	TH	753	740	633	526
01/24/92	FR	754	741	634	527
01/25/92	SA	755	742	635	NO WORK
01/26/92	SU	756	743	NO WORK	NO WORK
01/27/92	MO	757	744	636	528
01/28/92	TU	758	745	637	529
01/29/92	WE	759	746	638	530
01/30/92	TH	760	747	639	531
01/31/92	FR	761	748	640	532
02/01/92	SA	762	749	641	NO WORK
02/02/92	SU	763	750	NO WORK	NO WORK
02/03/92	MO	764	751	642	533
02/04/92	TU	765	752	643	534
02/05/92	WE	766	753	644	535
02/06/92	TH	767	754	645	536
02/07/92	FR	768	755	646	537
02/08/92	SA	769	756	647	NO WORK
02/09/92	SU	770	757	NO WORK	NO WORK
02/10/92	MO	771	758	648	538
02/11/92	TU	772	759	649	539
02/12/92	WE	773	760	650	540
02/13/92	TH	774	761	651	541
02/14/92	FR	775	762	652	542
02/15/92	SA	776	763	653	NO WORK
02/16/92	SU	777	764	NO WORK	NO WORK
02/17/92	MO	778	765	654	543
02/18/92	TU	779	766	655	544
02/19/92	WE	780	767	656	545
02/20/92	TH	781	768	657	546
02/21/92	FR	782	769	658	547
02/22/92	SA	783	770	659	NO WORK
02/23/92	SU	784	771	NO WORK	NO WORK
02/24/92	MO	785	772	660	548
02/25/92	TU	786	773	661	549
02/26/92	WE	787	774	662	550
02/27/92	TH	788	775	663	551
02/28/92	FR	789	776	664	552
02/29/92	SA	790	777	665	NO WORK
03/01/92	SU	791	778	NO WORK	NO WORK
03/02/92	MO	792	779	666	553
03/03/92	TU	793	780	667	554
03/04/92	WE	794	781	668	555
03/05/92	TH	795	782	669	556
03/06/92	FR	796	783	670	557
03/07/92	SA	797	784	671	NO WORK
03/08/92	SU	798	785	NO WORK	NO WORK
03/09/92	MO	799	786	672	558
03/10/92	TU	800	787	673	559
03/11/92	WE	801	788	674	560
03/12/92	TH	802	789	675	561
03/13/92	FR	803	790	676	562
03/14/92	SA	804	791	677	NO WORK
03/15/92	SU	805	792	NO WORK	NO WORK
03/16/92	MO	806	793	678	563
03/17/92	TU	807	794	679	564
03/18/92	WE	808	795	680	565
03/19/92	TH	809	796	681	566
03/20/92	FR	810	797	682	567
03/21/92	SA	811	798	683	NO WORK
03/22/92	SU	812	799	NO WORK	NO WORK
03/23/92	MO	813	800	684	568
03/24/92	TU	814	801	685	569
03/25/92	WE	815	802	686	570
03/26/92	TH	816	803	687	571
03/27/92	FR	817	804	688	572
03/28/92	SA	818	805	689	NO WORK
03/29/92	SU	819	806	NO WORK	NO WORK
03/30/92	MO	820	807	690	573
03/31/92	TU	821	808	691	574
04/01/92	WE	822	809	692	575
04/02/92	TH	823	810	693	576
04/03/92	FR	824	811	694	577
04/04/92	SA	825	812	695	NO WORK
04/05/92	SU	826	813	NO WORK	NO WORK
04/06/92	MO	827	814	696	578
04/07/92	TU	828	815	697	579
04/08/92	WE	829	816	698	580
04/09/92	TH	830	817	699	581
04/10/92	FR	831	818	700	582
04/11/92	SA	832	819	701	NO WORK
04/12/92	SU	833	820	NO WORK	NO WORK
04/13/92	MO	834	821	702	583
04/14/92	TU	835	822	703	584
04/15/92	WE	836	823	704	585

DATE	DAY OF THE WEEK	CALENDAR DAY	7-DAY WORK WEEK W/HOLIDAYS	6-DAY WORK WEEK W/HOLIDAYS	5-DAY WORK WEEK W/HOLIDAYS
04/16/92	TH	837	824	705	586
04/17/92	FR	838	825	706	587
04/18/92	SA	839	826	707	NO WORK
04/19/92	SU	840	827	NO WORK	NO WORK
04/20/92	MO	841	828	708	588
04/21/92	TU	842	829	709	589
04/22/92	WE	843	830	710	590
04/23/92	TH	844	831	711	591
04/24/92	FR	845	832	712	592
04/25/92	SA	846	833	713	NO WORK
04/26/92	SU	847	834	NO WORK	NO WORK
04/27/92	MO	848	835	714	593
04/28/92	TU	849	836	715	594
04/29/92	WE	850	837	716	595
04/30/92	TH	851	838	717	596
05/01/92	FR	852	839	718	597
05/02/92	SA	853	840	719	NO WORK
05/03/92	SU	854	841	NO WORK	NO WORK
05/04/92	MO	855	842	720	598
05/05/92	TU	856	843	721	599
05/06/92	WE	857	844	722	600
05/07/92	TH	858	845	723	601
05/08/92	FR	859	846	724	602
05/09/92	SA	860	847	725	NO WORK
05/10/92	SU	861	848	NO WORK	NO WORK
05/11/92	MO	862	849	726	603
05/12/92	TU	863	850	727	604
05/13/92	WE	864	851	728	605
05/14/92	TH	865	852	729	606
05/15/92	FR	866	853	730	607
05/16/92	SA	867	854	731	NO WORK
05/17/92	SU	868	855	NO WORK	NO WORK
05/18/92	MO	869	856	732	608
05/19/92	TU	870	857	733	609
05/20/92	WE	871	858	734	610
05/21/92	TH	872	859	735	611
05/22/92	FR	873	860	736	612
05/23/92	SA	874	861	737	NO WORK
05/24/92	SU	875	862	NO WORK	NO WORK
05/25/92	MO	876	H	H	H
05/26/92	TU	877	863	738	613
05/27/92	WE	878	864	739	614
05/28/92	TH	879	865	740	615
05/29/92	FR	880	866	741	616
05/30/92	SA	881	867	742	NO WORK
05/31/92	SU	882	868	NO WORK	NO WORK
06/01/92	MO	883	869	743	617
06/02/92	TU	884	870	744	618
06/03/92	WE	885	871	745	619
06/04/92	TH	886	872	746	620
06/05/92	FR	887	873	747	621
06/06/92	SA	888	874	748	NO WORK
06/07/92	SU	889	875	NO WORK	NO WORK
06/08/92	MO	890	876	749	622
06/09/92	TU	891	877	750	623
06/10/92	WE	892	878	751	624
06/11/92	TH	893	879	752	625
06/12/92	FR	894	880	753	626
06/13/92	SA	895	881	754	NO WORK
06/14/92	SU	896	882	NO WORK	NO WORK
06/15/92	MO	897	883	755	627
06/16/92	TU	898	884	756	628
06/17/92	WE	899	885	757	629
06/18/92	TH	900	886	758	630
06/19/92	FR	901	887	759	631
06/20/92	SA	902	888	760	NO WORK
06/21/92	SU	903	889	NO WORK	NO WORK
06/22/92	MO	904	890	761	632
06/23/92	TU	905	891	762	633
06/24/92	WE	906	892	763	634
06/25/92	TH	907	893	764	635
06/26/92	FR	908	894	765	636
06/27/92	SA	909	895	766	NO WORK
06/28/92	SU	910	896	NO WORK	NO WORK
06/29/92	MO	911	897	767	637
06/30/92	TU	912	898	768	638
07/01/92	WE	913	899	769	639
07/02/92	TH	914	900	770	640
07/03/92	FR	915	901	771	H
07/04/92	SA	916	H	H	NO WORK
07/05/92	SU	917	902	NO WORK	NO WORK
07/06/92	MO	918	903	772	641
07/07/92	TU	919	904	773	642
07/08/92	WE	920	905	774	643
07/09/92	TH	921	906	775	644
07/10/92	FR	922	907	776	645
07/11/92	SA	923	908	777	NO WORK
07/12/92	SU	924	909	NO WORK	NO WORK
07/13/92	MO	925	910	778	646
07/14/92	TU	926	911	779	647
07/15/92	WE	927	912	780	648
07/16/92	TH	928	913	781	649
07/17/92	FR	929	914	782	650
07/18/92	SA	930	915	783	NO WORK
07/19/92	SU	931	916	NO WORK	NO WORK
07/20/92	MO	932	917	784	651
07/21/92	TU	933	918	785	652
07/22/92	WE	934	919	786	653
07/23/92	TH	935	920	787	654
07/24/92	FR	936	921	788	655
07/25/92	SA	937	922	789	NO WORK
07/26/92	SU	938	923	NO WORK	NO WORK
07/27/92	MO	939	924	790	656
07/28/92	TU	940	925	791	657
07/29/92	WE	941	926	792	658
07/30/92	TH	942	927	793	659

DATE	DAY OF THE WEEK	CALENDAR DAY	7-DAY WORK WEEK W/HOLIDAYS	6-DAY WORK WEEK W/HOLIDAYS	5-DAY WORK WEEK W/HOLIDAYS
07/31/92	FR	943	928	794	660
08/01/92	SA	944	929	795	NO WORK
08/02/92	SU	945	930	NO WORK	NO WORK
08/03/92	MO	946	931	796	661
08/04/92	TU	947	932	797	662
08/05/92	WE	948	933	798	663
08/06/92	TH	949	934	799	664
08/07/92	FR	950	935	800	665
08/08/92	SA	951	936	801	NO WORK
08/09/92	SU	952	937	NO WORK	NO WORK
08/10/92	MO	953	938	802	666
08/11/92	TU	954	939	803	667
08/12/92	WE	955	940	804	668
08/13/92	TH	956	941	805	669
08/14/92	FR	957	942	806	670
08/15/92	SA	958	943	807	NO WORK
08/16/92	SU	959	944	NO WORK	NO WORK
08/17/92	MO	960	945	808	671
08/18/92	TU	961	946	809	672
08/19/92	WE	962	947	810	673
08/20/92	TH	963	948	811	674
08/21/92	FR	964	949	812	675
08/22/92	SA	965	950	813	NO WORK
08/23/92	SU	966	951	NO WORK	NO WORK
08/24/92	MO	967	952	814	676
08/25/92	TU	968	953	815	677
08/26/92	WE	969	954	816	678
08/27/92	TH	970	955	817	679
08/28/92	FR	971	956	818	680
08/29/92	SA	972	957	819	NO WORK
08/30/92	SU	973	958	NO WORK	NO WORK
08/31/92	MO	974	959	820	681
09/01/92	TU	975	960	821	682
09/02/92	WE	976	961	822	683
09/03/92	TH	977	962	823	684
09/04/92	FR	978	963	824	685
09/05/92	SA	979	964	825	NO WORK
09/06/92	SU	980	965	NO WORK	NO WORK
09/07/92	MO	981	H	H	H
09/08/92	TU	982	966	826	686
09/09/92	WE	983	967	827	687
09/10/92	TH	984	968	828	688
09/11/92	FR	985	969	829	689
09/12/92	SA	986	970	830	NO WORK
09/13/92	SU	987	971	NO WORK	NO WORK
09/14/92	MO	988	972	831	690
09/15/92	TU	989	973	832	691
09/16/92	WE	990	974	833	692
09/17/92	TH	991	975	834	693
09/18/92	FR	992	976	835	694
09/19/92	SA	993	977	836	NO WORK
09/20/92	SU	994	978	NO WORK	NO WORK
09/21/92	MO	995	979	837	695

DATE	DAY OF THE WEEK	CALENDAR DAY	7-DAY WORK WEEK W/HOLIDAYS	6-DAY WORK WEEK W/HOLIDAYS	5-DAY WORK WEEK W/HOLIDAYS
09/22/92	TU	996	980	838	696
09/23/92	WE	997	981	839	697
09/24/92	TH	998	982	840	698
09/25/92	FR	999	983	841	699
09/26/92	SA	1000	984	842	NO WORK
09/27/92	SU	1001	985	NO WORK	NO WORK
09/28/92	MO	1002	986	843	700
09/29/92	TU	1003	987	844	701
09/30/92	WE	1004	988	845	702
10/01/92	TH	1005	989	846	703
10/02/92	FR	1006	990	847	704
10/03/92	SA	1007	991	848	NO WORK
10/04/92	SU	1008	992	NO WORK	NO WORK
10/05/92	MO	1009	993	849	705
10/06/92	TU	1010	994	850	706
10/07/92	WE	1011	995	851	707
10/08/92	TH	1012	996	852	708
10/09/92	FR	1013	997	853	709
10/10/92	SA	1014	998	854	NO WORK
10/11/92	SU	1015	999	NO WORK	NO WORK
10/12/92	MO	1016	1000	855	710
10/13/92	TU	1017	1001	856	711
10/14/92	WE	1018	1002	857	712
10/15/92	TH	1019	1003	858	713
10/16/92	FR	1020	1004	859	714
10/17/92	SA	1021	1005	860	NO WORK
10/18/92	SU	1022	1006	NO WORK	NO WORK
10/19/92	MO	1023	1007	861	715
10/20/92	TU	1024	1008	862	716
10/21/92	WE	1025	1009	863	717
10/22/92	TH	1026	1010	864	718
10/23/92	FR	1027	1011	865	719
10/24/92	SA	1028	1012	866	NO WORK
10/25/92	SU	1029	1013	NO WORK	NO WORK
10/26/92	MO	1030	1014	867	720
10/27/92	TU	1031	1015	868	721
10/28/92	WE	1032	1016	869	722
10/29/92	TH	1033	1017	870	723
10/30/92	FR	1034	1018	871	724
10/31/92	SA	1035	1019	872	NO WORK
11/01/92	SU	1036	1020	NO WORK	NO WORK
11/02/92	MO	1037	1021	873	725
11/03/92	TU	1038	1022	874	726
11/04/92	WE	1039	1023	875	727
11/05/92	TH	1040	1024	876	728
11/06/92	FR	1041	1025	877	729
11/07/92	SA	1042	1026	878	NO WORK
11/08/92	SU	1043	1027	NO WORK	NO WORK
11/09/92	MO	1044	1028	879	730
11/10/92	TU	1045	1029	880	731
11/11/92	WE	1046	1030	881	732
11/12/92	TH	1047	1031	882	733
11/13/92	FR	1048	1032	883	734

DATE	DAY OF THE WEEK	CALENDAR DAY	7-DAY WORK WEEK W/HOLIDAYS	6-DAY WORK WEEK W/HOLIDAYS	5-DAY WORK WEEK W/HOLIDAYS
11/14/92	SA	1049	1033	884	NO WORK
11/15/92	SU	1050	1034	NO WORK	NO WORK
11/16/92	MO	1051	1035	885	735
11/17/92	TU	1052	1036	886	736
11/18/92	WE	1053	1037	887	737
11/19/92	TH	1054	1038	888	738
11/20/92	FR	1055	1039	889	739
11/21/92	SA	1056	1040	890	NO WORK
11/22/92	SU	1057	1041	NO WORK	NO WORK
11/23/92	MO	1058	1042	891	740
11/24/92	TU	1059	1043	892	741
11/25/92	WE	1060	1044	893	742
11/26/92	TH	1061	H	H	H
11/27/92	FR	1062	1045	894	743
11/28/92	SA	1063	1046	895	NO WORK
11/29/92	SU	1064	1047	NO WORK	NO WORK
11/30/92	MO	1065	1048	896	744
12/01/92	TU	1066	1049	897	745
12/02/92	WE	1067	1050	898	746
12/03/92	TH	1068	1051	899	747
12/04/92	FR	1069	1052	900	748
12/05/92	SA	1070	1053	901	NO WORK
12/06/92	SU	1071	1054	NO WORK	NO WORK
12/07/92	MO	1072	1055	902	749
12/08/92	TU	1073	1056	903	750
12/09/92	WE	1074	1057	904	751
12/10/92	TH	1075	1058	905	752
12/11/92	FR	1076	1059	906	753
12/12/92	SA	1077	1060	907	NO WORK
12/13/92	SU	1078	1061	NO WORK	NO WORK
12/14/92	MO	1079	1062	908	754
12/15/92	TU	1080	1063	909	755
12/16/92	WE	1081	1064	910	756
12/17/92	TH	1082	1065	911	757
12/18/92	FR	1083	1066	912	758
12/19/92	SA	1084	1067	913	NO WORK
12/20/92	SU	1085	1068	NO WORK	NO WORK
12/21/92	MO	1086	1069	914	759
12/22/92	TU	1087	1070	915	760
12/23/92	WE	1088	1071	916	761
12/24/92	TH	1089	1072	917	762
12/25/92	FR	1090	H	H	H
12/26/92	SA	1091	1073	918	NO WORK
12/27/92	SU	1092	1074	NO WORK	NO WORK
12/28/92	MO	1093	1075	919	763
12/29/92	TU	1094	1076	920	764
12/30/92	WE	1095	1077	921	765
12/31/92	TH	1096	1078	922	766
01/01/93	FR	1097	H	H	H
01/02/93	SA	1098	1079	923	NO WORK
01/03/93	SU	1099	1080	NO WORK	NO WORK
01/04/93	MO	1100	1081	924	767
01/05/93	TU	1101	1082	925	768
01/06/93	WE	1102	1083	926	769
01/07/93	TH	1103	1084	927	770
01/08/93	FR	1104	1085	928	771
01/09/93	SA	1105	1086	929	NO WORK
01/10/93	SU	1106	1087	NO WORK	NO WORK
01/11/93	MO	1107	1088	930	772
01/12/93	TU	1108	1089	931	773
01/13/93	WE	1109	1090	932	774
01/14/93	TH	1110	1091	933	775
01/15/93	FR	1111	1092	934	776
01/16/93	SA	1112	1093	935	NO WORK
01/17/93	SU	1113	1094	NO WORK	NO WORK
01/18/93	MO	1114	1095	936	777
01/19/93	TU	1115	1096	937	778
01/20/93	WE	1116	1097	938	779
01/21/93	TH	1117	1098	939	780
01/22/93	FR	1118	1099	940	781
01/23/93	SA	1119	1100	941	NO WORK
01/24/93	SU	1120	1101	NO WORK	NO WORK
01/25/93	MO	1121	1102	942	782
01/26/93	TU	1122	1103	943	783
01/27/93	WE	1123	1104	944	784
01/28/93	TH	1124	1105	945	785
01/29/93	FR	1125	1106	946	786
01/30/93	SA	1126	1107	947	NO WORK
01/31/93	SU	1127	1108	NO WORK	NO WORK
02/01/93	MO	1128	1109	948	787
02/02/93	TU	1129	1110	949	788
02/03/93	WE	1130	1111	950	789
02/04/93	TH	1131	1112	951	790
02/05/93	FR	1132	1113	952	791
02/06/93	SA	1133	1114	953	NO WORK
02/07/93	SU	1134	1115	NO WORK	NO WORK
02/08/93	MO	1135	1116	954	792
02/09/93	TU	1136	1117	955	793
02/10/93	WE	1137	1118	956	794
02/11/93	TH	1138	1119	957	795
02/12/93	FR	1139	1120	958	796
02/13/93	SA	1140	1121	959	NO WORK
02/14/93	SU	1141	1122	NO WORK	NO WORK
02/15/93	MO	1142	1123	960	797
02/16/93	TU	1143	1124	961	798
02/17/93	WE	1144	1125	962	799
02/18/93	TH	1145	1126	963	800
02/19/93	FR	1146	1127	964	801
02/20/93	SA	1147	1128	965	NO WORK
02/21/93	SU	1148	1129	NO WORK	NO WORK
02/22/93	MO	1149	1130	966	802

DATE	DAY OF THE WEEK	CALENDAR DAY	7-DAY WORK WEEK W/HOLIDAYS	6-DAY WORK WEEK W/HOLIDAYS	5-DAY WORK WEEK W/HOLIDAYS
02/23/93	TU	1150	1131	967	803
02/24/93	WE	1151	1132	968	804
02/25/93	TH	1152	1133	969	805
02/26/93	FR	1153	1134	970	806
02/27/93	SA	1154	1135	971	NO WORK
02/28/93	SU	1155	1136	NO WORK	NO WORK
03/01/93	MO	1156	1137	972	807
03/02/93	TU	1157	1138	973	808
03/03/93	WE	1158	1139	974	809
03/04/93	TH	1159	1140	975	810
03/05/93	FR	1160	1141	976	811
03/06/93	SA	1161	1142	977	NO WORK
03/07/93	SU	1162	1143	NO WORK	NO WORK
03/08/93	MO	1163	1144	978	812
03/09/93	TU	1164	1145	979	813
03/10/93	WE	1165	1146	980	814
03/11/93	TH	1166	1147	981	815
03/12/93	FR	1167	1148	982	816
03/13/93	SA	1168	1149	983	NO WORK
03/14/93	SU	1169	1150	NO WORK	NO WORK
03/15/93	MO	1170	1151	984	817
03/16/93	TU	1171	1152	985	818
03/17/93	WE	1172	1153	986	819
03/18/93	TH	1173	1154	987	820
03/19/93	FR	1174	1155	988	821
03/20/93	SA	1175	1156	989	NO WORK
03/21/93	SU	1176	1157	NO WORK	NO WORK
03/22/93	MO	1177	1158	990	822
03/23/93	TU	1178	1159	991	823
03/24/93	WE	1179	1160	992	824
03/25/93	TH	1180	1161	993	825
03/26/93	FR	1181	1162	994	826
03/27/93	SA	1182	1163	995	NO WORK
03/28/93	SU	1183	1164	NO WORK	NO WORK
03/29/93	MO	1184	1165	996	827
03/30/93	TU	1185	1166	997	828
03/31/93	WE	1186	1167	998	829
04/01/93	TH	1187	1168	999	830
04/02/93	FR	1188	1169	1000	831
04/03/93	SA	1189	1170	1001	NO WORK
04/04/93	SU	1190	1171	NO WORK	NO WORK
04/05/93	MO	1191	1172	1002	832
04/06/93	TU	1192	1173	1003	833
04/07/93	WE	1193	1174	1004	834
04/08/93	TH	1194	1175	1005	835
04/09/93	FR	1195	1176	1006	836
04/10/93	SA	1196	1177	1007	NO WORK
04/11/93	SU	1197	1178	NO WORK	NO WORK
04/12/93	MO	1198	1179	1008	837
04/13/93	TU	1199	1180	1009	838
04/14/93	WE	1200	1181	1010	839
04/15/93	TH	1201	1182	1011	840
04/16/93	FR	1202	1183	1012	841

DATE	DAY OF THE WEEK	CALENDAR DAY	7-DAY WORK WEEK W/HOLIDAYS	6-DAY WORK WEEK W/HOLIDAYS	5-DAY WORK WEEK W/HOLIDAYS
04/17/93	SA	1203	1184	1013	NO WORK
04/18/93	SU	1204	1185	NO WORK	NO WORK
04/19/93	MO	1205	1186	1014	842
04/20/93	TU	1206	1187	1015	843
04/21/93	WE	1207	1188	1016	844
04/22/93	TH	1208	1189	1017	845
04/23/93	FR	1209	1190	1018	846
04/24/93	SA	1210	1191	1019	NO WORK
04/25/93	SU	1211	1192	NO WORK	NO WORK
04/26/93	MO	1212	1193	1020	847
04/27/93	TU	1213	1194	1021	848
04/28/93	WE	1214	1195	1022	849
04/29/93	TH	1215	1196	1023	850
04/30/93	FR	1216	1197	1024	851
05/01/93	SA	1217	1198	1025	NO WORK
05/02/93	SU	1218	1199	NO WORK	NO WORK
05/03/93	MO	1219	1200	1026	852
05/04/93	TU	1220	1201	1027	853
05/05/93	WE	1221	1202	1028	854
05/06/93	TH	1222	1203	1029	855
05/07/93	FR	1223	1204	1030	856
05/08/93	SA	1224	1205	1031	NO WORK
05/09/93	SU	1225	1206	NO WORK	NO WORK
05/10/93	MO	1226	1207	1032	857
05/11/93	TU	1227	1208	1033	858
05/12/93	WE	1228	1209	1034	859
05/13/93	TH	1229	1210	1035	860
05/14/93	FR	1230	1211	1036	861
05/15/93	SA	1231	1212	1037	NO WORK
05/16/93	SU	1232	1213	NO WORK	NO WORK
05/17/93	MO	1233	1214	1038	862
05/18/93	TU	1234	1215	1039	863
05/19/93	WE	1235	1216	1040	864
05/20/93	TH	1236	1217	1041	865
05/21/93	FR	1237	1218	1042	866
05/22/93	SA	1238	1219	1043	NO WORK
05/23/93	SU	1239	1220	NO WORK	NO WORK
05/24/93	MO	1240	1221	1044	867
05/25/93	TU	1241	1222	1045	868
05/26/93	WE	1242	1223	1046	869
05/27/93	TH	1243	1224	1047	870
05/28/93	FR	1244	1225	1048	871
05/29/93	SA	1245	1226	1049	NO WORK
05/30/93	SU	1246	1227	NO WORK	NO WORK
05/31/93	MO	1247	H	H	H
06/01/93	TU	1248	1228	1050	872
06/02/93	WE	1249	1229	1051	873
06/03/93	TH	1250	1230	1052	874
06/04/93	FR	1251	1231	1053	875
06/05/93	SA	1252	1232	1054	NO WORK
06/06/93	SU	1253	1233	NO WORK	NO WORK
06/07/93	MO	1254	1234	1055	876
06/08/93	TU	1255	1235	1056	877

DATE	DAY OF THE WEEK	CALENDAR DAY	7-DAY WORK WEEK W/HOLIDAYS	6-DAY WORK WEEK W/HOLIDAYS	5-DAY WORK WEEK W/HOLIDAYS
06/09/93	WE	1256	1236	1057	878
06/10/93	TH	1257	1237	1058	879
06/11/93	FR	1258	1238	1059	880
06/12/93	SA	1259	1239	1060	NO WORK
06/13/93	SU	1260	1240	NO WORK	NO WORK
06/14/93	MO	1261	1241	1061	881
06/15/93	TU	1262	1242	1062	882
06/16/93	WE	1263	1243	1063	883
06/17/93	TH	1264	1244	1064	884
06/18/93	FR	1265	1245	1065	885
06/19/93	SA	1266	1246	1066	NO WORK
06/20/93	SU	1267	1247	NO WORK	NO WORK
06/21/93	MO	1268	1248	1067	886
06/22/93	TU	1269	1249	1068	887
06/23/93	WE	1270	1250	1069	888
06/24/93	TH	1271	1251	1070	889
06/25/93	FR	1272	1252	1071	890
06/26/93	SA	1273	1253	1072	NO WORK
06/27/93	SU	1274	1254	NO WORK	NO WORK
06/28/93	MO	1275	1255	1073	891
06/29/93	TU	1276	1256	1074	892
06/30/93	WE	1277	1257	1075	893
07/01/93	TH	1278	1258	1076	894
07/02/93	FR	1279	1259	1077	895
07/03/93	SA	1280	1260	1078	NO WORK
07/04/93	SU	1281	H	NO WORK	NO WORK
07/05/93	MO	1282	1261	H	H
07/06/93	TU	1283	1262	1079	896
07/07/93	WE	1284	1263	1080	897
07/08/93	TH	1285	1264	1081	898
07/09/93	FR	1286	1265	1082	899
07/10/93	SA	1287	1266	1083	NO WORK
07/11/93	SU	1288	1267	NO WORK	NO WORK
07/12/93	MO	1289	1268	1084	900
07/13/93	TU	1290	1269	1085	901
07/14/93	WE	1291	1270	1086	902
07/15/93	TH	1292	1271	1087	903
07/16/93	FR	1293	1272	1088	904
07/17/93	SA	1294	1273	1089	NO WORK
07/18/93	SU	1295	1274	NO WORK	NO WORK
07/19/93	MO	1296	1275	1090	905
07/20/93	TU	1297	1276	1091	906
07/21/93	WE	1298	1277	1092	907
07/22/93	TH	1299	1278	1093	908
07/23/93	FR	1300	1279	1094	909
07/24/93	SA	1301	1280	1095	NO WORK
07/25/93	SU	1302	1281	NO WORK	NO WORK
07/26/93	MO	1303	1282	1096	910
07/27/93	TU	1304	1283	1097	911
07/28/93	WE	1305	1284	1098	912
07/29/93	TH	1306	1285	1099	913
07/30/93	FR	1307	1286	1100	914
07/31/93	SA	1308	1287	1101	NO WORK

DATE	DAY OF THE WEEK	CALENDAR DAY	7-DAY WORK WEEK W/HOLIDAYS	6-DAY WORK WEEK W/HOLIDAYS	5-DAY WORK WEEK W/HOLIDAYS
08/01/93	SU	1309	1288	NO WORK	NO WORK
08/02/93	MO	1310	1289	1102	915
08/03/93	TU	1311	1290	1103	916
08/04/93	WE	1312	1291	1104	917
08/05/93	TH	1313	1292	1105	918
08/06/93	FR	1314	1293	1106	919
08/07/93	SA	1315	1294	1107	NO WORK
08/08/93	SU	1316	1295	NO WORK	NO WORK
08/09/93	MO	1317	1296	1108	920
08/10/93	TU	1318	1297	1109	921
08/11/93	WE	1319	1298	1110	922
08/12/93	TH	1320	1299	1111	923
08/13/93	FR	1321	1300	1112	924
08/14/93	SA	1322	1301	1113	NO WORK
08/15/93	SU	1323	1302	NO WORK	NO WORK
08/16/93	MO	1324	1303	1114	925
08/17/93	TU	1325	1304	1115	926
08/18/93	WE	1326	1305	1116	927
08/19/93	TH	1327	1306	1117	928
08/20/93	FR	1328	1307	1118	929
08/21/93	SA	1329	1308	1119	NO WORK
08/22/93	SU	1330	1309	NO WORK	NO WORK
08/23/93	MO	1331	1310	1120	930
08/24/93	TU	1332	1311	1121	931
08/25/93	WE	1333	1312	1122	932
08/26/93	TH	1334	1313	1123	933
08/27/93	FR	1335	1314	1124	934
08/28/93	SA	1336	1315	1125	NO WORK
08/29/93	SU	1337	1316	NO WORK	NO WORK
08/30/93	MO	1338	1317	1126	935
08/31/93	TU	1339	1318	1127	936
09/01/93	WE	1340	1319	1128	937
09/02/93	TH	1341	1320	1129	938
09/03/93	FR	1342	1321	1130	939
09/04/93	SA	1343	1322	1131	NO WORK
09/05/93	SU	1344	1323	NO WORK	NO WORK
09/06/93	MO	1345	H	H	H
09/07/93	TU	1346	1324	1132	940
09/08/93	WE	1347	1325	1133	941
09/09/93	TH	1348	1326	1134	942
09/10/93	FR	1349	1327	1135	943
09/11/93	SA	1350	1328	1136	NO WORK
09/12/93	SU	1351	1329	NO WORK	NO WORK
09/13/93	MO	1352	1330	1137	944
09/14/93	TU	1353	1331	1138	945
09/15/93	WE	1354	1332	1139	946
09/16/93	TH	1355	1333	1140	947
09/17/93	FR	1356	1334	1141	948
09/18/93	SA	1357	1335	1142	NO WORK
09/19/93	SU	1358	1336	NO WORK	NO WORK
09/20/93	MO	1359	1337	1143	949
09/21/93	TU	1360	1338	1144	950
09/22/93	WE	1361	1339	1145	951

DATE	DAY OF THE WEEK	CALENDAR DAY	7-DAY WORK WEEK W/HOLIDAYS	6-DAY WORK WEEK W/HOLIDAYS	5-DAY WORK WEEK W/HOLIDAYS
09/23/93	TH	1362	1340	1146	952
09/24/93	FR	1363	1341	1147	953
09/25/93	SA	1364	1342	1148	NO WORK
09/26/93	SU	1365	1343	NO WORK	NO WORK
09/27/93	MO	1366	1344	1149	954
09/28/93	TU	1367	1345	1150	955
09/29/93	WE	1368	1346	1151	956
09/30/93	TH	1369	1347	1152	957
10/01/93	FR	1370	1348	1153	958
10/02/93	SA	1371	1349	1154	NO WORK
10/03/93	SU	1372	1350	NO WORK	NO WORK
10/04/93	MO	1373	1351	1155	959
10/05/93	TU	1374	1352	1156	960
10/06/93	WE	1375	1353	1157	961
10/07/93	TH	1376	1354	1158	962
10/08/93	FR	1377	1355	1159	963
10/09/93	SA	1378	1356	1160	NO WORK
10/10/93	SU	1379	1357	NO WORK	NO WORK
10/11/93	MO	1380	1358	1161	964
10/12/93	TU	1381	1359	1162	965
10/13/93	WE	1382	1360	1163	966
10/14/93	TH	1383	1361	1164	967
10/15/93	FR	1384	1362	1165	968
10/16/93	SA	1385	1363	1166	NO WORK
10/17/93	SU	1386	1364	NO WORK	NO WORK
10/18/93	MO	1387	1365	1167	969
10/19/93	TU	1388	1366	1168	970
10/20/93	WE	1389	1367	1169	971
10/21/93	TH	1390	1368	1170	972
10/22/93	FR	1391	1369	1171	973
10/23/93	SA	1392	1370	1172	NO WORK
10/24/93	SU	1393	1371	NO WORK	NO WORK
10/25/93	MO	1394	1372	1173	974
10/26/93	TU	1395	1373	1174	975
10/27/93	WE	1396	1374	1175	976
10/28/93	TH	1397	1375	1176	977
10/29/93	FR	1398	1376	1177	978
10/30/93	SA	1399	1377	1178	NO WORK
10/31/93	SU	1400	1378	NO WORK	NO WORK
11/01/93	MO	1401	1379	1179	979
11/02/93	TU	1402	1380	1180	980
11/03/93	WE	1403	1381	1181	981
11/04/93	TH	1404	1382	1182	982
11/05/93	FR	1405	1383	1183	983
11/06/93	SA	1406	1384	1184	NO WORK
11/07/93	SU	1407	1385	NO WORK	NO WORK
11/08/93	MO	1408	1386	1185	984
11/09/93	TU	1409	1387	1186	985
11/10/93	WE	1410	1388	1187	986
11/11/93	TH	1411	1389	1188	987
11/12/93	FR	1412	1390	1189	988
11/13/93	SA	1413	1391	1190	NO WORK
11/14/93	SU	1414	1392	NO WORK	NO WORK
11/15/93	MO	1415	1393	1191	989
11/16/93	TU	1416	1394	1192	990
11/17/93	WE	1417	1395	1193	991
11/18/93	TH	1418	1396	1194	992
11/19/93	FR	1419	1397	1195	993
11/20/93	SA	1420	1398	1196	NO WORK
11/21/93	SU	1421	1399	NO WORK	NO WORK
11/22/93	MO	1422	1400	1197	994
11/23/93	TU	1423	1401	1198	995
11/24/93	WE	1424	1402	1199	996
11/25/93	TH	1425	H	H	H
11/26/93	FR	1426	1403	1200	997
11/27/93	SA	1427	1404	1201	NO WORK
11/28/93	SU	1428	1405	NO WORK	NO WORK
11/29/93	MO	1429	1406	1202	998
11/30/93	TU	1430	1407	1203	999
12/01/93	WE	1431	1408	1204	1000
12/02/93	TH	1432	1409	1205	1001
12/03/93	FR	1433	1410	1206	1002
12/04/93	SA	1434	1411	1207	NO WORK
12/05/93	SU	1435	1412	NO WORK	NO WORK
12/06/93	MO	1436	1413	1208	1003
12/07/93	TU	1437	1414	1209	1004
12/08/93	WE	1438	1415	1210	1005
12/09/93	TH	1439	1416	1211	1006
12/10/93	FR	1440	1417	1212	1007
12/11/93	SA	1441	1418	1213	NO WORK
12/12/93	SU	1442	1419	NO WORK	NO WORK
12/13/93	MO	1443	1420	1214	1008
12/14/93	TU	1444	1421	1215	1009
12/15/93	WE	1445	1422	1216	1010
12/16/93	TH	1446	1423	1217	1011
12/17/93	FR	1447	1424	1218	1012
12/18/93	SA	1448	1425	1219	NO WORK
12/19/93	SU	1449	1426	NO WORK	NO WORK
12/20/93	MO	1450	1427	1220	1013
12/21/93	TU	1451	1428	1221	1014
12/22/93	WE	1452	1429	1222	1015
12/23/93	TH	1453	1430	1223	1016
12/24/93	FR	1454	1431	1224	H
12/25/93	SA	1455	H	H	NO WORK
12/26/93	SU	1456	1432	NO WORK	NO WORK
12/27/93	MO	1457	1433	1225	1017
12/28/93	TU	1458	1434	1226	1018
12/29/93	WE	1459	1435	1227	1019
12/30/93	TH	1460	1436	1228	1020
12/31/93	FR	1461	1437	1229	H

DATE	DAY OF THE WEEK	CALENDAR DAY	7-DAY WORK WEEK W/HOLIDAYS	6-DAY WORK WEEK W/HOLIDAYS	5-DAY WORK WEEK W/HOLIDAYS
01/01/94	SA	1462	H	H	NO WORK
01/02/94	SU	1463	1438	NO WORK	NO WORK
01/03/94	MO	1464	1439	1230	1021
01/04/94	TU	1465	1440	1231	1022
01/05/94	WE	1466	1441	1232	1023
01/06/94	TH	1467	1442	1233	1024
01/07/94	FR	1468	1443	1234	1025
01/08/94	SA	1469	1444	1235	NO WORK
01/09/94	SU	1470	1445	NO WORK	NO WORK
01/10/94	MO	1471	1446	1236	1026
01/11/94	TU	1472	1447	1237	1027
01/12/94	WE	1473	1448	1238	1028
01/13/94	TH	1474	1449	1239	1029
01/14/94	FR	1475	1450	1240	1030
01/15/94	SA	1476	1451	1241	NO WORK
01/16/94	SU	1477	1452	NO WORK	NO WORK
01/17/94	MO	1478	1453	1242	1031
01/18/94	TU	1479	1454	1243	1032
01/19/94	WE	1480	1455	1244	1033
01/20/94	TH	1481	1456	1245	1034
01/21/94	FR	1482	1457	1246	1035
01/22/94	SA	1483	1458	1247	NO WORK
01/23/94	SU	1484	1459	NO WORK	NO WORK
01/24/94	MO	1485	1460	1248	1036
01/25/94	TU	1486	1461	1249	1037
01/26/94	WE	1487	1462	1250	1038
01/27/94	TH	1488	1463	1251	1039
01/28/94	FR	1489	1464	1252	1040
01/29/94	SA	1490	1465	1253	NO WORK
01/30/94	SU	1491	1466	NO WORK	NO WORK
01/31/94	MO	1492	1467	1254	1041
02/01/94	TU	1493	1468	1255	1042
02/02/94	WE	1494	1469	1256	1043
02/03/94	TH	1495	1470	1257	1044
02/04/94	FR	1496	1471	1258	1045
02/05/94	SA	1497	1472	1259	NO WORK
02/06/94	SU	1498	1473	NO WORK	NO WORK
02/07/94	MO	1499	1474	1260	1046
02/08/94	TU	1500	1475	1261	1047
02/09/94	WE	1501	1476	1262	1048
02/10/94	TH	1502	1477	1263	1049
02/11/94	FR	1503	1478	1264	1050
02/12/94	SA	1504	1479	1265	NO WORK
02/13/94	SU	1505	1480	NO WORK	NO WORK
02/14/94	MO	1506	1481	1266	1051
02/15/94	TU	1507	1482	1267	1052
02/16/94	WE	1508	1483	1268	1053
02/17/94	TH	1509	1484	1269	1054
02/18/94	FR	1510	1485	1270	1055
02/19/94	SA	1511	1486	1271	NO WORK
02/20/94	SU	1512	1487	NO WORK	NO WORK
02/21/94	MO	1513	1488	1272	1056
02/22/94	TU	1514	1489	1273	1057

DATE	DAY OF THE WEEK	CALENDAR DAY	7-DAY WORK WEEK W/HOLIDAYS	6-DAY WORK WEEK W/HOLIDAYS	5-DAY WORK WEEK W/HOLIDAYS
02/23/94	WE	1515	1490	1274	1058
02/24/94	TH	1516	1491	1275	1059
02/25/94	FR	1517	1492	1276	1060
02/26/94	SA	1518	1493	1277	NO WORK
02/27/94	SU	1519	1494	NO WORK	NO WORK
02/28/94	MO	1520	1495	1278	1061
03/01/94	TU	1521	1496	1279	1062
03/02/94	WE	1522	1497	1280	1063
03/03/94	TH	1523	1498	1281	1064
03/04/94	FR	1524	1499	1282	1065
03/05/94	SA	1525	1500	1283	NO WORK
03/06/94	SU	1526	1501	NO WORK	NO WORK
03/07/94	MO	1527	1502	1284	1066
03/08/94	TU	1528	1503	1285	1067
03/09/94	WE	1529	1504	1286	1068
03/10/94	TH	1530	1505	1287	1069
03/11/94	FR	1531	1506	1288	1070
03/12/94	SA	1532	1507	1289	NO WORK
03/13/94	SU	1533	1508	NO WORK	NO WORK
03/14/94	MO	1534	1509	1290	1071
03/15/94	TU	1535	1510	1291	1072
03/16/94	WE	1536	1511	1292	1073
03/17/94	TH	1537	1512	1293	1074
03/18/94	FR	1538	1513	1294	1075
03/19/94	SA	1539	1514	1295	NO WORK
03/20/94	SU	1540	1515	NO WORK	NO WORK
03/21/94	MO	1541	1516	1296	1076
03/22/94	TU	1542	1517	1297	1077
03/23/94	WE	1543	1518	1298	1078
03/24/94	TH	1544	1519	1299	1079
03/25/94	FR	1545	1520	1300	1080
03/26/94	SA	1546	1521	1301	NO WORK
03/27/94	SU	1547	1522	NO WORK	NO WORK
03/28/94	MO	1548	1523	1302	1081
03/29/94	TU	1549	1524	1303	1082
03/30/94	WE	1550	1525	1304	1083
03/31/94	TH	1551	1526	1305	1084
04/01/94	FR	1552	1527	1306	1085
04/02/94	SA	1553	1528	1307	NO WORK
04/03/94	SU	1554	1529	NO WORK	NO WORK
04/04/94	MO	1555	1530	1308	1086
04/05/94	TU	1556	1531	1309	1087
04/06/94	WE	1557	1532	1310	1088
04/07/94	TH	1558	1533	1311	1089
04/08/94	FR	1559	1534	1312	1090
04/09/94	SA	1560	1535	1313	NO WORK
04/10/94	SU	1561	1536	NO WORK	NO WORK
04/11/94	MO	1562	1537	1314	1091
04/12/94	TU	1563	1538	1315	1092
04/13/94	WE	1564	1539	1316	1093
04/14/94	TH	1565	1540	1317	1094
04/15/94	FR	1566	1541	1318	1095
04/16/94	SA	1567	1542	1319	NO WORK

DATE	DAY OF THE WEEK	CALENDAR DAY	7-DAY WORK WEEK W/HOLIDAYS	6-DAY WORK WEEK W/HOLIDAYS	5-DAY WORK WEEK W/HOLIDAYS
04/17/94	SU	1568	1543	NO WORK	NO WORK
04/18/94	MO	1569	1544	1320	1096
04/19/94	TU	1570	1545	1321	1097
04/20/94	WE	1571	1546	1322	1098
04/21/94	TH	1572	1547	1323	1099
04/22/94	FR	1573	1548	1324	1100
04/23/94	SA	1574	1549	1325	NO WORK
04/24/94	SU	1575	1550	NO WORK	NO WORK
04/25/94	MO	1576	1551	1326	1101
04/26/94	TU	1577	1552	1327	1102
04/27/94	WE	1578	1553	1328	1103
04/28/94	TH	1579	1554	1329	1104
04/29/94	FR	1580	1555	1330	1105
04/30/94	SA	1581	1556	1331	NO WORK
05/01/94	SU	1582	1557	NO WORK	NO WORK
05/02/94	MO	1583	1558	1332	1106
05/03/94	TU	1584	1559	1333	1107
05/04/94	WE	1585	1560	1334	1108
05/05/94	TH	1586	1561	1335	1109
05/06/94	FR	1587	1562	1336	1110
05/07/94	SA	1588	1563	1337	NO WORK
05/08/94	SU	1589	1564	NO WORK	NO WORK
05/09/94	MO	1590	1565	1338	1111
05/10/94	TU	1591	1566	1339	1112
05/11/94	WE	1592	1567	1340	1113
05/12/94	TH	1593	1568	1341	1114
05/13/94	FR	1594	1569	1342	1115
05/14/94	SA	1595	1570	1343	NO WORK
05/15/94	SU	1596	1571	NO WORK	NO WORK
05/16/94	MO	1597	1572	1344	1116
05/17/94	TU	1598	1573	1345	1117
05/18/94	WE	1599	1574	1346	1118
05/19/94	TH	1600	1575	1347	1119
05/20/94	FR	1601	1576	1348	1120
05/21/94	SA	1602	1577	1349	NO WORK
05/22/94	SU	1603	1578	NO WORK	NO WORK
05/23/94	MO	1604	1579	1350	1121
05/24/94	TU	1605	1580	1351	1122
05/25/94	WE	1606	1581	1352	1123
05/26/94	TH	1607	1582	1353	1124
05/27/94	FR	1608	1583	1354	1125
05/28/94	SA	1609	1584	1355	NO WORK
05/29/94	SU	1610	1585	NO WORK	NO WORK
05/30/94	MO	1611	H	H	H
05/31/94	TU	1612	1586	1356	1126
06/01/94	WE	1613	1587	1357	1127
06/02/94	TH	1614	1588	1358	1128
06/03/94	FR	1615	1589	1359	1129
06/04/94	SA	1616	1590	1360	NO WORK
06/05/94	SU	1617	1591	NO WORK	NO WORK
06/06/94	MO	1618	1592	1361	1130
06/07/94	TU	1619	1593	1362	1131
06/08/94	WE	1620	1594	1363	1132
06/09/94	TH	1621	1595	1364	1133
06/10/94	FR	1622	1596	1365	1134
06/11/94	SA	1623	1597	1366	NO WORK
06/12/94	SU	1624	1598	NO WORK	NO WORK
06/13/94	MO	1625	1599	1367	1135
06/14/94	TU	1626	1600	1368	1136
06/15/94	WE	1627	1601	1369	1137
06/16/94	TH	1628	1602	1370	1138
06/17/94	FR	1629	1603	1371	1139
06/18/94	SA	1630	1604	1372	NO WORK
06/19/94	SU	1631	1605	NO WORK	NO WORK
06/20/94	MO	1632	1606	1373	1140
06/21/94	TU	1633	1607	1374	1141
06/22/94	WE	1634	1608	1375	1142
06/23/94	TH	1635	1609	1376	1143
06/24/94	FR	1636	1610	1377	1144
06/25/94	SA	1637	1611	1378	NO WORK
06/26/94	SU	1638	1612	NO WORK	NO WORK
06/27/94	MO	1639	1613	1379	1145
06/28/94	TU	1640	1614	1380	1146
06/29/94	WE	1641	1615	1381	1147
06/30/94	TH	1642	1616	1382	1148
07/01/94	FR	1643	1617	1383	1149
07/02/94	SA	1644	1618	1384	NO WORK
07/03/94	SU	1645	1619	NO WORK	NO WORK
07/04/94	MO	1646	H	H	H
07/05/94	TU	1647	1620	1385	1150
07/06/94	WE	1648	1621	1386	1151
07/07/94	TH	1649	1622	1387	1152
07/08/94	FR	1650	1623	1388	1153
07/09/94	SA	1651	1624	1389	NO WORK
07/10/94	SU	1652	1625	NO WORK	NO WORK
07/11/94	MO	1653	1626	1390	1154
07/12/94	TU	1654	1627	1391	1155
07/13/94	WE	1655	1628	1392	1156
07/14/94	TH	1656	1629	1393	1157
07/15/94	FR	1657	1630	1394	1158
07/16/94	SA	1658	1631	1395	NO WORK
07/17/94	SU	1659	1632	NO WORK	NO WORK
07/18/94	MO	1660	1633	1396	1159
07/19/94	TU	1661	1634	1397	1160
07/20/94	WE	1662	1635	1398	1161
07/21/94	TH	1663	1636	1399	1162
07/22/94	FR	1664	1637	1400	1163
07/23/94	SA	1665	1638	1401	NO WORK
07/24/94	SU	1666	1639	NO WORK	NO WORK
07/25/94	MO	1667	1640	1402	1164
07/26/94	TU	1668	1641	1403	1165
07/27/94	WE	1669	1642	1404	1166
07/28/94	TH	1670	1643	1405	1167
07/29/94	FR	1671	1644	1406	1168
07/30/94	SA	1672	1645	1407	NO WORK
07/31/94	SU	1673	1646	NO WORK	NO WORK

DATE	DAY OF THE WEEK	CALENDAR DAY	7-DAY WORK WEEK W/HOLIDAYS	6-DAY WORK WEEK W/HOLIDAYS	5-DAY WORK WEEK W/HOLIDAYS
08/01/94	MO	1674	1647	1408	1169
08/02/94	TU	1675	1648	1409	1170
08/03/94	WE	1676	1649	1410	1171
08/04/94	TH	1677	1650	1411	1172
08/05/94	FR	1678	1651	1412	1173
08/06/94	SA	1679	1652	1413	NO WORK
08/07/94	SU	1680	1653	NO WORK	NO WORK
08/08/94	MO	1681	1654	1414	1174
08/09/94	TU	1682	1655	1415	1175
08/10/94	WE	1683	1656	1416	1176
08/11/94	TH	1684	1657	1417	1177
08/12/94	FR	1685	1658	1418	1178
08/13/94	SA	1686	1659	1419	NO WORK
08/14/94	SU	1687	1660	NO WORK	NO WORK
08/15/94	MO	1688	1661	1420	1179
08/16/94	TU	1689	1662	1421	1180
08/17/94	WE	1690	1663	1422	1181
08/18/94	TH	1691	1664	1423	1182
08/19/94	FR	1692	1665	1424	1183
08/20/94	SA	1693	1666	1425	NO WORK
08/21/94	SU	1694	1667	NO WORK	NO WORK
08/22/94	MO	1695	1668	1426	1184
08/23/94	TU	1696	1669	1427	1185
08/24/94	WE	1697	1670	1428	1186
08/25/94	TH	1698	1671	1429	1187
08/26/94	FR	1699	1672	1430	1188
08/27/94	SA	1700	1673	1431	NO WORK
08/28/94	SU	1701	1674	NO WORK	NO WORK
08/29/94	MO	1702	1675	1432	1189
08/30/94	TU	1703	1676	1433	1190
08/31/94	WE	1704	1677	1434	1191
09/01/94	TH	1705	1678	1435	1192
09/02/94	FR	1706	1679	1436	1193
09/03/94	SA	1707	1680	1437	NO WORK
09/04/94	SU	1708	1681	NO WORK	NO WORK
09/05/94	MO	1709	H	H	H
09/06/94	TU	1710	1682	1438	1194
09/07/94	WE	1711	1683	1439	1195
09/08/94	TH	1712	1684	1440	1196
09/09/94	FR	1713	1685	1441	1197
09/10/94	SA	1714	1686	1442	NO WORK
09/11/94	SU	1715	1687	NO WORK	NO WORK
09/12/94	MO	1716	1688	1443	1198
09/13/94	TU	1717	1689	1444	1199
09/14/94	WE	1718	1690	1445	1200
09/15/94	TH	1719	1691	1446	1201
09/16/94	FR	1720	1692	1447	1202
09/17/94	SA	1721	1693	1448	NO WORK
09/18/94	SU	1722	1694	NO WORK	NO WORK
09/19/94	MO	1723	1695	1449	1203
09/20/94	TU	1724	1696	1450	1204
09/21/94	WE	1725	1697	1451	1205
09/22/94	TH	1726	1698	1452	1206

DATE	DAY OF THE WEEK	CALENDAR DAY	7-DAY WORK WEEK W/HOLIDAYS	6-DAY WORK WEEK W/HOLIDAYS	5-DAY WORK WEEK W/HOLIDAYS
09/23/94	FR	1727	1699	1453	1207
09/24/94	SA	1728	1700	1454	NO WORK
09/25/94	SU	1729	1701	NO WORK	NO WORK
09/26/94	MO	1730	1702	1455	1208
09/27/94	TU	1731	1703	1456	1209
09/28/94	WE	1732	1704	1457	1210
09/29/94	TH	1733	1705	1458	1211
09/30/94	FR	1734	1706	1459	1212
10/01/94	SA	1735	1707	1460	NO WORK
10/02/94	SU	1736	1708	NO WORK	NO WORK
10/03/94	MO	1737	1709	1461	1213
10/04/94	TU	1738	1710	1462	1214
10/05/94	WE	1739	1711	1463	1215
10/06/94	TH	1740	1712	1464	1216
10/07/94	FR	1741	1713	1465	1217
10/08/94	SA	1742	1714	1466	NO WORK
10/09/94	SU	1743	1715	NO WORK	NO WORK
10/10/94	MO	1744	1716	1467	1218
10/11/94	TU	1745	1717	1468	1219
10/12/94	WE	1746	1718	1469	1220
10/13/94	TH	1747	1719	1470	1221
10/14/94	FR	1748	1720	1471	1222
10/15/94	SA	1749	1721	1472	NO WORK
10/16/94	SU	1750	1722	NO WORK	NO WORK
10/17/94	MO	1751	1723	1473	1223
10/18/94	TU	1752	1724	1474	1224
10/19/94	WE	1753	1725	1475	1225
10/20/94	TH	1754	1726	1476	1226
10/21/94	FR	1755	1727	1477	1227
10/22/94	SA	1756	1728	1478	NO WORK
10/23/94	SU	1757	1729	NO WORK	NO WORK
10/24/94	MO	1758	1730	1479	1228
10/25/94	TU	1759	1731	1480	1229
10/26/94	WE	1760	1732	1481	1230
10/27/94	TH	1761	1733	1482	1231
10/28/94	FR	1762	1734	1483	1232
10/29/94	SA	1763	1735	1484	NO WORK
10/30/94	SU	1764	1736	NO WORK	NO WORK
10/31/94	MO	1765	1737	1485	1233
11/01/94	TU	1766	1738	1486	1234
11/02/94	WE	1767	1739	1487	1235
11/03/94	TH	1768	1740	1488	1236
11/04/94	FR	1769	1741	1489	1237
11/05/94	SA	1770	1742	1490	NO WORK
11/06/94	SU	1771	1743	NO WORK	NO WORK
11/07/94	MO	1772	1744	1491	1238
11/08/94	TU	1773	1745	1492	1239
11/09/94	WE	1774	1746	1493	1240
11/10/94	TH	1775	1747	1494	1241
11/11/94	FR	1776	1748	1495	1242
11/12/94	SA	1777	1749	1496	NO WORK
11/13/94	SU	1778	1750	NO WORK	NO WORK
11/14/94	MO	1779	1751	1497	1243

DATE	DAY OF THE WEEK	CALENDAR DAY	7-DAY WORK WEEK W/HOLIDAYS	6-DAY WORK WEEK W/HOLIDAYS	5-DAY WORK WEEK W/HOLIDAYS
11/15/94	TU	1780	1752	1498	1244
11/16/94	WE	1781	1753	1499	1245
11/17/94	TH	1782	1754	1500	1246
11/18/94	FR	1783	1755	1501	1247
11/19/94	SA	1784	1756	1502	NO WORK
11/20/94	SU	1785	1757	NO WORK	NO WORK
11/21/94	MO	1786	1758	1503	1248
11/22/94	TU	1787	1759	1504	1249
11/23/94	WE	1788	1760	1505	1250
11/24/94	TH	1789	H	H	H
11/25/94	FR	1790	1761	1506	1251
11/26/94	SA	1791	1762	1507	NO WORK
11/27/94	SU	1792	1763	NO WORK	NO WORK
11/28/94	MO	1793	1764	1508	1252
11/29/94	TU	1794	1765	1509	1253
11/30/94	WE	1795	1766	1510	1254
12/01/94	TH	1796	1767	1511	1255
12/02/94	FR	1797	1768	1512	1256
12/03/94	SA	1798	1769	1513	NO WORK
12/04/94	SU	1799	1770	NO WORK	NO WORK
12/05/94	MO	1800	1771	1514	1257
12/06/94	TU	1801	1772	1515	1258
12/07/94	WE	1802	1773	1516	1259
12/08/94	TH	1803	1774	1517	1260
12/09/94	FR	1804	1775	1518	1261
12/10/94	SA	1805	1776	1519	NO WORK
12/11/94	SU	1806	1777	NO WORK	NO WORK
12/12/94	MO	1807	1778	1520	1262
12/13/94	TU	1808	1779	1521	1263
12/14/94	WE	1809	1780	1522	1264
12/15/94	TH	1810	1781	1523	1265
12/16/94	FR	1811	1782	1524	1266
12/17/94	SA	1812	1783	1525	NO WORK
12/18/94	SU	1813	1784	NO WORK	NO WORK
12/19/94	MO	1814	1785	1526	1267
12/20/94	TU	1815	1786	1527	1268
12/21/94	WE	1816	1787	1528	1269
12/22/94	TH	1817	1788	1529	1270
12/23/94	FR	1818	1789	1530	1271
12/24/94	SA	1819	1790	1531	NO WORK
12/25/94	SU	1820	H	NO WORK	NO WORK
12/26/94	MO	1821	1791	H	H
12/27/94	TU	1822	1792	1532	1272
12/28/94	WE	1823	1793	1533	1273
12/29/94	TH	1824	1794	1534	1274
12/30/94	FR	1825	1795	1535	1275
12/31/94	SA	1826	1796	1536	NO WORK

DATE	DAY OF THE WEEK	CALENDAR DAY	7-DAY WORK WEEK W/HOLIDAYS	6-DAY WORK WEEK W/HOLIDAYS	5-DAY WORK WEEK W/HOLIDAYS
01/01/95	SU	1827	H	NO WORK	NO WORK
01/02/95	MO	1828	1797	H	H
01/03/95	TU	1829	1798	1537	1276
01/04/95	WE	1830	1799	1538	1277
01/05/95	TH	1831	1800	1539	1278
01/06/95	FR	1832	1801	1540	1279
01/07/95	SA	1833	1802	1541	NO WORK
01/08/95	SU	1834	1803	NO WORK	NO WORK
01/09/95	MO	1835	1804	1542	1280
01/10/95	TU	1836	1805	1543	1281
01/11/95	WE	1837	1806	1544	1282
01/12/95	TH	1838	1807	1545	1283
01/13/95	FR	1839	1808	1546	1284
01/14/95	SA	1840	1809	1547	NO WORK
01/15/95	SU	1841	1810	NO WORK	NO WORK
01/16/95	MO	1842	1811	1548	1285
01/17/95	TU	1843	1812	1549	1286
01/18/95	WE	1844	1813	1550	1287
01/19/95	TH	1845	1814	1551	1288
01/20/95	FR	1846	1815	1552	1289
01/21/95	SA	1847	1816	1553	NO WORK
01/22/95	SU	1848	1817	NO WORK	NO WORK
01/23/95	MO	1849	1818	1554	1290
01/24/95	TU	1850	1819	1555	1291
01/25/95	WE	1851	1820	1556	1292
01/26/95	TH	1852	1821	1557	1293
01/27/95	FR	1853	1822	1558	1294
01/28/95	SA	1854	1823	1559	NO WORK
01/29/95	SU	1855	1824	NO WORK	NO WORK
01/30/95	MO	1856	1825	1560	1295
01/31/95	TU	1857	1826	1561	1296
02/01/95	WE	1858	1827	1562	1297
02/02/95	TH	1859	1828	1563	1298
02/03/95	FR	1860	1829	1564	1299
02/04/95	SA	1861	1830	1565	NO WORK
02/05/95	SU	1862	1831	NO WORK	NO WORK
02/06/95	MO	1863	1832	1566	1300
02/07/95	TU	1864	1833	1567	1301
02/08/95	WE	1865	1834	1568	1302
02/09/95	TH	1866	1835	1569	1303
02/10/95	FR	1867	1836	1570	1304
02/11/95	SA	1868	1837	1571	NO WORK
02/12/95	SU	1869	1838	NO WORK	NO WORK
02/13/95	MO	1870	1839	1572	1305
02/14/95	TU	1871	1840	1573	1306
02/15/95	WE	1872	1841	1574	1307
02/16/95	TH	1873	1842	1575	1308
02/17/95	FR	1874	1843	1576	1309
02/18/95	SA	1875	1844	1577	NO WORK
02/19/95	SU	1876	1845	NO WORK	NO WORK
02/20/95	MO	1877	1846	1578	1310
02/21/95	TU	1878	1847	1579	1311
02/22/95	WE	1879	1848	1580	1312

DATE	DAY OF THE WEEK	CALENDAR DAY	7-DAY WORK WEEK W/HOLIDAYS	6-DAY WORK WEEK W/HOLIDAYS	5-DAY WORK WEEK W/HOLIDAYS
02/23/95	TH	1880	1849	1581	1313
02/24/95	FR	1881	1850	1582	1314
02/25/95	SA	1882	1851	1583	NO WORK
02/26/95	SU	1883	1852	NO WORK	NO WORK
02/27/95	MO	1884	1853	1584	1315
02/28/95	TU	1885	1854	1585	1316
03/01/95	WE	1886	1855	1586	1317
03/02/95	TH	1887	1856	1587	1318
03/03/95	FR	1888	1857	1588	1319
03/04/95	SA	1889	1858	1589	NO WORK
03/05/95	SU	1890	1859	NO WORK	NO WORK
03/06/95	MO	1891	1860	1590	1320
03/07/95	TU	1892	1861	1591	1321
03/08/95	WE	1893	1862	1592	1322
03/09/95	TH	1894	1863	1593	1323
03/10/95	FR	1895	1864	1594	1324
03/11/95	SA	1896	1865	1595	NO WORK
03/12/95	SU	1897	1866	NO WORK	NO WORK
03/13/95	MO	1898	1867	1596	1325
03/14/95	TU	1899	1868	1597	1326
03/15/95	WE	1900	1869	1598	1327
03/16/95	TH	1901	1870	1599	1328
03/17/95	FR	1902	1871	1600	1329
03/18/95	SA	1903	1872	1601	NO WORK
03/19/95	SU	1904	1873	NO WORK	NO WORK
03/20/95	MO	1905	1874	1602	1330
03/21/95	TU	1906	1875	1603	1331
03/22/95	WE	1907	1876	1604	1332
03/23/95	TH	1908	1877	1605	1333
03/24/95	FR	1909	1878	1606	1334
03/25/95	SA	1910	1879	1607	NO WORK
03/26/95	SU	1911	1880	NO WORK	NO WORK
03/27/95	MO	1912	1881	1608	1335
03/28/95	TU	1913	1882	1609	1336
03/29/95	WE	1914	1883	1610	1337
03/30/95	TH	1915	1884	1611	1338
03/31/95	FR	1916	1885	1612	1339
04/01/95	SA	1917	1886	1613	NO WORK
04/02/95	SU	1918	1887	NO WORK	NO WORK
04/03/95	MO	1919	1888	1614	1340
04/04/95	TU	1920	1889	1615	1341
04/05/95	WE	1921	1890	1616	1342
04/06/95	TH	1922	1891	1617	1343
04/07/95	FR	1923	1892	1618	1344
04/08/95	SA	1924	1893	1619	NO WORK
04/09/95	SU	1925	1894	NO WORK	NO WORK
04/10/95	MO	1926	1895	1620	1345
04/11/95	TU	1927	1896	1621	1346
04/12/95	WE	1928	1897	1622	1347
04/13/95	TH	1929	1898	1623	1348
04/14/95	FR	1930	1899	1624	1349
04/15/95	SA	1931	1900	1625	NO WORK
04/16/95	SU	1932	1901	NO WORK	NO WORK
04/17/95	MO	1933	1902	1626	1350
04/18/95	TU	1934	1903	1627	1351
04/19/95	WE	1935	1904	1628	1352
04/20/95	TH	1936	1905	1629	1353
04/21/95	FR	1937	1906	1630	1354
04/22/95	SA	1938	1907	1631	NO WORK
04/23/95	SU	1939	1908	NO WORK	NO WORK
04/24/95	MO	1940	1909	1632	1355
04/25/95	TU	1941	1910	1633	1356
04/26/95	WE	1942	1911	1634	1357
04/27/95	TH	1943	1912	1635	1358
04/28/95	FR	1944	1913	1636	1359
04/29/95	SA	1945	1914	1637	NO WORK
04/30/95	SU	1946	1915	NO WORK	NO WORK
05/01/95	MO	1947	1916	1638	1360
05/02/95	TU	1948	1917	1639	1361
05/03/95	WE	1949	1918	1640	1362
05/04/95	TH	1950	1919	1641	1363
05/05/95	FR	1951	1920	1642	1364
05/06/95	SA	1952	1921	1643	NO WORK
05/07/95	SU	1953	1922	NO WORK	NO WORK
05/08/95	MO	1954	1923	1644	1365
05/09/95	TU	1955	1924	1645	1366
05/10/95	WE	1956	1925	1646	1367
05/11/95	TH	1957	1926	1647	1368
05/12/95	FR	1958	1927	1648	1369
05/13/95	SA	1959	1928	1649	NO WORK
05/14/95	SU	1960	1929	NO WORK	NO WORK
05/15/95	MO	1961	1930	1650	1370
05/16/95	TU	1962	1931	1651	1371
05/17/95	WE	1963	1932	1652	1372
05/18/95	TH	1964	1933	1653	1373
05/19/95	FR	1965	1934	1654	1374
05/20/95	SA	1966	1935	1655	NO WORK
05/21/95	SU	1967	1936	NO WORK	NO WORK
05/22/95	MO	1968	1937	1656	1375
05/23/95	TU	1969	1938	1657	1376
05/24/95	WE	1970	1939	1658	1377
05/25/95	TH	1971	1940	1659	1378
05/26/95	FR	1972	1941	1660	1379
05/27/95	SA	1973	1942	1661	NO WORK
05/28/95	SU	1974	1943	NO WORK	NO WORK
05/29/95	MO	1975	H	H	H
05/30/95	TU	1976	1944	1662	1380
05/31/95	WE	1977	1945	1663	1381
06/01/95	TH	1978	1946	1664	1382
06/02/95	FR	1979	1947	1665	1383
06/03/95	SA	1980	1948	1666	NO WORK
06/04/95	SU	1981	1949	NO WORK	NO WORK
06/05/95	MO	1982	1950	1667	1384
06/06/95	TU	1983	1951	1668	1385
06/07/95	WE	1984	1952	1669	1386
06/08/95	TH	1985	1953	1670	1387

411

DATE	DAY OF THE WEEK	CALENDAR DAY	7-DAY WORK WEEK W/HOLIDAYS	6-DAY WORK WEEK W/HOLIDAYS	5-DAY WORK WEEK W/HOLIDAYS
06/09/95	FR	1986	1954	1671	1388
06/10/95	SA	1987	1955	1672	NO WORK
06/11/95	SU	1988	1956	NO WORK	NO WORK
06/12/95	MO	1989	1957	1673	1389
06/13/95	TU	1990	1958	1674	1390
06/14/95	WE	1991	1959	1675	1391
06/15/95	TH	1992	1960	1676	1392
06/16/95	FR	1993	1961	1677	1393
06/17/95	SA	1994	1962	1678	NO WORK
06/18/95	SU	1995	1963	NO WORK	NO WORK
06/19/95	MO	1996	1964	1679	1394
06/20/95	TU	1997	1965	1680	1395
06/21/95	WE	1998	1966	1681	1396
06/22/95	TH	1999	1967	1682	1397
06/23/95	FR	2000	1968	1683	1398
06/24/95	SA	2001	1969	1684	NO WORK
06/25/95	SU	2002	1970	NO WORK	NO WORK
06/26/95	MO	2003	1971	1685	1399
06/27/95	TU	2004	1972	1686	1400
06/28/95	WE	2005	1973	1687	1401
06/29/95	TH	2006	1974	1688	1402
06/30/95	FR	2007	1975	1689	1403
07/01/95	SA	2008	1976	1690	NO WORK
07/02/95	SU	2009	1977	NO WORK	NO WORK
07/03/95	MO	2010	1978	1691	1404
07/04/95	TU	2011	H	H	H
07/05/95	WE	2012	1979	1692	1405
07/06/95	TH	2013	1980	1693	1406
07/07/95	FR	2014	1981	1694	1407
07/08/95	SA	2015	1982	1695	NO WORK
07/09/95	SU	2016	1983	NO WORK	NO WORK
07/10/95	MO	2017	1984	1696	1408
07/11/95	TU	2018	1985	1697	1409
07/12/95	WE	2019	1986	1698	1410
07/13/95	TH	2020	1987	1699	1411
07/14/95	FR	2021	1988	1700	1412
07/15/95	SA	2022	1989	1701	NO WORK
07/16/95	SU	2023	1990	NO WORK	NO WORK
07/17/95	MO	2024	1991	1702	1413
07/18/95	TU	2025	1992	1703	1414
07/19/95	WE	2026	1993	1704	1415
07/20/95	TH	2027	1994	1705	1416
07/21/95	FR	2028	1995	1706	1417
07/22/95	SA	2029	1996	1707	NO WORK
07/23/95	SU	2030	1997	NO WORK	NO WORK
07/24/95	MO	2031	1998	1708	1418
07/25/95	TU	2032	1999	1709	1419
07/26/95	WE	2033	2000	1710	1420
07/27/95	TH	2034	2001	1711	1421
07/28/95	FR	2035	2002	1712	1422
07/29/95	SA	2036	2003	1713	NO WORK
07/30/95	SU	2037	2004	NO WORK	NO WORK
07/31/95	MO	2038	2005	1714	1423

DATE	DAY OF THE WEEK	CALENDAR DAY	7-DAY WORK WEEK W/HOLIDAYS	6-DAY WORK WEEK W/HOLIDAYS	5-DAY WORK WEEK W/HOLIDAYS
08/01/95	TU	2039	2006	1715	1424
08/02/95	WE	2040	2007	1716	1425
08/03/95	TH	2041	2008	1717	1426
08/04/95	FR	2042	2009	1718	1427
08/05/95	SA	2043	2010	1719	NO WORK
08/06/95	SU	2044	2011	NO WORK	NO WORK
08/07/95	MO	2045	2012	1720	1428
08/08/95	TU	2046	2013	1721	1429
08/09/95	WE	2047	2014	1722	1430
08/10/95	TH	2048	2015	1723	1431
08/11/95	FR	2049	2016	1724	1432
08/12/95	SA	2050	2017	1725	NO WORK
08/13/95	SU	2051	2018	NO WORK	NO WORK
08/14/95	MO	2052	2019	1726	1433
08/15/95	TU	2053	2020	1727	1434
08/16/95	WE	2054	2021	1728	1435
08/17/95	TH	2055	2022	1729	1436
08/18/95	FR	2056	2023	1730	1437
08/19/95	SA	2057	2024	1731	NO WORK
08/20/95	SU	2058	2025	NO WORK	NO WORK
08/21/95	MO	2059	2026	1732	1438
08/22/95	TU	2060	2027	1733	1439
08/23/95	WE	2061	2028	1734	1440
08/24/95	TH	2062	2029	1735	1441
08/25/95	FR	2063	2030	1736	1442
08/26/95	SA	2064	2031	1737	NO WORK
08/27/95	SU	2065	2032	NO WORK	NO WORK
08/28/95	MO	2066	2033	1738	1443
08/29/95	TU	2067	2034	1739	1444
08/30/95	WE	2068	2035	1740	1445
08/31/95	TH	2069	2036	1741	1446
09/01/95	FR	2070	2037	1742	1447
09/02/95	SA	2071	2038	1743	NO WORK
09/03/95	SU	2072	2039	NO WORK	NO WORK
09/04/95	MO	2073	H	H	H
09/05/95	TU	2074	2040	1744	1448
09/06/95	WE	2075	2041	1745	1449
09/07/95	TH	2076	2042	1746	1450
09/08/95	FR	2077	2043	1747	1451
09/09/95	SA	2078	2044	1748	NO WORK
09/10/95	SU	2079	2045	NO WORK	NO WORK
09/11/95	MO	2080	2046	1749	1452
09/12/95	TU	2081	2047	1750	1453
09/13/95	WE	2082	2048	1751	1454
09/14/95	TH	2083	2049	1752	1455
09/15/95	FR	2084	2050	1753	1456
09/16/95	SA	2085	2051	1754	NO WORK
09/17/95	SU	2086	2052	NO WORK	NO WORK
09/18/95	MO	2087	2053	1755	1457
09/19/95	TU	2088	2054	1756	1458
09/20/95	WE	2089	2055	1757	1459
09/21/95	TH	2090	2056	1758	1460
09/22/95	FR	2091	2057	1759	1461

DATE	DAY OF THE WEEK	CALENDAR DAY	7-DAY WORK WEEK W/HOLIDAYS	6-DAY WORK WEEK W/HOLIDAYS	5-DAY WORK WEEK W/HOLIDAYS
09/23/95	SA	2092	2058	1760	NO WORK
09/24/95	SU	2093	2059	NO WORK	NO WORK
09/25/95	MO	2094	2060	1761	1462
09/26/95	TU	2095	2061	1762	1463
09/27/95	WE	2096	2062	1763	1464
09/28/95	TH	2097	2063	1764	1465
09/29/95	FR	2098	2064	1765	1466
09/30/95	SA	2099	2065	1766	NO WORK
10/01/95	SU	2100	2066	NO WORK	NO WORK
10/02/95	MO	2101	2067	1767	1467
10/03/95	TU	2102	2068	1768	1468
10/04/95	WE	2103	2069	1769	1469
10/05/95	TH	2104	2070	1770	1470
10/06/95	FR	2105	2071	1771	1471
10/07/95	SA	2106	2072	1772	NO WORK
10/08/95	SU	2107	2073	NO WORK	NO WORK
10/09/95	MO	2108	2074	1773	1472
10/10/95	TU	2109	2075	1774	1473
10/11/95	WE	2110	2076	1775	1474
10/12/95	TH	2111	2077	1776	1475
10/13/95	FR	2112	2078	1777	1476
10/14/95	SA	2113	2079	1778	NO WORK
10/15/95	SU	2114	2080	NO WORK	NO WORK
10/16/95	MO	2115	2081	1779	1477
10/17/95	TU	2116	2082	1780	1478
10/18/95	WE	2117	2083	1781	1479
10/19/95	TH	2118	2084	1782	1480
10/20/95	FR	2119	2085	1783	1481
10/21/95	SA	2120	2086	1784	NO WORK
10/22/95	SU	2121	2087	NO WORK	NO WORK
10/23/95	MO	2122	2088	1785	1482
10/24/95	TU	2123	2089	1786	1483
10/25/95	WE	2124	2090	1787	1484
10/26/95	TH	2125	2091	1788	1485
10/27/95	FR	2126	2092	1789	1486
10/28/95	SA	2127	2093	1790	NO WORK
10/29/95	SU	2128	2094	NO WORK	NO WORK
10/30/95	MO	2129	2095	1791	1487
10/31/95	TU	2130	2096	1792	1488
11/01/95	WE	2131	2097	1793	1489
11/02/95	TH	2132	2098	1794	1490
11/03/95	FR	2133	2099	1795	1491
11/04/95	SA	2134	2100	1796	NO WORK
11/05/95	SU	2135	2101	NO WORK	NO WORK
11/06/95	MO	2136	2102	1797	1492
11/07/95	TU	2137	2103	1798	1493
11/08/95	WE	2138	2104	1799	1494
11/09/95	TH	2139	2105	1800	1495
11/10/95	FR	2140	2106	1801	1496
11/11/95	SA	2141	2107	1802	NO WORK
11/12/95	SU	2142	2108	NO WORK	NO WORK
11/13/95	MO	2143	2109	1803	1497
11/14/95	TU	2144	2110	1804	1498

DATE	DAY OF THE WEEK	CALENDAR DAY	7-DAY WORK WEEK W/HOLIDAYS	6-DAY WORK WEEK W/HOLIDAYS	5-DAY WORK WEEK W/HOLIDAYS
11/15/95	WE	2145	2111	1805	1499
11/16/95	TH	2146	2112	1806	1500
11/17/95	FR	2147	2113	1807	1501
11/18/95	SA	2148	2114	1808	NO WORK
11/19/95	SU	2149	2115	NO WORK	NO WORK
11/20/95	MO	2150	2116	1809	1502
11/21/95	TU	2151	2117	1810	1503
11/22/95	WE	2152	2118	1811	1504
11/23/95	TH	2153	H	H	H
11/24/95	FR	2154	2119	1812	1505
11/25/95	SA	2155	2120	1813	NO WORK
11/26/95	SU	2156	2121	NO WORK	NO WORK
11/27/95	MO	2157	2122	1814	1506
11/28/95	TU	2158	2123	1815	1507
11/29/95	WE	2159	2124	1816	1508
11/30/95	TH	2160	2125	1817	1509
12/01/95	FR	2161	2126	1818	1510
12/02/95	SA	2162	2127	1819	NO WORK
12/03/95	SU	2163	2128	NO WORK	NO WORK
12/04/95	MO	2164	2129	1820	1511
12/05/95	TU	2165	2130	1821	1512
12/06/95	WE	2166	2131	1822	1513
12/07/95	TH	2167	2132	1823	1514
12/08/95	FR	2168	2133	1824	1515
12/09/95	SA	2169	2134	1825	NO WORK
12/10/95	SU	2170	2135	NO WORK	NO WORK
12/11/95	MO	2171	2136	1826	1516
12/12/95	TU	2172	2137	1827	1517
12/13/95	WE	2173	2138	1828	1518
12/14/95	TH	2174	2139	1829	1519
12/15/95	FR	2175	2140	1830	1520
12/16/95	SA	2176	2141	1831	NO WORK
12/17/95	SU	2177	2142	NO WORK	NO WORK
12/18/95	MO	2178	2143	1832	1521
12/19/95	TU	2179	2144	1833	1522
12/20/95	WE	2180	2145	1834	1523
12/21/95	TH	2181	2146	1835	1524
12/22/95	FR	2182	2147	1836	1525
12/23/95	SA	2183	2148	1837	NO WORK
12/24/95	SU	2184	2149	NO WORK	NO WORK
12/25/95	MO	2185	H	H	H
12/26/95	TU	2186	2150	1838	1526
12/27/95	WE	2187	2151	1839	1527
12/28/95	TH	2188	2152	1840	1528
12/29/95	FR	2189	2153	1841	1529
12/30/95	SA	2190	2154	1842	NO WORK
12/31/95	SU	2191	2155	NO WORK	NO WORK

413

DATE	DAY OF THE WEEK	CALENDAR DAY	7-DAY WORK WEEK W/HOLIDAYS	6-DAY WORK WEEK W/HOLIDAYS	5-DAY WORK WEEK W/HOLIDAYS
01/01/96	MO	2192	H	H	H
01/02/96	TU	2193	2156	1843	1530
01/03/96	WE	2194	2157	1844	1531
01/04/96	TH	2195	2158	1845	1532
01/05/96	FR	2196	2159	1846	1533
01/06/96	SA	2197	2160	1847	NO WORK
01/07/96	SU	2198	2161	NO WORK	NO WORK
01/08/96	MO	2199	2162	1848	1534
01/09/96	TU	2200	2163	1849	1535
01/10/96	WE	2201	2164	1850	1536
01/11/96	TH	2202	2165	1851	1537
01/12/96	FR	2203	2166	1852	1538
01/13/96	SA	2204	2167	1853	NO WORK
01/14/96	SU	2205	2168	NO WORK	NO WORK
01/15/96	MO	2206	2169	1854	1539
01/16/96	TU	2207	2170	1855	1540
01/17/96	WE	2208	2171	1856	1541
01/18/96	TH	2209	2172	1857	1542
01/19/96	FR	2210	2173	1858	1543
01/20/96	SA	2211	2174	1859	NO WORK
01/21/96	SU	2212	2175	NO WORK	NO WORK
01/22/96	MO	2213	2176	1860	1544
01/23/96	TU	2214	2177	1861	1545
01/24/96	WE	2215	2178	1862	1546
01/25/96	TH	2216	2179	1863	1547
01/26/96	FR	2217	2180	1864	1548
01/27/96	SA	2218	2181	1865	NO WORK
01/28/96	SU	2219	2182	NO WORK	NO WORK
01/29/96	MO	2220	2183	1866	1549
01/30/96	TU	2221	2184	1867	1550
01/31/96	WE	2222	2185	1868	1551
02/01/96	TH	2223	2186	1869	1552
02/02/96	FR	2224	2187	1870	1553
02/03/96	SA	2225	2188	1871	NO WORK
02/04/96	SU	2226	2189	NO WORK	NO WORK
02/05/96	MO	2227	2190	1872	1554
02/06/96	TU	2228	2191	1873	1555
02/07/96	WE	2229	2192	1874	1556
02/08/96	TH	2230	2193	1875	1557
02/09/96	FR	2231	2194	1876	1558
02/10/96	SA	2232	2195	1877	NO WORK
02/11/96	SU	2233	2196	NO WORK	NO WORK
02/12/96	MO	2234	2197	1878	1559
02/13/96	TU	2235	2198	1879	1560
02/14/96	WE	2236	2199	1880	1561
02/15/96	TH	2237	2200	1881	1562
02/16/96	FR	2238	2201	1882	1563
02/17/96	SA	2239	2202	1883	NO WORK
02/18/96	SU	2240	2203	NO WORK	NO WORK
02/19/96	MO	2241	2204	1884	1564
02/20/96	TU	2242	2205	1885	1565
02/21/96	WE	2243	2206	1886	1566
02/22/96	TH	2244	2207	1887	1567

DATE	DAY OF THE WEEK	CALENDAR DAY	7-DAY WORK WEEK W/HOLIDAYS	6-DAY WORK WEEK W/HOLIDAYS	5-DAY WORK WEEK W/HOLIDAYS
02/23/96	FR	2245	2208	1888	1568
02/24/96	SA	2246	2209	1889	NO WORK
02/25/96	SU	2247	2210	NO WORK	NO WORK
02/26/96	MO	2248	2211	1890	1569
02/27/96	TU	2249	2212	1891	1570
02/28/96	WE	2250	2213	1892	1571
02/29/96	TH	2251	2214	1893	1572
03/01/96	FR	2252	2215	1894	1573
03/02/96	SA	2253	2216	1895	NO WORK
03/03/96	SU	2254	2217	NO WORK	NO WORK
03/04/96	MO	2255	2218	1896	1574
03/05/96	TU	2256	2219	1897	1575
03/06/96	WE	2257	2220	1898	1576
03/07/96	TH	2258	2221	1899	1577
03/08/96	FR	2259	2222	1900	1578
03/09/96	SA	2260	2223	1901	NO WORK
03/10/96	SU	2261	2224	NO WORK	NO WORK
03/11/96	MO	2262	2225	1902	1579
03/12/96	TU	2263	2226	1903	1580
03/13/96	WE	2264	2227	1904	1581
03/14/96	TH	2265	2228	1905	1582
03/15/96	FR	2266	2229	1906	1583
03/16/96	SA	2267	2230	1907	NO WORK
03/17/96	SU	2268	2231	NO WORK	NO WORK
03/18/96	MO	2269	2232	1908	1584
03/19/96	TU	2270	2233	1909	1585
03/20/96	WE	2271	2234	1910	1586
03/21/96	TH	2272	2235	1911	1587
03/22/96	FR	2273	2236	1912	1588
03/23/96	SA	2274	2237	1913	NO WORK
03/24/96	SU	2275	2238	NO WORK	NO WORK
03/25/96	MO	2276	2239	1914	1589
03/26/96	TU	2277	2240	1915	1590
03/27/96	WE	2278	2241	1916	1591
03/28/96	TH	2279	2242	1917	1592
03/29/96	FR	2280	2243	1918	1593
03/30/96	SA	2281	2244	1919	NO WORK
03/31/96	SU	2282	2245	NO WORK	NO WORK
04/01/96	MO	2283	2246	1920	1594
04/02/96	TU	2284	2247	1921	1595
04/03/96	WE	2285	2248	1922	1596
04/04/96	TH	2286	2249	1923	1597
04/05/96	FR	2287	2250	1924	1598
04/06/96	SA	2288	2251	1925	NO WORK
04/07/96	SU	2289	2252	NO WORK	NO WORK
04/08/96	MO	2290	2253	1926	1599
04/09/96	TU	2291	2254	1927	1600
04/10/96	WE	2292	2255	1928	1601
04/11/96	TH	2293	2256	1929	1602
04/12/96	FR	2294	2257	1930	1603
04/13/96	SA	2295	2258	1931	NO WORK
04/14/96	SU	2296	2259	NO WORK	NO WORK
04/15/96	MO	2297	2260	1932	1604

DATE	DAY OF THE WEEK	CALENDAR DAY	7-DAY WORK WEEK W/HOLIDAYS	6-DAY WORK WEEK W/HOLIDAYS	5-DAY WORK WEEK W/HOLIDAYS
04/16/96	TU	2298	2261	1933	1605
04/17/96	WE	2299	2262	1934	1606
04/18/96	TH	2300	2263	1935	1607
04/19/96	FR	2301	2264	1936	1608
04/20/96	SA	2302	2265	1937	NO WORK
04/21/96	SU	2303	2266	NO WORK	NO WORK
04/22/96	MO	2304	2267	1938	1609
04/23/96	TU	2305	2268	1939	1610
04/24/96	WE	2306	2269	1940	1611
04/25/96	TH	2307	2270	1941	1612
04/26/96	FR	2308	2271	1942	1613
04/27/96	SA	2309	2272	1943	NO WORK
04/28/96	SU	2310	2273	NO WORK	NO WORK
04/29/96	MO	2311	2274	1944	1614
04/30/96	TU	2312	2275	1945	1615
05/01/96	WE	2313	2276	1946	1616
05/02/96	TH	2314	2277	1947	1617
05/03/96	FR	2315	2278	1948	1618
05/04/96	SA	2316	2279	1949	NO WORK
05/05/96	SU	2317	2280	NO WORK	NO WORK
05/06/96	MO	2318	2281	1950	1619
05/07/96	TU	2319	2282	1951	1620
05/08/96	WE	2320	2283	1952	1621
05/09/96	TH	2321	2284	1953	1622
05/10/96	FR	2322	2285	1954	1623
05/11/96	SA	2323	2286	1955	NO WORK
05/12/96	SU	2324	2287	NO WORK	NO WORK
05/13/96	MO	2325	2288	1956	1624
05/14/96	TU	2326	2289	1957	1625
05/15/96	WE	2327	2290	1958	1626
05/16/96	TH	2328	2291	1959	1627
05/17/96	FR	2329	2292	1960	1628
05/18/96	SA	2330	2293	1961	NO WORK
05/19/96	SU	2331	2294	NO WORK	NO WORK
05/20/96	MO	2332	2295	1962	1629
05/21/96	TU	2333	2296	1963	1630
05/22/96	WE	2334	2297	1964	1631
05/23/96	TH	2335	2298	1965	1632
05/24/96	FR	2336	2299	1966	1633
05/25/96	SA	2337	2300	1967	NO WORK
05/26/96	SU	2338	2301	NO WORK	NO WORK
05/27/96	MO	2339	H	H	H
05/28/96	TU	2340	2302	1968	1634
05/29/96	WE	2341	2303	1969	1635
05/30/96	TH	2342	2304	1970	1636
05/31/96	FR	2343	2305	1971	1637
06/01/96	SA	2344	2306	1972	NO WORK
06/02/96	SU	2345	2307	NO WORK	NO WORK
06/03/96	MO	2346	2308	1973	1638
06/04/96	TU	2347	2309	1974	1639
06/05/96	WE	2348	2310	1975	1640
06/06/96	TH	2349	2311	1976	1641
06/07/96	FR	2350	2312	1977	1642

DATE	DAY OF THE WEEK	CALENDAR DAY	7-DAY WORK WEEK W/HOLIDAYS	6-DAY WORK WEEK W/HOLIDAYS	5-DAY WORK WEEK W/HOLIDAYS
06/08/96	SA	2351	2313	1978	NO WORK
06/09/96	SU	2352	2314	NO WORK	NO WORK
06/10/96	MO	2353	2315	1979	1643
06/11/96	TU	2354	2316	1980	1644
06/12/96	WE	2355	2317	1981	1645
06/13/96	TH	2356	2318	1982	1646
06/14/96	FR	2357	2319	1983	1647
06/15/96	SA	2358	2320	1984	NO WORK
06/16/96	SU	2359	2321	NO WORK	NO WORK
06/17/96	MO	2360	2322	1985	1648
06/18/96	TU	2361	2323	1986	1649
06/19/96	WE	2362	2324	1987	1650
06/20/96	TH	2363	2325	1988	1651
06/21/96	FR	2364	2326	1989	1652
06/22/96	SA	2365	2327	1990	NO WORK
06/23/96	SU	2366	2328	NO WORK	NO WORK
06/24/96	MO	2367	2329	1991	1653
06/25/96	TU	2368	2330	1992	1654
06/26/96	WE	2369	2331	1993	1655
06/27/96	TH	2370	2332	1994	1656
06/28/96	FR	2371	2333	1995	1657
06/29/96	SA	2372	2334	1996	NO WORK
06/30/96	SU	2373	2335	NO WORK	NO WORK
07/01/96	MO	2374	2336	1997	1658
07/02/96	TU	2375	2337	1998	1659
07/03/96	WE	2376	2338	1999	1660
07/04/96	TH	2377	H	H	H
07/05/96	FR	2378	2339	2000	1661
07/06/96	SA	2379	2340	2001	NO WORK
07/07/96	SU	2380	2341	NO WORK	NO WORK
07/08/96	MO	2381	2342	2002	1662
07/09/96	TU	2382	2343	2003	1663
07/10/96	WE	2383	2344	2004	1664
07/11/96	TH	2384	2345	2005	1665
07/12/96	FR	2385	2346	2006	1666
07/13/96	SA	2386	2347	2007	NO WORK
07/14/96	SU	2387	2348	NO WORK	NO WORK
07/15/96	MO	2388	2349	2008	1667
07/16/96	TU	2389	2350	2009	1668
07/17/96	WE	2390	2351	2010	1669
07/18/96	TH	2391	2352	2011	1670
07/19/96	FR	2392	2353	2012	1671
07/20/96	SA	2393	2354	2013	NO WORK
07/21/96	SU	2394	2355	NO WORK	NO WORK
07/22/96	MO	2395	2356	2014	1672
07/23/96	TU	2396	2357	2015	1673
07/24/96	WE	2397	2358	2016	1674
07/25/96	TH	2398	2359	2017	1675
07/26/96	FR	2399	2360	2018	1676
07/27/96	SA	2400	2361	2019	NO WORK
07/28/96	SU	2401	2362	NO WORK	NO WORK
07/29/96	MO	2402	2363	2020	1677
07/30/96	TU	2403	2364	2021	1678

DATE	DAY OF THE WEEK	CALENDAR DAY	7-DAY WORK WEEK W/HOLIDAYS	6-DAY WORK WEEK W/HOLIDAYS	5-DAY WORK WEEK W/HOLIDAYS
07/31/96	WE	2404	2365	2022	1679
08/01/96	TH	2405	2366	2023	1680
08/02/96	FR	2406	2367	2024	1681
08/03/96	SA	2407	2368	2025	NO WORK
08/04/96	SU	2408	2369	NO WORK	NO WORK
08/05/96	MO	2409	2370	2026	1682
08/06/96	TU	2410	2371	2027	1683
08/07/96	WE	2411	2372	2028	1684
08/08/96	TH	2412	2373	2029	1685
08/09/96	FR	2413	2374	2030	1686
08/10/96	SA	2414	2375	2031	NO WORK
08/11/96	SU	2415	2376	NO WORK	NO WORK
08/12/96	MO	2416	2377	2032	1687
08/13/96	TU	2417	2378	2033	1688
08/14/96	WE	2418	2379	2034	1689
08/15/96	TH	2419	2380	2035	1690
08/16/96	FR	2420	2381	2036	1691
08/17/96	SA	2421	2382	2037	NO WORK
08/18/96	SU	2422	2383	NO WORK	NO WORK
08/19/96	MO	2423	2384	2038	1692
08/20/96	TU	2424	2385	2039	1693
08/21/96	WE	2425	2386	2040	1694
08/22/96	TH	2426	2387	2041	1695
08/23/96	FR	2427	2388	2042	1696
08/24/96	SA	2428	2389	2043	NO WORK
08/25/96	SU	2429	2390	NO WORK	NO WORK
08/26/96	MO	2430	2391	2044	1697
08/27/96	TU	2431	2392	2045	1698
08/28/96	WE	2432	2393	2046	1699
08/29/96	TH	2433	2394	2047	1700
08/30/96	FR	2434	2395	2048	1701
08/31/96	SA	2435	2396	2049	NO WORK
09/01/96	SU	2436	2397	NO WORK	NO WORK
09/02/96	MO	2437	H	H	H
09/03/96	TU	2438	2398	2050	1702
09/04/96	WE	2439	2399	2051	1703
09/05/96	TH	2440	2400	2052	1704
09/06/96	FR	2441	2401	2053	1705
09/07/96	SA	2442	2402	2054	NO WORK
09/08/96	SU	2443	2403	NO WORK	NO WORK
09/09/96	MO	2444	2404	2055	1706
09/10/96	TU	2445	2405	2056	1707
09/11/96	WE	2446	2406	2057	1708
09/12/96	TH	2447	2407	2058	1709
09/13/96	FR	2448	2408	2059	1710
09/14/96	SA	2449	2409	2060	NO WORK
09/15/96	SU	2450	2410	NO WORK	NO WORK
09/16/96	MO	2451	2411	2061	1711
09/17/96	TU	2452	2412	2062	1712
09/18/96	WE	2453	2413	2063	1713
09/19/96	TH	2454	2414	2064	1714
09/20/96	FR	2455	2415	2065	1715
09/21/96	SA	2456	2416	2066	NO WORK
09/22/96	SU	2457	2417	NO WORK	NO WORK
09/23/96	MO	2458	2418	2067	1716
09/24/96	TU	2459	2419	2068	1717
09/25/96	WE	2460	2420	2069	1718
09/26/96	TH	2461	2421	2070	1719
09/27/96	FR	2462	2422	2071	1720
09/28/96	SA	2463	2423	2072	NO WORK
09/29/96	SU	2464	2424	NO WORK	NO WORK
09/30/96	MO	2465	2425	2073	1721
10/01/96	TU	2466	2426	2074	1722
10/02/96	WE	2467	2427	2075	1723
10/03/96	TH	2468	2428	2076	1724
10/04/96	FR	2469	2429	2077	1725
10/05/96	SA	2470	2430	2078	NO WORK
10/06/96	SU	2471	2431	NO WORK	NO WORK
10/07/96	MO	2472	2432	2079	1726
10/08/96	TU	2473	2433	2080	1727
10/09/96	WE	2474	2434	2081	1728
10/10/96	TH	2475	2435	2082	1729
10/11/96	FR	2476	2436	2083	1730
10/12/96	SA	2477	2437	2084	NO WORK
10/13/96	SU	2478	2438	NO WORK	NO WORK
10/14/96	MO	2479	2439	2085	1731
10/15/96	TU	2480	2440	2086	1732
10/16/96	WE	2481	2441	2087	1733
10/17/96	TH	2482	2442	2088	1734
10/18/96	FR	2483	2443	2089	1735
10/19/96	SA	2484	2444	2090	NO WORK
10/20/96	SU	2485	2445	NO WORK	NO WORK
10/21/96	MO	2486	2446	2091	1736
10/22/96	TU	2487	2447	2092	1737
10/23/96	WE	2488	2448	2093	1738
10/24/96	TH	2489	2449	2094	1739
10/25/96	FR	2490	2450	2095	1740
10/26/96	SA	2491	2451	2096	NO WORK
10/27/96	SU	2492	2452	NO WORK	NO WORK
10/28/96	MO	2493	2453	2097	1741
10/29/96	TU	2494	2454	2098	1742
10/30/96	WE	2495	2455	2099	1743
10/31/96	TH	2496	2456	2100	1744
11/01/96	FR	2497	2457	2101	1745
11/02/96	SA	2498	2458	2102	NO WORK
11/03/96	SU	2499	2459	NO WORK	NO WORK
11/04/96	MO	2500	2460	2103	1746
11/05/96	TU	2501	2461	2104	1747
11/06/96	WE	2502	2462	2105	1748
11/07/96	TH	2503	2463	2106	1749
11/08/96	FR	2504	2464	2107	1750
11/09/96	SA	2505	2465	2108	NO WORK
11/10/96	SU	2506	2466	NO WORK	NO WORK
11/11/96	MO	2507	2467	2109	1751
11/12/96	TU	2508	2468	2110	1752
11/13/96	WE	2509	2469	2111	1753

DATE	DAY OF THE WEEK	CALENDAR DAY	7-DAY WORK WEEK W/HOLIDAYS	6-DAY WORK WEEK W/HOLIDAYS	5-DAY WORK WEEK W/HOLIDAYS
11/14/96	TH	2510	2470	2112	1754
11/15/96	FR	2511	2471	2113	1755
11/16/96	SA	2512	2472	2114	NO WORK
11/17/96	SU	2513	2473	NO WORK	NO WORK
11/18/96	MO	2514	2474	2115	1756
11/19/96	TU	2515	2475	2116	1757
11/20/96	WE	2516	2476	2117	1758
11/21/96	TH	2517	2477	2118	1759
11/22/96	FR	2518	2478	2119	1760
11/23/96	SA	2519	2479	2120	NO WORK
11/24/96	SU	2520	2480	NO WORK	NO WORK
11/25/96	MO	2521	2481	2121	1761
11/26/96	TU	2522	2482	2122	1762
11/27/96	WE	2523	2483	2123	1763
11/28/96	TH	2524	H	H	H
11/29/96	FR	2525	2484	2124	1764
11/30/96	SA	2526	2485	2125	NO WORK
12/01/96	SU	2527	2486	NO WORK	NO WORK
12/02/96	MO	2528	2487	2126	1765
12/03/96	TU	2529	2488	2127	1766
12/04/96	WE	2530	2489	2128	1767
12/05/96	TH	2531	2490	2129	1768
12/06/96	FR	2532	2491	2130	1769
12/07/96	SA	2533	2492	2131	NO WORK
12/08/96	SU	2534	2493	NO WORK	NO WORK
12/09/96	MO	2535	2494	2132	1770
12/10/96	TU	2536	2495	2133	1771
12/11/96	WE	2537	2496	2134	1772
12/12/96	TH	2538	2497	2135	1773
12/13/96	FR	2539	2498	2136	1774
12/14/96	SA	2540	2499	2137	NO WORK
12/15/96	SU	2541	2500	NO WORK	NO WORK
12/16/96	MO	2542	2501	2138	1775
12/17/96	TU	2543	2502	2139	1776
12/18/96	WE	2544	2503	2140	1777
12/19/96	TH	2545	2504	2141	1778
12/20/96	FR	2546	2505	2142	1779
12/21/96	SA	2547	2506	2143	NO WORK
12/22/96	SU	2548	2507	NO WORK	NO WORK
12/23/96	MO	2549	2508	2144	1780
12/24/96	TU	2550	2509	2145	1781
12/25/96	WE	2551	H	H	H
12/26/96	TH	2552	2510	2146	1782
12/27/96	FR	2553	2511	2147	1783
12/28/96	SA	2554	2512	2148	NO WORK
12/29/96	SU	2555	2513	NO WORK	NO WORK
12/30/96	MO	2556	2514	2149	1784
12/31/96	TU	2557	2515	2150	1785
01/01/97	WE	2558	H	H	H
01/02/97	TH	2559	2516	2151	1786
01/03/97	FR	2560	2517	2152	1787
01/04/97	SA	2561	2518	2153	NO WORK
01/05/97	SU	2562	2519	NO WORK	NO WORK
01/06/97	MO	2563	2520	2154	1788
01/07/97	TU	2564	2521	2155	1789
01/08/97	WE	2565	2522	2156	1790
01/09/97	TH	2566	2523	2157	1791
01/10/97	FR	2567	2524	2158	1792
01/11/97	SA	2568	2525	2159	NO WORK
01/12/97	SU	2569	2526	NO WORK	NO WORK
01/13/97	MO	2570	2527	2160	1793
01/14/97	TU	2571	2528	2161	1794
01/15/97	WE	2572	2529	2162	1795
01/16/97	TH	2573	2530	2163	1796
01/17/97	FR	2574	2531	2164	1797
01/18/97	SA	2575	2532	2165	NO WORK
01/19/97	SU	2576	2533	NO WORK	NO WORK
01/20/97	MO	2577	2534	2166	1798
01/21/97	TU	2578	2535	2167	1799
01/22/97	WE	2579	2536	2168	1800
01/23/97	TH	2580	2537	2169	1801
01/24/97	FR	2581	2538	2170	1802
01/25/97	SA	2582	2539	2171	NO WORK
01/26/97	SU	2583	2540	NO WORK	NO WORK
01/27/97	MO	2584	2541	2172	1803
01/28/97	TU	2585	2542	2173	1804
01/29/97	WE	2586	2543	2174	1805
01/30/97	TH	2587	2544	2175	1806
01/31/97	FR	2588	2545	2176	1807
02/01/97	SA	2589	2546	2177	NO WORK
02/02/97	SU	2590	2547	NO WORK	NO WORK
02/03/97	MO	2591	2548	2178	1808
02/04/97	TU	2592	2549	2179	1809
02/05/97	WE	2593	2550	2180	1810
02/06/97	TH	2594	2551	2181	1811
02/07/97	FR	2595	2552	2182	1812
02/08/97	SA	2596	2553	2183	NO WORK
02/09/97	SU	2597	2554	NO WORK	NO WORK
02/10/97	MO	2598	2555	2184	1813
02/11/97	TU	2599	2556	2185	1814
02/12/97	WE	2600	2557	2186	1815
02/13/97	TH	2601	2558	2187	1816
02/14/97	FR	2602	2559	2188	1817
02/15/97	SA	2603	2560	2189	NO WORK
02/16/97	SU	2604	2561	NO WORK	NO WORK
02/17/97	MO	2605	2562	2190	1818
02/18/97	TU	2606	2563	2191	1819
02/19/97	WE	2607	2564	2192	1820
02/20/97	TH	2608	2565	2193	1821
02/21/97	FR	2609	2566	2194	1822
02/22/97	SA	2610	2567	2195	NO WORK

DATE	DAY OF THE WEEK	CALENDAR DAY	7-DAY WORK WEEK W/HOLIDAYS	6-DAY WORK WEEK W/HOLIDAYS	5-DAY WORK WEEK W/HOLIDAYS
02/23/97	SU	2611	2568	NO WORK	NO WORK
02/24/97	MO	2612	2569	2196	1823
02/25/97	TU	2613	2570	2197	1824
02/26/97	WE	2614	2571	2198	1825
02/27/97	TH	2615	2572	2199	1826
02/28/97	FR	2616	2573	2200	1827
03/01/97	SA	2617	2574	2201	NO WORK
03/02/97	SU	2618	2575	NO WORK	NO WORK
03/03/97	MO	2619	2576	2202	1828
03/04/97	TU	2620	2577	2203	1829
03/05/97	WE	2621	2578	2204	1830
03/06/97	TH	2622	2579	2205	1831
03/07/97	FR	2623	2580	2206	1832
03/08/97	SA	2624	2581	2207	NO WORK
03/09/97	SU	2625	2582	NO WORK	NO WORK
03/10/97	MO	2626	2583	2208	1833
03/11/97	TU	2627	2584	2209	1834
03/12/97	WE	2628	2585	2210	1835
03/13/97	TH	2629	2586	2211	1836
03/14/97	FR	2630	2587	2212	1837
03/15/97	SA	2631	2588	2213	NO WORK
03/16/97	SU	2632	2589	NO WORK	NO WORK
03/17/97	MO	2633	2590	2214	1838
03/18/97	TU	2634	2591	2215	1839
03/19/97	WE	2635	2592	2216	1840
03/20/97	TH	2636	2593	2217	1841
03/21/97	FR	2637	2594	2218	1842
03/22/97	SA	2638	2595	2219	NO WORK
03/23/97	SU	2639	2596	NO WORK	NO WORK
03/24/97	MO	2640	2597	2220	1843
03/25/97	TU	2641	2598	2221	1844
03/26/97	WE	2642	2599	2222	1845
03/27/97	TH	2643	2600	2223	1846
03/28/97	FR	2644	2601	2224	1847
03/29/97	SA	2645	2602	2225	NO WORK
03/30/97	SU	2646	2603	NO WORK	NO WORK
03/31/97	MO	2647	2604	2226	1848
04/01/97	TU	2648	2605	2227	1849
04/02/97	WE	2649	2606	2228	1850
04/03/97	TH	2650	2607	2229	1851
04/04/97	FR	2651	2608	2230	1852
04/05/97	SA	2652	2609	2231	NO WORK
04/06/97	SU	2653	2610	NO WORK	NO WORK
04/07/97	MO	2654	2611	2232	1853
04/08/97	TU	2655	2612	2233	1854
04/09/97	WE	2656	2613	2234	1855
04/10/97	TH	2657	2614	2235	1856
04/11/97	FR	2658	2615	2236	1857
04/12/97	SA	2659	2616	2237	NO WORK
04/13/97	SU	2660	2617	NO WORK	NO WORK
04/14/97	MO	2661	2618	2238	1858
04/15/97	TU	2662	2619	2239	1859
04/16/97	WE	2663	2620	2240	1860

DATE	DAY OF THE WEEK	CALENDAR DAY	7-DAY WORK WEEK W/HOLIDAYS	6-DAY WORK WEEK W/HOLIDAYS	5-DAY WORK WEEK W/HOLIDAYS
04/17/97	TH	2664	2621	2241	1861
04/18/97	FR	2665	2622	2242	1862
04/19/97	SA	2666	2623	2243	NO WORK
04/20/97	SU	2667	2624	NO WORK	NO WORK
04/21/97	MO	2668	2625	2244	1863
04/22/97	TU	2669	2626	2245	1864
04/23/97	WE	2670	2627	2246	1865
04/24/97	TH	2671	2628	2247	1866
04/25/97	FR	2672	2629	2248	1867
04/26/97	SA	2673	2630	2249	NO WORK
04/27/97	SU	2674	2631	NO WORK	NO WORK
04/28/97	MO	2675	2632	2250	1868
04/29/97	TU	2676	2633	2251	1869
04/30/97	WE	2677	2634	2252	1870
05/01/97	TH	2678	2635	2253	1871
05/02/97	FR	2679	2636	2254	1872
05/03/97	SA	2680	2637	2255	NO WORK
05/04/97	SU	2681	2638	NO WORK	NO WORK
05/05/97	MO	2682	2639	2256	1873
05/06/97	TU	2683	2640	2257	1874
05/07/97	WE	2684	2641	2258	1875
05/08/97	TH	2685	2642	2259	1876
05/09/97	FR	2686	2643	2260	1877
05/10/97	SA	2687	2644	2261	NO WORK
05/11/97	SU	2688	2645	NO WORK	NO WORK
05/12/97	MO	2689	2646	2262	1878
05/13/97	TU	2690	2647	2263	1879
05/14/97	WE	2691	2648	2264	1880
05/15/97	TH	2692	2649	2265	1881
05/16/97	FR	2693	2650	2266	1882
05/17/97	SA	2694	2651	2267	NO WORK
05/18/97	SU	2695	2652	NO WORK	NO WORK
05/19/97	MO	2696	2653	2268	1883
05/20/97	TU	2697	2654	2269	1884
05/21/97	WE	2698	2655	2270	1885
05/22/97	TH	2699	2656	2271	1886
05/23/97	FR	2700	2657	2272	1887
05/24/97	SA	2701	2658	2273	NO WORK
05/25/97	SU	2702	2659	NO WORK	NO WORK
05/26/97	MO	2703	H	H	H
05/27/97	TU	2704	2660	2274	1888
05/28/97	WE	2705	2661	2275	1889
05/29/97	TH	2706	2662	2276	1890
05/30/97	FR	2707	2663	2277	1891
05/31/97	SA	2708	2664	2278	NO WORK
06/01/97	SU	2709	2665	NO WORK	NO WORK
06/02/97	MO	2710	2666	2279	1892
06/03/97	TU	2711	2667	2280	1893
06/04/97	WE	2712	2668	2281	1894
06/05/97	TH	2713	2669	2282	1895
06/06/97	FR	2714	2670	2283	1896
06/07/97	SA	2715	2671	2284	NO WORK
06/08/97	SU	2716	2672	NO WORK	NO WORK

DATE	DAY OF THE WEEK	CALENDAR DAY	7-DAY WORK WEEK W/HOLIDAYS	6-DAY WORK WEEK W/HOLIDAYS	5-DAY WORK WEEK W/HOLIDAYS
06/09/97	MO	2717	2673	2285	1897
06/10/97	TU	2718	2674	2286	1898
06/11/97	WE	2719	2675	2287	1899
06/12/97	TH	2720	2676	2288	1900
06/13/97	FR	2721	2677	2289	1901
06/14/97	SA	2722	2678	2290	NO WORK
06/15/97	SU	2723	2679	NO WORK	NO WORK
06/16/97	MO	2724	2680	2291	1902
06/17/97	TU	2725	2681	2292	1903
06/18/97	WE	2726	2682	2293	1904
06/19/97	TH	2727	2683	2294	1905
06/20/97	FR	2728	2684	2295	1906
06/21/97	SA	2729	2685	2296	NO WORK
06/22/97	SU	2730	2686	NO WORK	NO WORK
06/23/97	MO	2731	2687	2297	1907
06/24/97	TU	2732	2688	2298	1908
06/25/97	WE	2733	2689	2299	1909
06/26/97	TH	2734	2690	2300	1910
06/27/97	FR	2735	2691	2301	1911
06/28/97	SA	2736	2692	2302	NO WORK
06/29/97	SU	2737	2693	NO WORK	NO WORK
06/30/97	MO	2738	2694	2303	1912
07/01/97	TU	2739	2695	2304	1913
07/02/97	WE	2740	2696	2305	1914
07/03/97	TH	2741	2697	2306	1915
07/04/97	FR	2742	H	H	H
07/05/97	SA	2743	2698	2307	NO WORK
07/06/97	SU	2744	2699	NO WORK	NO WORK
07/07/97	MO	2745	2700	2308	1916
07/08/97	TU	2746	2701	2309	1917
07/09/97	WE	2747	2702	2310	1918
07/10/97	TH	2748	2703	2311	1919
07/11/97	FR	2749	2704	2312	1920
07/12/97	SA	2750	2705	2313	NO WORK
07/13/97	SU	2751	2706	NO WORK	NO WORK
07/14/97	MO	2752	2707	2314	1921
07/15/97	TU	2753	2708	2315	1922
07/16/97	WE	2754	2709	2316	1923
07/17/97	TH	2755	2710	2317	1924
07/18/97	FR	2756	2711	2318	1925
07/19/97	SA	2757	2712	2319	NO WORK
07/20/97	SU	2758	2713	NO WORK	NO WORK
07/21/97	MO	2759	2714	2320	1926
07/22/97	TU	2760	2715	2321	1927
07/23/97	WE	2761	2716	2322	1928
07/24/97	TH	2762	2717	2323	1929
07/25/97	FR	2763	2718	2324	1930
07/26/97	SA	2764	2719	2325	NO WORK
07/27/97	SU	2765	2720	NO WORK	NO WORK
07/28/97	MO	2766	2721	2326	1931
07/29/97	TU	2767	2722	2327	1932
07/30/97	WE	2768	2723	2328	1933
07/31/97	TH	2769	2724	2329	1934

DATE	DAY OF THE WEEK	CALENDAR DAY	7-DAY WORK WEEK W/HOLIDAYS	6-DAY WORK WEEK W/HOLIDAYS	5-DAY WORK WEEK W/HOLIDAYS
08/01/97	FR	2770	2725	2330	1935
08/02/97	SA	2771	2726	2331	NO WORK
08/03/97	SU	2772	2727	NO WORK	NO WORK
08/04/97	MO	2773	2728	2332	1936
08/05/97	TU	2774	2729	2333	1937
08/06/97	WE	2775	2730	2334	1938
08/07/97	TH	2776	2731	2335	1939
08/08/97	FR	2777	2732	2336	1940
08/09/97	SA	2778	2733	2337	NO WORK
08/10/97	SU	2779	2734	NO WORK	NO WORK
08/11/97	MO	2780	2735	2338	1941
08/12/97	TU	2781	2736	2339	1942
08/13/97	WE	2782	2737	2340	1943
08/14/97	TH	2783	2738	2341	1944
08/15/97	FR	2784	2739	2342	1945
08/16/97	SA	2785	2740	2343	NO WORK
08/17/97	SU	2786	2741	NO WORK	NO WORK
08/18/97	MO	2787	2742	2344	1946
08/19/97	TU	2788	2743	2345	1947
08/20/97	WE	2789	2744	2346	1948
08/21/97	TH	2790	2745	2347	1949
08/22/97	FR	2791	2746	2348	1950
08/23/97	SA	2792	2747	2349	NO WORK
08/24/97	SU	2793	2748	NO WORK	NO WORK
08/25/97	MO	2794	2749	2350	1951
08/26/97	TU	2795	2750	2351	1952
08/27/97	WE	2796	2751	2352	1953
08/28/97	TH	2797	2752	2353	1954
08/29/97	FR	2798	2753	2354	1955
08/30/97	SA	2799	2754	2355	NO WORK
08/31/97	SU	2800	2755	NO WORK	NO WORK
09/01/97	MO	2801	H	H	H
09/02/97	TU	2802	2756	2356	1956
09/03/97	WE	2803	2757	2357	1957
09/04/97	TH	2804	2758	2358	1958
09/05/97	FR	2805	2759	2359	1959
09/06/97	SA	2806	2760	2360	NO WORK
09/07/97	SU	2807	2761	NO WORK	NO WORK
09/08/97	MO	2808	2762	2361	1960
09/09/97	TU	2809	2763	2362	1961
09/10/97	WE	2810	2764	2363	1962
09/11/97	TH	2811	2765	2364	1963
09/12/97	FR	2812	2766	2365	1964
09/13/97	SA	2813	2767	2366	NO WORK
09/14/97	SU	2814	2768	NO WORK	NO WORK
09/15/97	MO	2815	2769	2367	1965
09/16/97	TU	2816	2770	2368	1966
09/17/97	WE	2817	2771	2369	1967
09/18/97	TH	2818	2772	2370	1968
09/19/97	FR	2819	2773	2371	1969
09/20/97	SA	2820	2774	2372	NO WORK
09/21/97	SU	2821	2775	NO WORK	NO WORK
09/22/97	MO	2822	2776	2373	1970

DATE	DAY OF THE WEEK	CALENDAR DAY	7-DAY WORK WEEK W/HOLIDAYS	6-DAY WORK WEEK W/HOLIDAYS	5-DAY WORK WEEK W/HOLIDAYS
09/23/97	TU	2823	2777	2374	1971
09/24/97	WE	2824	2778	2375	1972
09/25/97	TH	2825	2779	2376	1973
09/26/97	FR	2826	2780	2377	1974
09/27/97	SA	2827	2781	2378	NO WORK
09/28/97	SU	2828	2782	NO WORK	NO WORK
09/29/97	MO	2829	2783	2379	1975
09/30/97	TU	2830	2784	2380	1976
10/01/97	WE	2831	2785	2381	1977
10/02/97	TH	2832	2786	2382	1978
10/03/97	FR	2833	2787	2383	1979
10/04/97	SA	2834	2788	2384	NO WORK
10/05/97	SU	2835	2789	NO WORK	NO WORK
10/06/97	MO	2836	2790	2385	1980
10/07/97	TU	2837	2791	2386	1981
10/08/97	WE	2838	2792	2387	1982
10/09/97	TH	2839	2793	2388	1983
10/10/97	FR	2840	2794	2389	1984
10/11/97	SA	2841	2795	2390	NO WORK
10/12/97	SU	2842	2796	NO WORK	NO WORK
10/13/97	MO	2843	2797	2391	1985
10/14/97	TU	2844	2798	2392	1986
10/15/97	WE	2845	2799	2393	1987
10/16/97	TH	2846	2800	2394	1988
10/17/97	FR	2847	2801	2395	1989
10/18/97	SA	2848	2802	2396	NO WORK
10/19/97	SU	2849	2803	NO WORK	NO WORK
10/20/97	MO	2850	2804	2397	1990
10/21/97	TU	2851	2805	2398	1991
10/22/97	WE	2852	2806	2399	1992
10/23/97	TH	2853	2807	2400	1993
10/24/97	FR	2854	2808	2401	1994
10/25/97	SA	2855	2809	2402	NO WORK
10/26/97	SU	2856	2810	NO WORK	NO WORK
10/27/97	MO	2857	2811	2403	1995
10/28/97	TU	2858	2812	2404	1996
10/29/97	WE	2859	2813	2405	1997
10/30/97	TH	2860	2814	2406	1998
10/31/97	FR	2861	2815	2407	1999
11/01/97	SA	2862	2816	2408	NO WORK
11/02/97	SU	2863	2817	NO WORK	NO WORK
11/03/97	MO	2864	2818	2409	2000
11/04/97	TU	2865	2819	2410	2001
11/05/97	WE	2866	2820	2411	2002
11/06/97	TH	2867	2821	2412	2003
11/07/97	FR	2868	2822	2413	2004
11/08/97	SA	2869	2823	2414	NO WORK
11/09/97	SU	2870	2824	NO WORK	NO WORK
11/10/97	MO	2871	2825	2415	2005
11/11/97	TU	2872	2826	2416	2006
11/12/97	WE	2873	2827	2417	2007
11/13/97	TH	2874	2828	2418	2008
11/14/97	FR	2875	2829	2419	2009
11/15/97	SA	2876	2830	2420	NO WORK
11/16/97	SU	2877	2831	NO WORK	NO WORK
11/17/97	MO	2878	2832	2421	2010
11/18/97	TU	2879	2833	2422	2011
11/19/97	WE	2880	2834	2423	2012
11/20/97	TH	2881	2835	2424	2013
11/21/97	FR	2882	2836	2425	2014
11/22/97	SA	2883	2837	2426	NO WORK
11/23/97	SU	2884	2838	NO WORK	NO WORK
11/24/97	MO	2885	2839	2427	2015
11/25/97	TU	2886	2840	2428	2016
11/26/97	WE	2887	2841	2429	2017
11/27/97	TH	2888	H	H	H
11/28/97	FR	2889	2842	2430	2018
11/29/97	SA	2890	2843	2431	NO WORK
11/30/97	SU	2891	2844	NO WORK	NO WORK
12/01/97	MO	2892	2845	2432	2019
12/02/97	TU	2893	2846	2433	2020
12/03/97	WE	2894	2847	2434	2021
12/04/97	TH	2895	2848	2435	2022
12/05/97	FR	2896	2849	2436	2023
12/06/97	SA	2897	2850	2437	NO WORK
12/07/97	SU	2898	2851	NO WORK	NO WORK
12/08/97	MO	2899	2852	2438	2024
12/09/97	TU	2900	2853	2439	2025
12/10/97	WE	2901	2854	2440	2026
12/11/97	TH	2902	2855	2441	2027
12/12/97	FR	2903	2856	2442	2028
12/13/97	SA	2904	2857	2443	NO WORK
12/14/97	SU	2905	2858	NO WORK	NO WORK
12/15/97	MO	2906	2859	2444	2029
12/16/97	TU	2907	2860	2445	2030
12/17/97	WE	2908	2861	2446	2031
12/18/97	TH	2909	2862	2447	2032
12/19/97	FR	2910	2863	2448	2033
12/20/97	SA	2911	2864	2449	NO WORK
12/21/97	SU	2912	2865	NO WORK	NO WORK
12/22/97	MO	2913	2866	2450	2034
12/23/97	TU	2914	2867	2451	2035
12/24/97	WE	2915	2868	2452	2036
12/25/97	TH	2916	H	H	H
12/26/97	FR	2917	2869	2453	2037
12/27/97	SA	2918	2870	2454	NO WORK
12/28/97	SU	2919	2871	NO WORK	NO WORK
12/29/97	MO	2920	2872	2455	2038
12/30/97	TU	2921	2873	2456	2039
12/31/97	WE	2922	2874	2457	2040

DATE	DAY OF THE WEEK	CALENDAR DAY	7-DAY WORK WEEK W/HOLIDAYS	6-DAY WORK WEEK W/HOLIDAYS	5-DAY WORK WEEK W/HOLIDAYS
01/01/98	TH	2923	H	H	H
01/02/98	FR	2924	2875	2458	2041
01/03/98	SA	2925	2876	2459	NO WORK
01/04/98	SU	2926	2877	NO WORK	NO WORK
01/05/98	MO	2927	2878	2460	2042
01/06/98	TU	2928	2879	2461	2043
01/07/98	WE	2929	2880	2462	2044
01/08/98	TH	2930	2881	2463	2045
01/09/98	FR	2931	2882	2464	2046
01/10/98	SA	2932	2883	2465	NO WORK
01/11/98	SU	2933	2884	NO WORK	NO WORK
01/12/98	MO	2934	2885	2466	2047
01/13/98	TU	2935	2886	2467	2048
01/14/98	WE	2936	2887	2468	2049
01/15/98	TH	2937	2888	2469	2050
01/16/98	FR	2938	2889	2470	2051
01/17/98	SA	2939	2890	2471	NO WORK
01/18/98	SU	2940	2891	NO WORK	NO WORK
01/19/98	MO	2941	2892	2472	2052
01/20/98	TU	2942	2893	2473	2053
01/21/98	WE	2943	2894	2474	2054
01/22/98	TH	2944	2895	2475	2055
01/23/98	FR	2945	2896	2476	2056
01/24/98	SA	2946	2897	2477	NO WORK
01/25/98	SU	2947	2898	NO WORK	NO WORK
01/26/98	MO	2948	2899	2478	2057
01/27/98	TU	2949	2900	2479	2058
01/28/98	WE	2950	2901	2480	2059
01/29/98	TH	2951	2902	2481	2060
01/30/98	FR	2952	2903	2482	2061
01/31/98	SA	2953	2904	2483	NO WORK
02/01/98	SU	2954	2905	NO WORK	NO WORK
02/02/98	MO	2955	2906	2484	2062
02/03/98	TU	2956	2907	2485	2063
02/04/98	WE	2957	2908	2486	2064
02/05/98	TH	2958	2909	2487	2065
02/06/98	FR	2959	2910	2488	2066
02/07/98	SA	2960	2911	2489	NO WORK
02/08/98	SU	2961	2912	NO WORK	NO WORK
02/09/98	MO	2962	2913	2490	2067
02/10/98	TU	2963	2914	2491	2068
02/11/98	WE	2964	2915	2492	2069
02/12/98	TH	2965	2916	2493	2070
02/13/98	FR	2966	2917	2494	2071
02/14/98	SA	2967	2918	2495	NO WORK
02/15/98	SU	2968	2919	NO WORK	NO WORK
02/16/98	MO	2969	2920	2496	2072
02/17/98	TU	2970	2921	2497	2073
02/18/98	WE	2971	2922	2498	2074
02/19/98	TH	2972	2923	2499	2075
02/20/98	FR	2973	2924	2500	2076
02/21/98	SA	2974	2925	2501	NO WORK
02/22/98	SU	2975	2926	NO WORK	NO WORK
02/23/98	MO	2976	2927	2502	2077
02/24/98	TU	2977	2928	2503	2078
02/25/98	WE	2978	2929	2504	2079
02/26/98	TH	2979	2930	2505	2080
02/27/98	FR	2980	2931	2506	2081
02/28/98	SA	2981	2932	2507	NO WORK
03/01/98	SU	2982	2933	NO WORK	NO WORK
03/02/98	MO	2983	2934	2508	2082
03/03/98	TU	2984	2935	2509	2083
03/04/98	WE	2985	2936	2510	2084
03/05/98	TH	2986	2937	2511	2085
03/06/98	FR	2987	2938	2512	2086
03/07/98	SA	2988	2939	2513	NO WORK
03/08/98	SU	2989	2940	NO WORK	NO WORK
03/09/98	MO	2990	2941	2514	2087
03/10/98	TU	2991	2942	2515	2088
03/11/98	WE	2992	2943	2516	2089
03/12/98	TH	2993	2944	2517	2090
03/13/98	FR	2994	2945	2518	2091
03/14/98	SA	2995	2946	2519	NO WORK
03/15/98	SU	2996	2947	NO WORK	NO WORK
03/16/98	MO	2997	2948	2520	2092
03/17/98	TU	2998	2949	2521	2093
03/18/98	WE	2999	2950	2522	2094
03/19/98	TH	3000	2951	2523	2095
03/20/98	FR	3001	2952	2524	2096
03/21/98	SA	3002	2953	2525	NO WORK
03/22/98	SU	3003	2954	NO WORK	NO WORK
03/23/98	MO	3004	2955	2526	2097
03/24/98	TU	3005	2956	2527	2098
03/25/98	WE	3006	2957	2528	2099
03/26/98	TH	3007	2958	2529	2100
03/27/98	FR	3008	2959	2530	2101
03/28/98	SA	3009	2960	2531	NO WORK
03/29/98	SU	3010	2961	NO WORK	NO WORK
03/30/98	MO	3011	2962	2532	2102
03/31/98	TU	3012	2963	2533	2103
04/01/98	WE	3013	2964	2534	2104
04/02/98	TH	3014	2965	2535	2105
04/03/98	FR	3015	2966	2536	2106
04/04/98	SA	3016	2967	2537	NO WORK
04/05/98	SU	3017	2968	NO WORK	NO WORK
04/06/98	MO	3018	2969	2538	2107
04/07/98	TU	3019	2970	2539	2108
04/08/98	WE	3020	2971	2540	2109
04/09/98	TH	3021	2972	2541	2110
04/10/98	FR	3022	2973	2542	2111
04/11/98	SA	3023	2974	2543	NO WORK
04/12/98	SU	3024	2975	NO WORK	NO WORK
04/13/98	MO	3025	2976	2544	2112
04/14/98	TU	3026	2977	2545	2113
04/15/98	WE	3027	2978	2546	2114
04/16/98	TH	3028	2979	2547	2115

DATE	DAY OF THE WEEK	CALENDAR DAY	7-DAY WORK WEEK W/HOLIDAYS	6-DAY WORK WEEK W/HOLIDAYS	5-DAY WORK WEEK W/HOLIDAYS
04/17/98	FR	3029	2980	2548	2116
04/18/98	SA	3030	2981	2549	NO WORK
04/19/98	SU	3031	2982	NO WORK	NO WORK
04/20/98	MO	3032	2983	2550	2117
04/21/98	TU	3033	2984	2551	2118
04/22/98	WE	3034	2985	2552	2119
04/23/98	TH	3035	2986	2553	2120
04/24/98	FR	3036	2987	2554	2121
04/25/98	SA	3037	2988	2555	NO WORK
04/26/98	SU	3038	2989	NO WORK	NO WORK
04/27/98	MO	3039	2990	2556	2122
04/28/98	TU	3040	2991	2557	2123
04/29/98	WE	3041	2992	2558	2124
04/30/98	TH	3042	2993	2559	2125
05/01/98	FR	3043	2994	2560	2126
05/02/98	SA	3044	2995	2561	NO WORK
05/03/98	SU	3045	2996	NO WORK	NO WORK
05/04/98	MO	3046	2997	2562	2127
05/05/98	TU	3047	2998	2563	2128
05/06/98	WE	3048	2999	2564	2129
05/07/98	TH	3049	3000	2565	2130
05/08/98	FR	3050	3001	2566	2131
05/09/98	SA	3051	3002	2567	NO WORK
05/10/98	SU	3052	3003	NO WORK	NO WORK
05/11/98	MO	3053	3004	2568	2132
05/12/98	TU	3054	3005	2569	2133
05/13/98	WE	3055	3006	2570	2134
05/14/98	TH	3056	3007	2571	2135
05/15/98	FR	3057	3008	2572	2136
05/16/98	SA	3058	3009	2573	NO WORK
05/17/98	SU	3059	3010	NO WORK	NO WORK
05/18/98	MO	3060	3011	2574	2137
05/19/98	TU	3061	3012	2575	2138
05/20/98	WE	3062	3013	2576	2139
05/21/98	TH	3063	3014	2577	2140
05/22/98	FR	3064	3015	2578	2141
05/23/98	SA	3065	3016	2579	NO WORK
05/24/98	SU	3066	3017	NO WORK	NO WORK
05/25/98	MO	3067	H	H	H
05/26/98	TU	3068	3018	2580	2142
05/27/98	WE	3069	3019	2581	2143
05/28/98	TH	3070	3020	2582	2144
05/29/98	FR	3071	3021	2583	2145
05/30/98	SA	3072	3022	2584	NO WORK
05/31/98	SU	3073	3023	NO WORK	NO WORK
06/01/98	MO	3074	3024	2585	2146
06/02/98	TU	3075	3025	2586	2147
06/03/98	WE	3076	3026	2587	2148
06/04/98	TH	3077	3027	2588	2149
06/05/98	FR	3078	3028	2589	2150
06/06/98	SA	3079	3029	2590	NO WORK
06/07/98	SU	3080	3030	NO WORK	NO WORK
06/08/98	MO	3081	3031	2591	2151

DATE	DAY OF THE WEEK	CALENDAR DAY	7-DAY WORK WEEK W/HOLIDAYS	6-DAY WORK WEEK W/HOLIDAYS	5-DAY WORK WEEK W/HOLIDAYS
06/09/98	TU	3082	3032	2592	2152
06/10/98	WE	3083	3033	2593	2153
06/11/98	TH	3084	3034	2594	2154
06/12/98	FR	3085	3035	2595	2155
06/13/98	SA	3086	3036	2596	NO WORK
06/14/98	SU	3087	3037	NO WORK	NO WORK
06/15/98	MO	3088	3038	2597	2156
06/16/98	TU	3089	3039	2598	2157
06/17/98	WE	3090	3040	2599	2158
06/18/98	TH	3091	3041	2600	2159
06/19/98	FR	3092	3042	2601	2160
06/20/98	SA	3093	3043	2602	NO WORK
06/21/98	SU	3094	3044	NO WORK	NO WORK
06/22/98	MO	3095	3045	2603	2161
06/23/98	TU	3096	3046	2604	2162
06/24/98	WE	3097	3047	2605	2163
06/25/98	TH	3098	3048	2606	2164
06/26/98	FR	3099	3049	2607	2165
06/27/98	SA	3100	3050	2608	NO WORK
06/28/98	SU	3101	3051	NO WORK	NO WORK
06/29/98	MO	3102	3052	2609	2166
06/30/98	TU	3103	3053	2610	2167
07/01/98	WE	3104	3054	2611	2168
07/02/98	TH	3105	3055	2612	2169
07/03/98	FR	3106	3056	2613	H
07/04/98	SA	3107	H	H	NO WORK
07/05/98	SU	3108	3057	NO WORK	NO WORK
07/06/98	MO	3109	3058	2614	2170
07/07/98	TU	3110	3059	2615	2171
07/08/98	WE	3111	3060	2616	2172
07/09/98	TH	3112	3061	2617	2173
07/10/98	FR	3113	3062	2618	2174
07/11/98	SA	3114	3063	2619	NO WORK
07/12/98	SU	3115	3064	NO WORK	NO WORK
07/13/98	MO	3116	3065	2620	2175
07/14/98	TU	3117	3066	2621	2176
07/15/98	WE	3118	3067	2622	2177
07/16/98	TH	3119	3068	2623	2178
07/17/98	FR	3120	3069	2624	2179
07/18/98	SA	3121	3070	2625	NO WORK
07/19/98	SU	3122	3071	NO WORK	NO WORK
07/20/98	MO	3123	3072	2626	2180
07/21/98	TU	3124	3073	2627	2181
07/22/98	WE	3125	3074	2628	2182
07/23/98	TH	3126	3075	2629	2183
07/24/98	FR	3127	3076	2630	2184
07/25/98	SA	3128	3077	2631	NO WORK
07/26/98	SU	3129	3078	NO WORK	NO WORK
07/27/98	MO	3130	3079	2632	2185
07/28/98	TU	3131	3080	2633	2186
07/29/98	WE	3132	3081	2634	2187
07/30/98	TH	3133	3082	2635	2188
07/31/98	FR	3134	3083	2636	2189

DATE	DAY OF THE WEEK	CALENDAR DAY	7-DAY WORK WEEK W/HOLIDAYS	6-DAY WORK WEEK W/HOLIDAYS	5-DAY WORK WEEK W/HOLIDAYS
08/01/98	SA	3135	3084	2637	NO WORK
08/02/98	SU	3136	3085	NO WORK	NO WORK
08/03/98	MO	3137	3086	2638	2190
08/04/98	TU	3138	3087	2639	2191
08/05/98	WE	3139	3088	2640	2192
08/06/98	TH	3140	3089	2641	2193
08/07/98	FR	3141	3090	2642	2194
08/08/98	SA	3142	3091	2643	NO WORK
08/09/98	SU	3143	3092	NO WORK	NO WORK
08/10/98	MO	3144	3093	2644	2195
08/11/98	TU	3145	3094	2645	2196
08/12/98	WE	3146	3095	2646	2197
08/13/98	TH	3147	3096	2647	2198
08/14/98	FR	3148	3097	2648	2199
08/15/98	SA	3149	3098	2649	NO WORK
08/16/98	SU	3150	3099	NO WORK	NO WORK
08/17/98	MO	3151	3100	2650	2200
08/18/98	TU	3152	3101	2651	2201
08/19/98	WE	3153	3102	2652	2202
08/20/98	TH	3154	3103	2653	2203
08/21/98	FR	3155	3104	2654	2204
08/22/98	SA	3156	3105	2655	NO WORK
08/23/98	SU	3157	3106	NO WORK	NO WORK
08/24/98	MO	3158	3107	2656	2205
08/25/98	TU	3159	3108	2657	2206
08/26/98	WE	3160	3109	2658	2207
08/27/98	TH	3161	3110	2659	2208
08/28/98	FR	3162	3111	2660	2209
08/29/98	SA	3163	3112	2661	NO WORK
08/30/98	SU	3164	3113	NO WORK	NO WORK
08/31/98	MO	3165	3114	2662	2210
09/01/98	TU	3166	3115	2663	2211
09/02/98	WE	3167	3116	2664	2212
09/03/98	TH	3168	3117	2665	2213
09/04/98	FR	3169	3118	2666	2214
09/05/98	SA	3170	3119	2667	NO WORK
09/06/98	SU	3171	3120	NO WORK	NO WORK
09/07/98	MO	3172	H	H	H
09/08/98	TU	3173	3121	2668	2215
09/09/98	WE	3174	3122	2669	2216
09/10/98	TH	3175	3123	2670	2217
09/11/98	FR	3176	3124	2671	2218
09/12/98	SA	3177	3125	2672	NO WORK
09/13/98	SU	3178	3126	NO WORK	NO WORK
09/14/98	MO	3179	3127	2673	2219
09/15/98	TU	3180	3128	2674	2220
09/16/98	WE	3181	3129	2675	2221
09/17/98	TH	3182	3130	2676	2222
09/18/98	FR	3183	3131	2677	2223
09/19/98	SA	3184	3132	2678	NO WORK
09/20/98	SU	3185	3133	NO WORK	NO WORK
09/21/98	MO	3186	3134	2679	2224
09/22/98	TU	3187	3135	2680	2225
09/23/98	WE	3188	3136	2681	2226
09/24/98	TH	3189	3137	2682	2227
09/25/98	FR	3190	3138	2683	2228
09/26/98	SA	3191	3139	2684	NO WORK
09/27/98	SU	3192	3140	NO WORK	NO WORK
09/28/98	MO	3193	3141	2685	2229
09/29/98	TU	3194	3142	2686	2230
09/30/98	WE	3195	3143	2687	2231
10/01/98	TH	3196	3144	2688	2232
10/02/98	FR	3197	3145	2689	2233
10/03/98	SA	3198	3146	2690	NO WORK
10/04/98	SU	3199	3147	NO WORK	NO WORK
10/05/98	MO	3200	3148	2691	2234
10/06/98	TU	3201	3149	2692	2235
10/07/98	WE	3202	3150	2693	2236
10/08/98	TH	3203	3151	2694	2237
10/09/98	FR	3204	3152	2695	2238
10/10/98	SA	3205	3153	2696	NO WORK
10/11/98	SU	3206	3154	NO WORK	NO WORK
10/12/98	MO	3207	3155	2697	2239
10/13/98	TU	3208	3156	2698	2240
10/14/98	WE	3209	3157	2699	2241
10/15/98	TH	3210	3158	2700	2242
10/16/98	FR	3211	3159	2701	2243
10/17/98	SA	3212	3160	2702	NO WORK
10/18/98	SU	3213	3161	NO WORK	NO WORK
10/19/98	MO	3214	3162	2703	2244
10/20/98	TU	3215	3163	2704	2245
10/21/98	WE	3216	3164	2705	2246
10/22/98	TH	3217	3165	2706	2247
10/23/98	FR	3218	3166	2707	2248
10/24/98	SA	3219	3167	2708	NO WORK
10/25/98	SU	3220	3168	NO WORK	NO WORK
10/26/98	MO	3221	3169	2709	2249
10/27/98	TU	3222	3170	2710	2250
10/28/98	WE	3223	3171	2711	2251
10/29/98	TH	3224	3172	2712	2252
10/30/98	FR	3225	3173	2713	2253
10/31/98	SA	3226	3174	2714	NO WORK
11/01/98	SU	3227	3175	NO WORK	NO WORK
11/02/98	MO	3228	3176	2715	2254
11/03/98	TU	3229	3177	2716	2255
11/04/98	WE	3230	3178	2717	2256
11/05/98	TH	3231	3179	2718	2257
11/06/98	FR	3232	3180	2719	2258
11/07/98	SA	3233	3181	2720	NO WORK
11/08/98	SU	3234	3182	NO WORK	NO WORK
11/09/98	MO	3235	3183	2721	2259
11/10/98	TU	3236	3184	2722	2260
11/11/98	WE	3237	3185	2723	2261
11/12/98	TH	3238	3186	2724	2262
11/13/98	FR	3239	3187	2725	2263
11/14/98	SA	3240	3188	2726	NO WORK

DATE	DAY OF THE WEEK	CALENDAR DAY	7-DAY WORK WEEK W/HOLIDAYS	6-DAY WORK WEEK W/HOLIDAYS	5-DAY WORK WEEK W/HOLIDAYS
11/15/98	SU	3241	3189	NO WORK	NO WORK
11/16/98	MO	3242	3190	2727	2264
11/17/98	TU	3243	3191	2728	2265
11/18/98	WE	3244	3192	2729	2266
11/19/98	TH	3245	3193	2730	2267
11/20/98	FR	3246	3194	2731	2268
11/21/98	SA	3247	3195	2732	NO WORK
11/22/98	SU	3248	3196	NO WORK	NO WORK
11/23/98	MO	3249	3197	2733	2269
11/24/98	TU	3250	3198	2734	2270
11/25/98	WE	3251	3199	2735	2271
11/26/98	TH	3252	H	H	H
11/27/98	FR	3253	3200	2736	2272
11/28/98	SA	3254	3201	2737	NO WORK
11/29/98	SU	3255	3202	NO WORK	NO WORK
11/30/98	MO	3256	3203	2738	2273
12/01/98	TU	3257	3204	2739	2274
12/02/98	WE	3258	3205	2740	2275
12/03/98	TH	3259	3206	2741	2276
12/04/98	FR	3260	3207	2742	2277
12/05/98	SA	3261	3208	2743	NO WORK
12/06/98	SU	3262	3209	NO WORK	NO WORK
12/07/98	MO	3263	3210	2744	2278
12/08/98	TU	3264	3211	2745	2279
12/09/98	WE	3265	3212	2746	2280
12/10/98	TH	3266	3213	2747	2281
12/11/98	FR	3267	3214	2748	2282
12/12/98	SA	3268	3215	2749	NO WORK
12/13/98	SU	3269	3216	NO WORK	NO WORK
12/14/98	MO	3270	3217	2750	2283
12/15/98	TU	3271	3218	2751	2284
12/16/98	WE	3272	3219	2752	2285
12/17/98	TH	3273	3220	2753	2286
12/18/98	FR	3274	3221	2754	2287
12/19/98	SA	3275	3222	2755	NO WORK
12/20/98	SU	3276	3223	NO WORK	NO WORK
12/21/98	MO	3277	3224	2756	2288
12/22/98	TU	3278	3225	2757	2289
12/23/98	WE	3279	3226	2758	2290
12/24/98	TH	3280	3227	2759	2291
12/25/98	FR	3281	H	H	H
12/26/98	SA	3282	3228	2760	NO WORK
12/27/98	SU	3283	3229	NO WORK	NO WORK
12/28/98	MO	3284	3230	2761	2292
12/29/98	TU	3285	3231	2762	2293
12/30/98	WE	3286	3232	2763	2294
12/31/98	TH	3287	3233	2764	2295

DATE	DAY OF THE WEEK	CALENDAR DAY	7-DAY WORK WEEK W/HOLIDAYS	6-DAY WORK WEEK W/HOLIDAYS	5-DAY WORK WEEK W/HOLIDAYS
01/01/99	FR	3288	H	H	H
01/02/99	SA	3289	3234	2765	NO WORK
01/03/99	SU	3290	3235	NO WORK	NO WORK
01/04/99	MO	3291	3236	2766	2296
01/05/99	TU	3292	3237	2767	2297
01/06/99	WE	3293	3238	2768	2298
01/07/99	TH	3294	3239	2769	2299
01/08/99	FR	3295	3240	2770	2300
01/09/99	SA	3296	3241	2771	NO WORK
01/10/99	SU	3297	3242	NO WORK	NO WORK
01/11/99	MO	3298	3243	2772	2301
01/12/99	TU	3299	3244	2773	2302
01/13/99	WE	3300	3245	2774	2303
01/14/99	TH	3301	3246	2775	2304
01/15/99	FR	3302	3247	2776	2305
01/16/99	SA	3303	3248	2777	NO WORK
01/17/99	SU	3304	3249	NO WORK	NO WORK
01/18/99	MO	3305	3250	2778	2306
01/19/99	TU	3306	3251	2779	2307
01/20/99	WE	3307	3252	2780	2308
01/21/99	TH	3308	3253	2781	2309
01/22/99	FR	3309	3254	2782	2310
01/23/99	SA	3310	3255	2783	NO WORK
01/24/99	SU	3311	3256	NO WORK	NO WORK
01/25/99	MO	3312	3257	2784	2311
01/26/99	TU	3313	3258	2785	2312
01/27/99	WE	3314	3259	2786	2313
01/28/99	TH	3315	3260	2787	2314
01/29/99	FR	3316	3261	2788	2315
01/30/99	SA	3317	3262	2789	NO WORK
01/31/99	SU	3318	3263	NO WORK	NO WORK
02/01/99	MO	3319	3264	2790	2316
02/02/99	TU	3320	3265	2791	2317
02/03/99	WE	3321	3266	2792	2318
02/04/99	TH	3322	3267	2793	2319
02/05/99	FR	3323	3268	2794	2320
02/06/99	SA	3324	3269	2795	NO WORK
02/07/99	SU	3325	3270	NO WORK	NO WORK
02/08/99	MO	3326	3271	2796	2321
02/09/99	TU	3327	3272	2797	2322
02/10/99	WE	3328	3273	2798	2323
02/11/99	TH	3329	3274	2799	2324
02/12/99	FR	3330	3275	2800	2325
02/13/99	SA	3331	3276	2801	NO WORK
02/14/99	SU	3332	3277	NO WORK	NO WORK
02/15/99	MO	3333	3278	2802	2326
02/16/99	TU	3334	3279	2803	2327
02/17/99	WE	3335	3280	2804	2328
02/18/99	TH	3336	3281	2805	2329
02/19/99	FR	3337	3282	2806	2330
02/20/99	SA	3338	3283	2807	NO WORK
02/21/99	SU	3339	3284	NO WORK	NO WORK
02/22/99	MO	3340	3285	2808	2331

DATE	DAY OF THE WEEK	CALENDAR DAY	7-DAY WORK WEEK W/HOLIDAYS	6-DAY WORK WEEK W/HOLIDAYS	5-DAY WORK WEEK W/HOLIDAYS
02/23/99	TU	3341	3286	2809	2332
02/24/99	WE	3342	3287	2810	2333
02/25/99	TH	3343	3288	2811	2334
02/26/99	FR	3344	3289	2812	2335
02/27/99	SA	3345	3290	2813	NO WORK
02/28/99	SU	3346	3291	NO WORK	NO WORK
03/01/99	MO	3347	3292	2814	2336
03/02/99	TU	3348	3293	2815	2337
03/03/99	WE	3349	3294	2816	2338
03/04/99	TH	3350	3295	2817	2339
03/05/99	FR	3351	3296	2818	2340
03/06/99	SA	3352	3297	2819	NO WORK
03/07/99	SU	3353	3298	NO WORK	NO WORK
03/08/99	MO	3354	3299	2820	2341
03/09/99	TU	3355	3300	2821	2342
03/10/99	WE	3356	3301	2822	2343
03/11/99	TH	3357	3302	2823	2344
03/12/99	FR	3358	3303	2824	2345
03/13/99	SA	3359	3304	2825	NO WORK
03/14/99	SU	3360	3305	NO WORK	NO WORK
03/15/99	MO	3361	3306	2826	2346
03/16/99	TU	3362	3307	2827	2347
03/17/99	WE	3363	3308	2828	2348
03/18/99	TH	3364	3309	2829	2349
03/19/99	FR	3365	3310	2830	2350
03/20/99	SA	3366	3311	2831	NO WORK
03/21/99	SU	3367	3312	NO WORK	NO WORK
03/22/99	MO	3368	3313	2832	2351
03/23/99	TU	3369	3314	2833	2352
03/24/99	WE	3370	3315	2834	2353
03/25/99	TH	3371	3316	2835	2354
03/26/99	FR	3372	3317	2836	2355
03/27/99	SA	3373	3318	2837	NO WORK
03/28/99	SU	3374	3319	NO WORK	NO WORK
03/29/99	MO	3375	3320	2838	2356
03/30/99	TU	3376	3321	2839	2357
03/31/99	WE	3377	3322	2840	2358
04/01/99	TH	3378	3323	2841	2359
04/02/99	FR	3379	3324	2842	2360
04/03/99	SA	3380	3325	2843	NO WORK
04/04/99	SU	3381	3326	NO WORK	NO WORK
04/05/99	MO	3382	3327	2844	2361
04/06/99	TU	3383	3328	2845	2362
04/07/99	WE	3384	3329	2846	2363
04/08/99	TH	3385	3330	2847	2364
04/09/99	FR	3386	3331	2848	2365
04/10/99	SA	3387	3332	2849	NO WORK
04/11/99	SU	3388	3333	NO WORK	NO WORK
04/12/99	MO	3389	3334	2850	2366
04/13/99	TU	3390	3335	2851	2367
04/14/99	WE	3391	3336	2852	2368
04/15/99	TH	3392	3337	2853	2369
04/16/99	FR	3393	3338	2854	2370
04/17/99	SA	3394	3339	2855	NO WORK
04/18/99	SU	3395	3340	NO WORK	NO WORK
04/19/99	MO	3396	3341	2856	2371
04/20/99	TU	3397	3342	2857	2372
04/21/99	WE	3398	3343	2858	2373
04/22/99	TH	3399	3344	2859	2374
04/23/99	FR	3400	3345	2860	2375
04/24/99	SA	3401	3346	2861	NO WORK
04/25/99	SU	3402	3347	NO WORK	NO WORK
04/26/99	MO	3403	3348	2862	2376
04/27/99	TU	3404	3349	2863	2377
04/28/99	WE	3405	3350	2864	2378
04/29/99	TH	3406	3351	2865	2379
04/30/99	FR	3407	3352	2866	2380
05/01/99	SA	3408	3353	2867	NO WORK
05/02/99	SU	3409	3354	NO WORK	NO WORK
05/03/99	MO	3410	3355	2868	2381
05/04/99	TU	3411	3356	2869	2382
05/05/99	WE	3412	3357	2870	2383
05/06/99	TH	3413	3358	2871	2384
05/07/99	FR	3414	3359	2872	2385
05/08/99	SA	3415	3360	2873	NO WORK
05/09/99	SU	3416	3361	NO WORK	NO WORK
05/10/99	MO	3417	3362	2874	2386
05/11/99	TU	3418	3363	2875	2387
05/12/99	WE	3419	3364	2876	2388
05/13/99	TH	3420	3365	2877	2389
05/14/99	FR	3421	3366	2878	2390
05/15/99	SA	3422	3367	2879	NO WORK
05/16/99	SU	3423	3368	NO WORK	NO WORK
05/17/99	MO	3424	3369	2880	2391
05/18/99	TU	3425	3370	2881	2392
05/19/99	WE	3426	3371	2882	2393
05/20/99	TH	3427	3372	2883	2394
05/21/99	FR	3428	3373	2884	2395
05/22/99	SA	3429	3374	2885	NO WORK
05/23/99	SU	3430	3375	NO WORK	NO WORK
05/24/99	MO	3431	3376	2886	2396
05/25/99	TU	3432	3377	2887	2397
05/26/99	WE	3433	3378	2888	2398
05/27/99	TH	3434	3379	2889	2399
05/28/99	FR	3435	3380	2890	2400
05/29/99	SA	3436	3381	2891	NO WORK
05/30/99	SU	3437	3382	NO WORK	NO WORK
05/31/99	MO	3438	H	H	H
06/01/99	TU	3439	3383	2892	2401
06/02/99	WE	3440	3384	2893	2402
06/03/99	TH	3441	3385	2894	2403
06/04/99	FR	3442	3386	2895	2404
06/05/99	SA	3443	3387	2896	NO WORK
06/06/99	SU	3444	3388	NO WORK	NO WORK
06/07/99	MO	3445	3389	2897	2405
06/08/99	TU	3446	3390	2898	2406

DATE	DAY OF THE WEEK	CALENDAR DAY	7-DAY WORK WEEK W/HOLIDAYS	6-DAY WORK WEEK W/HOLIDAYS	5-DAY WORK WEEK W/HOLIDAYS
06/09/99	WE	3447	3391	2899	2407
06/10/99	TH	3448	3392	2900	2408
06/11/99	FR	3449	3393	2901	2409
06/12/99	SA	3450	3394	2902	NO WORK
06/13/99	SU	3451	3395	NO WORK	NO WORK
06/14/99	MO	3452	3396	2903	2410
06/15/99	TU	3453	3397	2904	2411
06/16/99	WE	3454	3398	2905	2412
06/17/99	TH	3455	3399	2906	2413
06/18/99	FR	3456	3400	2907	2414
06/19/99	SA	3457	3401	2908	NO WORK
06/20/99	SU	3458	3402	NO WORK	NO WORK
06/21/99	MO	3459	3403	2909	2415
06/22/99	TU	3460	3404	2910	2416
06/23/99	WE	3461	3405	2911	2417
06/24/99	TH	3462	3406	2912	2418
06/25/99	FR	3463	3407	2913	2419
06/26/99	SA	3464	3408	2914	NO WORK
06/27/99	SU	3465	3409	NO WORK	NO WORK
06/28/99	MO	3466	3410	2915	2420
06/29/99	TU	3467	3411	2916	2421
06/30/99	WE	3468	3412	2917	2422
07/01/99	TH	3469	3413	2918	2423
07/02/99	FR	3470	3414	2919	2424
07/03/99	SA	3471	3415	2920	NO WORK
07/04/99	SU	3472	H	NO WORK	NO WORK
07/05/99	MO	3473	3416	H	H
07/06/99	TU	3474	3417	2921	2425
07/07/99	WE	3475	3418	2922	2426
07/08/99	TH	3476	3419	2923	2427
07/09/99	FR	3477	3420	2924	2428
07/10/99	SA	3478	3421	2925	NO WORK
07/11/99	SU	3479	3422	NO WORK	NO WORK
07/12/99	MO	3480	3423	2926	2429
07/13/99	TU	3481	3424	2927	2430
07/14/99	WE	3482	3425	2928	2431
07/15/99	TH	3483	3426	2929	2432
07/16/99	FR	3484	3427	2930	2433
07/17/99	SA	3485	3428	2931	NO WORK
07/18/99	SU	3486	3429	NO WORK	NO WORK
07/19/99	MO	3487	3430	2932	2434
07/20/99	TU	3488	3431	2933	2435
07/21/99	WE	3489	3432	2934	2436
07/22/99	TH	3490	3433	2935	2437
07/23/99	FR	3491	3434	2936	2438
07/24/99	SA	3492	3435	2937	NO WORK
07/25/99	SU	3493	3436	NO WORK	NO WORK
07/26/99	MO	3494	3437	2938	2439
07/27/99	TU	3495	3438	2939	2440
07/28/99	WE	3496	3439	2940	2441
07/29/99	TH	3497	3440	2941	2442
07/30/99	FR	3498	3441	2942	2443
07/31/99	SA	3499	3442	2943	NO WORK

DATE	DAY OF THE WEEK	CALENDAR DAY	7-DAY WORK WEEK W/HOLIDAYS	6-DAY WORK WEEK W/HOLIDAYS	5-DAY WORK WEEK W/HOLIDAYS
08/01/99	SU	3500	3443	NO WORK	NO WORK
08/02/99	MO	3501	3444	2944	2444
08/03/99	TU	3502	3445	2945	2445
08/04/99	WE	3503	3446	2946	2446
08/05/99	TH	3504	3447	2947	2447
08/06/99	FR	3505	3448	2948	2448
08/07/99	SA	3506	3449	2949	NO WORK
08/08/99	SU	3507	3450	NO WORK	NO WORK
08/09/99	MO	3508	3451	2950	2449
08/10/99	TU	3509	3452	2951	2450
08/11/99	WE	3510	3453	2952	2451
08/12/99	TH	3511	3454	2953	2452
08/13/99	FR	3512	3455	2954	2453
08/14/99	SA	3513	3456	2955	NO WORK
08/15/99	SU	3514	3457	NO WORK	NO WORK
08/16/99	MO	3515	3458	2956	2454
08/17/99	TU	3516	3459	2957	2455
08/18/99	WE	3517	3460	2958	2456
08/19/99	TH	3518	3461	2959	2457
08/20/99	FR	3519	3462	2960	2458
08/21/99	SA	3520	3463	2961	NO WORK
08/22/99	SU	3521	3464	NO WORK	NO WORK
08/23/99	MO	3522	3465	2962	2459
08/24/99	TU	3523	3466	2963	2460
08/25/99	WE	3524	3467	2964	2461
08/26/99	TH	3525	3468	2965	2462
08/27/99	FR	3526	3469	2966	2463
08/28/99	SA	3527	3470	2967	NO WORK
08/29/99	SU	3528	3471	NO WORK	NO WORK
08/30/99	MO	3529	3472	2968	2464
08/31/99	TU	3530	3473	2969	2465
09/01/99	WE	3531	3474	2970	2466
09/02/99	TH	3532	3475	2971	2467
09/03/99	FR	3533	3476	2972	2468
09/04/99	SA	3534	3477	2973	NO WORK
09/05/99	SU	3535	3478	NO WORK	NO WORK
09/06/99	MO	3536	H	H	H
09/07/99	TU	3537	3479	2974	2469
09/08/99	WE	3538	3480	2975	2470
09/09/99	TH	3539	3481	2976	2471
09/10/99	FR	3540	3482	2977	2472
09/11/99	SA	3541	3483	2978	NO WORK
09/12/99	SU	3542	3484	NO WORK	NO WORK
09/13/99	MO	3543	3485	2979	2473
09/14/99	TU	3544	3486	2980	2474
09/15/99	WE	3545	3487	2981	2475
09/16/99	TH	3546	3488	2982	2476
09/17/99	FR	3547	3489	2983	2477
09/18/99	SA	3548	3490	2984	NO WORK
09/19/99	SU	3549	3491	NO WORK	NO WORK
09/20/99	MO	3550	3492	2985	2478
09/21/99	TU	3551	3493	2986	2479
09/22/99	WE	3552	3494	2987	2480

DATE	DAY OF THE WEEK	CALENDAR DAY	7-DAY WORK WEEK W/HOLIDAYS	6-DAY WORK WEEK W/HOLIDAYS	5-DAY WORK WEEK W/HOLIDAYS
09/23/99	TH	3553	3495	2988	2481
09/24/99	FR	3554	3496	2989	2482
09/25/99	SA	3555	3497	2990	NO WORK
09/26/99	SU	3556	3498	NO WORK	NO WORK
09/27/99	MO	3557	3499	2991	2483
09/28/99	TU	3558	3500	2992	2484
09/29/99	WE	3559	3501	2993	2485
09/30/99	TH	3560	3502	2994	2486
10/01/99	FR	3561	3503	2995	2487
10/02/99	SA	3562	3504	2996	NO WORK
10/03/99	SU	3563	3505	NO WORK	NO WORK
10/04/99	MO	3564	3506	2997	2488
10/05/99	TU	3565	3507	2998	2489
10/06/99	WE	3566	3508	2999	2490
10/07/99	TH	3567	3509	3000	2491
10/08/99	FR	3568	3510	3001	2492
10/09/99	SA	3569	3511	3002	NO WORK
10/10/99	SU	3570	3512	NO WORK	NO WORK
10/11/99	MO	3571	3513	3003	2493
10/12/99	TU	3572	3514	3004	2494
10/13/99	WE	3573	3515	3005	2495
10/14/99	TH	3574	3516	3006	2496
10/15/99	FR	3575	3517	3007	2497
10/16/99	SA	3576	3518	3008	NO WORK
10/17/99	SU	3577	3519	NO WORK	NO WORK
10/18/99	MO	3578	3520	3009	2498
10/19/99	TU	3579	3521	3010	2499
10/20/99	WE	3580	3522	3011	2500
10/21/99	TH	3581	3523	3012	2501
10/22/99	FR	3582	3524	3013	2502
10/23/99	SA	3583	3525	3014	NO WORK
10/24/99	SU	3584	3526	NO WORK	NO WORK
10/25/99	MO	3585	3527	3015	2503
10/26/99	TU	3586	3528	3016	2504
10/27/99	WE	3587	3529	3017	2505
10/28/99	TH	3588	3530	3018	2506
10/29/99	FR	3589	3531	3019	2507
10/30/99	SA	3590	3532	3020	NO WORK
10/31/99	SU	3591	3533	NO WORK	NO WORK
11/01/99	MO	3592	3534	3021	2508
11/02/99	TU	3593	3535	3022	2509
11/03/99	WE	3594	3536	3023	2510
11/04/99	TH	3595	3537	3024	2511
11/05/99	FR	3596	3538	3025	2512
11/06/99	SA	3597	3539	3026	NO WORK
11/07/99	SU	3598	3540	NO WORK	NO WORK
11/08/99	MO	3599	3541	3027	2513
11/09/99	TU	3600	3542	3028	2514
11/10/99	WE	3601	3543	3029	2515
11/11/99	TH	3602	3544	3030	2516
11/12/99	FR	3603	3545	3031	2517
11/13/99	SA	3604	3546	3032	NO WORK
11/14/99	SU	3605	3547	NO WORK	NO WORK
11/15/99	MO	3606	3548	3033	2518
11/16/99	TU	3607	3549	3034	2519
11/17/99	WE	3608	3550	3035	2520
11/18/99	TH	3609	3551	3036	2521
11/19/99	FR	3610	3552	3037	2522
11/20/99	SA	3611	3553	3038	NO WORK
11/21/99	SU	3612	3554	NO WORK	NO WORK
11/22/99	MO	3613	3555	3039	2523
11/23/99	TU	3614	3556	3040	2524
11/24/99	WE	3615	3557	3041	2525
11/25/99	TH	3616	H	H	H
11/26/99	FR	3617	3558	3042	2526
11/27/99	SA	3618	3559	3043	NO WORK
11/28/99	SU	3619	3560	NO WORK	NO WORK
11/29/99	MO	3620	3561	3044	2527
11/30/99	TU	3621	3562	3045	2528
12/01/99	WE	3622	3563	3046	2529
12/02/99	TH	3623	3564	3047	2530
12/03/99	FR	3624	3565	3048	2531
12/04/99	SA	3625	3566	3049	NO WORK
12/05/99	SU	3626	3567	NO WORK	NO WORK
12/06/99	MO	3627	3568	3050	2532
12/07/99	TU	3628	3569	3051	2533
12/08/99	WE	3629	3570	3052	2534
12/09/99	TH	3630	3571	3053	2535
12/10/99	FR	3631	3572	3054	2536
12/11/99	SA	3632	3573	3055	NO WORK
12/12/99	SU	3633	3574	NO WORK	NO WORK
12/13/99	MO	3634	3575	3056	2537
12/14/99	TU	3635	3576	3057	2538
12/15/99	WE	3636	3577	3058	2539
12/16/99	TH	3637	3578	3059	2540
12/17/99	FR	3638	3579	3060	2541
12/18/99	SA	3639	3580	3061	NO WORK
12/19/99	SU	3640	3581	NO WORK	NO WORK
12/20/99	MO	3641	3582	3062	2542
12/21/99	TU	3642	3583	3063	2543
12/22/99	WE	3643	3584	3064	2544
12/23/99	TH	3644	3585	3065	2545
12/24/99	FR	3645	3586	3066	H
12/25/99	SA	3646	H	H	NO WORK
12/26/99	SU	3647	3587	NO WORK	NO WORK
12/27/99	MO	3648	3588	3067	2546
12/28/99	TU	3649	3589	3068	2547
12/29/99	WE	3650	3590	3069	2548
12/30/99	TH	3651	3591	3070	2549
12/31/99	FR	3652	3592	3071	H

DATE	DAY OF THE WEEK	CALENDAR DAY	7-DAY WORK WEEK W/HOLIDAYS	6-DAY WORK WEEK W/HOLIDAYS	5-DAY WORK WEEK W/HOLIDAYS
01/01/2000	SA	3653	H	H	NO WORK
01/02/2000	SU	3654	3593	NO WORK	NO WORK
01/03/2000	MO	3655	3594	3072	2550
01/04/2000	TU	3656	3595	3073	2551
01/05/2000	WE	3657	3596	3074	2552
01/06/2000	TH	3658	3597	3075	2553
01/07/2000	FR	3659	3598	3076	2554
01/08/2000	SA	3660	3599	3077	NO WORK
01/09/2000	SU	3661	3600	NO WORK	NO WORK
01/10/2000	MO	3662	3601	3078	2555
01/11/2000	TU	3663	3602	3079	2556
01/12/2000	WE	3664	3603	3080	2557
01/13/2000	TH	3665	3604	3081	2558
01/14/2000	FR	3666	3605	3082	2559
01/15/2000	SA	3667	3606	3083	NO WORK
01/16/2000	SU	3668	3607	NO WORK	NO WORK
01/17/2000	MO	3669	3608	3084	2560
01/18/2000	TU	3670	3609	3085	2561
01/19/2000	WE	3671	3610	3086	2562
01/20/2000	TH	3672	3611	3087	2563
01/21/2000	FR	3673	3612	3088	2564
01/22/2000	SA	3674	3613	3089	NO WORK
01/23/2000	SU	3675	3614	NO WORK	NO WORK
01/24/2000	MO	3676	3615	3090	2565
01/25/2000	TU	3677	3616	3091	2566
01/26/2000	WE	3678	3617	3092	2567
01/27/2000	TH	3679	3618	3093	2568
01/28/2000	FR	3680	3619	3094	2569
01/29/2000	SA	3681	3620	3095	NO WORK
01/30/2000	SU	3682	3621	NO WORK	NO WORK
01/31/2000	MO	3683	3622	3096	2570
02/01/2000	TU	3684	3623	3097	2571
02/02/2000	WE	3685	3624	3098	2572
02/03/2000	TH	3686	3625	3099	2573
02/04/2000	FR	3687	3626	3100	2574
02/05/2000	SA	3688	3627	3101	NO WORK
02/06/2000	SU	3689	3628	NO WORK	NO WORK
02/07/2000	MO	3690	3629	3102	2575
02/08/2000	TU	3691	3630	3103	2576
02/09/2000	WE	3692	3631	3104	2577
02/10/2000	TH	3693	3632	3105	2578
02/11/2000	FR	3694	3633	3106	2579
02/12/2000	SA	3695	3634	3107	NO WORK
02/13/2000	SU	3696	3635	NO WORK	NO WORK
02/14/2000	MO	3697	3636	3108	2580
02/15/2000	TU	3698	3637	3109	2581
02/16/2000	WE	3699	3638	3110	2582
02/17/2000	TH	3700	3639	3111	2583
02/18/2000	FR	3701	3640	3112	2584
02/19/2000	SA	3702	3641	3113	NO WORK
02/20/2000	SU	3703	3642	NO WORK	NO WORK
02/21/2000	MO	3704	3643	3114	2585
02/22/2000	TU	3705	3644	3115	2586

DATE	DAY OF THE WEEK	CALENDAR DAY	7-DAY WORK WEEK W/HOLIDAYS	6-DAY WORK WEEK W/HOLIDAYS	5-DAY WORK WEEK W/HOLIDAYS
02/23/2000	WE	3706	3645	3116	2587
02/24/2000	TH	3707	3646	3117	2588
02/25/2000	FR	3708	3647	3118	2589
02/26/2000	SA	3709	3648	3119	NO WORK
02/27/2000	SU	3710	3649	NO WORK	NO WORK
02/28/2000	MO	3711	3650	3120	2590
02/29/2000	TU	3712	3651	3121	2591
03/01/2000	WE	3713	3652	3122	2592
03/02/2000	TH	3714	3653	3123	2593
03/03/2000	FR	3715	3654	3124	2594
03/04/2000	SA	3716	3655	3125	NO WORK
03/05/2000	SU	3717	3656	NO WORK	NO WORK
03/06/2000	MO	3718	3657	3126	2595
03/07/2000	TU	3719	3658	3127	2596
03/08/2000	WE	3720	3659	3128	2597
03/09/2000	TH	3721	3660	3129	2598
03/10/2000	FR	3722	3661	3130	2599
03/11/2000	SA	3723	3662	3131	NO WORK
03/12/2000	SU	3724	3663	NO WORK	NO WORK
03/13/2000	MO	3725	3664	3132	2600
03/14/2000	TU	3726	3665	3133	2601
03/15/2000	WE	3727	3666	3134	2602
03/16/2000	TH	3728	3667	3135	2603
03/17/2000	FR	3729	3668	3136	2604
03/18/2000	SA	3730	3669	3137	NO WORK
03/19/2000	SU	3731	3670	NO WORK	NO WORK
03/20/2000	MO	3732	3671	3138	2605
03/21/2000	TU	3733	3672	3139	2606
03/22/2000	WE	3734	3673	3140	2607
03/23/2000	TH	3735	3674	3141	2608
03/24/2000	FR	3736	3675	3142	2609
03/25/2000	SA	3737	3676	3143	NO WORK
03/26/2000	SU	3738	3677	NO WORK	NO WORK
03/27/2000	MO	3739	3678	3144	2610
03/28/2000	TU	3740	3679	3145	2611
03/29/2000	WE	3741	3680	3146	2612
03/30/2000	TH	3742	3681	3147	2613
03/31/2000	FR	3743	3682	3148	2614
04/01/2000	SA	3744	3683	3149	NO WORK
04/02/2000	SU	3745	3684	NO WORK	NO WORK
04/03/2000	MO	3746	3685	3150	2615
04/04/2000	TU	3747	3686	3151	2616
04/05/2000	WE	3748	3687	3152	2617
04/06/2000	TH	3749	3688	3153	2618
04/07/2000	FR	3750	3689	3154	2619
04/08/2000	SA	3751	3690	3155	NO WORK
04/09/2000	SU	3752	3691	NO WORK	NO WORK
04/10/2000	MO	3753	3692	3156	2620
04/11/2000	TU	3754	3693	3157	2621
04/12/2000	WE	3755	3694	3158	2622
04/13/2000	TH	3756	3695	3159	2623
04/14/2000	FR	3757	3696	3160	2624
04/15/2000	SA	3758	3697	3161	NO WORK

DATE	DAY OF THE WEEK	CALENDAR DAY	7-DAY WORK WEEK W/HOLIDAYS	6-DAY WORK WEEK W/HOLIDAYS	5-DAY WORK WEEK W/HOLIDAYS
04/16/2000	SU	3759	3698	NO WORK	NO WORK
04/17/2000	MO	3760	3699	3162	2625
04/18/2000	TU	3761	3700	3163	2626
04/19/2000	WE	3762	3701	3164	2627
04/20/2000	TH	3763	3702	3165	2628
04/21/2000	FR	3764	3703	3166	2629
04/22/2000	SA	3765	3704	3167	NO WORK
04/23/2000	SU	3766	3705	NO WORK	NO WORK
04/24/2000	MO	3767	3706	3168	2630
04/25/2000	TU	3768	3707	3169	2631
04/26/2000	WE	3769	3708	3170	2632
04/27/2000	TH	3770	3709	3171	2633
04/28/2000	FR	3771	3710	3172	2634
04/29/2000	SA	3772	3711	3173	NO WORK
04/30/2000	SU	3773	3712	NO WORK	NO WORK
05/01/2000	MO	3774	3713	3174	2635
05/02/2000	TU	3775	3714	3175	2636
05/03/2000	WE	3776	3715	3176	2637
05/04/2000	TH	3777	3716	3177	2638
05/05/2000	FR	3778	3717	3178	2639
05/06/2000	SA	3779	3718	3179	NO WORK
05/07/2000	SU	3780	3719	NO WORK	NO WORK
05/08/2000	MO	3781	3720	3180	2640
05/09/2000	TU	3782	3721	3181	2641
05/10/2000	WE	3783	3722	3182	2642
05/11/2000	TH	3784	3723	3183	2643
05/12/2000	FR	3785	3724	3184	2644
05/13/2000	SA	3786	3725	3185	NO WORK
05/14/2000	SU	3787	3726	NO WORK	NO WORK
05/15/2000	MO	3788	3727	3186	2645
05/16/2000	TU	3789	3728	3187	2646
05/17/2000	WE	3790	3729	3188	2647
05/18/2000	TH	3791	3730	3189	2648
05/19/2000	FR	3792	3731	3190	2649
05/20/2000	SA	3793	3732	3191	NO WORK
05/21/2000	SU	3794	3733	NO WORK	NO WORK
05/22/2000	MO	3795	3734	3192	2650
05/23/2000	TU	3796	3735	3193	2651
05/24/2000	WE	3797	3736	3194	2652
05/25/2000	TH	3798	3737	3195	2653
05/26/2000	FR	3799	3738	3196	2654
05/27/2000	SA	3800	3739	3197	NO WORK
05/28/2000	SU	3801	3740	NO WORK	NO WORK
05/29/2000	MO	3802	H	H	H
05/30/2000	TU	3803	3741	3198	2655
05/31/2000	WE	3804	3742	3199	2656
06/01/2000	TH	3805	3743	3200	2657
06/02/2000	FR	3806	3744	3201	2658
06/03/2000	SA	3807	3745	3202	NO WORK
06/04/2000	SU	3808	3746	NO WORK	NO WORK
06/05/2000	MO	3809	3747	3203	2659
06/06/2000	TU	3810	3748	3204	2660
06/07/2000	WE	3811	3749	3205	2661

DATE	DAY OF THE WEEK	CALENDAR DAY	7-DAY WORK WEEK W/HOLIDAYS	6-DAY WORK WEEK W/HOLIDAYS	5-DAY WORK WEEK W/HOLIDAYS
06/08/2000	TH	3812	3750	3206	2662
06/09/2000	FR	3813	3751	3207	2663
06/10/2000	SA	3814	3752	3208	NO WORK
06/11/2000	SU	3815	3753	NO WORK	NO WORK
06/12/2000	MO	3816	3754	3209	2664
06/13/2000	TU	3817	3755	3210	2665
06/14/2000	WE	3818	3756	3211	2666
06/15/2000	TH	3819	3757	3212	2667
06/16/2000	FR	3820	3758	3213	2668
06/17/2000	SA	3821	3759	3214	NO WORK
06/18/2000	SU	3822	3760	NO WORK	NO WORK
06/19/2000	MO	3823	3761	3215	2669
06/20/2000	TU	3824	3762	3216	2670
06/21/2000	WE	3825	3763	3217	2671
06/22/2000	TH	3826	3764	3218	2672
06/23/2000	FR	3827	3765	3219	2673
06/24/2000	SA	3828	3766	3220	NO WORK
06/25/2000	SU	3829	3767	NO WORK	NO WORK
06/26/2000	MO	3830	3768	3221	2674
06/27/2000	TU	3831	3769	3222	2675
06/28/2000	WE	3832	3770	3223	2676
06/29/2000	TH	3833	3771	3224	2677
06/30/2000	FR	3834	3772	3225	2678
07/01/2000	SA	3835	3773	3226	NO WORK
07/02/2000	SU	3836	3774	NO WORK	NO WORK
07/03/2000	MO	3837	3775	3227	2679
07/04/2000	TU	3838	H	H	H
07/05/2000	WE	3839	3776	3228	2680
07/06/2000	TH	3840	3777	3229	2681
07/07/2000	FR	3841	3778	3230	2682
07/08/2000	SA	3842	3779	3231	NO WORK
07/09/2000	SU	3843	3780	NO WORK	NO WORK
07/10/2000	MO	3844	3781	3232	2683
07/11/2000	TU	3845	3782	3233	2684
07/12/2000	WE	3846	3783	3234	2685
07/13/2000	TH	3847	3784	3235	2686
07/14/2000	FR	3848	3785	3236	2687
07/15/2000	SA	3849	3786	3237	NO WORK
07/16/2000	SU	3850	3787	NO WORK	NO WORK
07/17/2000	MO	3851	3788	3238	2688
07/18/2000	TU	3852	3789	3239	2689
07/19/2000	WE	3853	3790	3240	2690
07/20/2000	TH	3854	3791	3241	2691
07/21/2000	FR	3855	3792	3242	2692
07/22/2000	SA	3856	3793	3243	NO WORK
07/23/2000	SU	3857	3794	NO WORK	NO WORK
07/24/2000	MO	3858	3795	3244	2693
07/25/2000	TU	3859	3796	3245	2694
07/26/2000	WE	3860	3797	3246	2695
07/27/2000	TH	3861	3798	3247	2696
07/28/2000	FR	3862	3799	3248	2697
07/29/2000	SA	3863	3800	3249	NO WORK
07/30/2000	SU	3864	3801	NO WORK	NO WORK

DATE	DAY OF THE WEEK	CALENDAR DAY	7-DAY WORK WEEK W/HOLIDAYS	6-DAY WORK WEEK W/HOLIDAYS	5-DAY WORK WEEK W/HOLIDAYS
07/31/2000	MO	3865	3802	3250	2698
08/01/2000	TU	3866	3803	3251	2699
08/02/2000	WE	3867	3804	3252	2700
08/03/2000	TH	3868	3805	3253	2701
08/04/2000	FR	3869	3806	3254	2702
08/05/2000	SA	3870	3807	3255	NO WORK
08/06/2000	SU	3871	3808	NO WORK	NO WORK
08/07/2000	MO	3872	3809	3256	2703
08/08/2000	TU	3873	3810	3257	2704
08/09/2000	WE	3874	3811	3258	2705
08/10/2000	TH	3875	3812	3259	2706
08/11/2000	FR	3876	3813	3260	2707
08/12/2000	SA	3877	3814	3261	NO WORK
08/13/2000	SU	3878	3815	NO WORK	NO WORK
08/14/2000	MO	3879	3816	3262	2708
08/15/2000	TU	3880	3817	3263	2709
08/16/2000	WE	3881	3818	3264	2710
08/17/2000	TH	3882	3819	3265	2711
08/18/2000	FR	3883	3820	3266	2712
08/19/2000	SA	3884	3821	3267	NO WORK
08/20/2000	SU	3885	3822	NO WORK	NO WORK
08/21/2000	MO	3886	3823	3268	2713
08/22/2000	TU	3887	3824	3269	2714
08/23/2000	WE	3888	3825	3270	2715
08/24/2000	TH	3889	3826	3271	2716
08/25/2000	FR	3890	3827	3272	2717
08/26/2000	SA	3891	3828	3273	NO WORK
08/27/2000	SU	3892	3829	NO WORK	NO WORK
08/28/2000	MO	3893	3830	3274	2718
08/29/2000	TU	3894	3831	3275	2719
08/30/2000	WE	3895	3832	3276	2720
08/31/2000	TH	3896	3833	3277	2721
09/01/2000	FR	3897	3834	3278	2722
09/02/2000	SA	3898	3835	3279	NO WORK
09/03/2000	SU	3899	3836	NO WORK	NO WORK
09/04/2000	MO	3900	H	H	H
09/05/2000	TU	3901	3837	3280	2723
09/06/2000	WE	3902	3838	3281	2724
09/07/2000	TH	3903	3839	3282	2725
09/08/2000	FR	3904	3840	3283	2726
09/09/2000	SA	3905	3841	3284	NO WORK
09/10/2000	SU	3906	3842	NO WORK	NO WORK
09/11/2000	MO	3907	3843	3285	2727
09/12/2000	TU	3908	3844	3286	2728
09/13/2000	WE	3909	3845	3287	2729
09/14/2000	TH	3910	3846	3288	2730
09/15/2000	FR	3911	3847	3289	2731
09/16/2000	SA	3912	3848	3290	NO WORK
09/17/2000	SU	3913	3849	NO WORK	NO WORK
09/18/2000	MO	3914	3850	3291	2732
09/19/2000	TU	3915	3851	3292	2733
09/20/2000	WE	3916	3852	3293	2734
09/21/2000	TH	3917	3853	3294	2735
09/22/2000	FR	3918	3854	3295	2736
09/23/2000	SA	3919	3855	3296	NO WORK
09/24/2000	SU	3920	3856	NO WORK	NO WORK
09/25/2000	MO	3921	3857	3297	2737
09/26/2000	TU	3922	3858	3298	2738
09/27/2000	WE	3923	3859	3299	2739
09/28/2000	TH	3924	3860	3300	2740
09/29/2000	FR	3925	3861	3301	2741
09/30/2000	SA	3926	3862	3302	NO WORK
10/01/2000	SU	3927	3863	NO WORK	NO WORK
10/02/2000	MO	3928	3864	3303	2742
10/03/2000	TU	3929	3865	3304	2743
10/04/2000	WE	3930	3866	3305	2744
10/05/2000	TH	3931	3867	3306	2745
10/06/2000	FR	3932	3868	3307	2746
10/07/2000	SA	3933	3869	3308	NO WORK
10/08/2000	SU	3934	3870	NO WORK	NO WORK
10/09/2000	MO	3935	3871	3309	2747
10/10/2000	TU	3936	3872	3310	2748
10/11/2000	WE	3937	3873	3311	2749
10/12/2000	TH	3938	3874	3312	2750
10/13/2000	FR	3939	3875	3313	2751
10/14/2000	SA	3940	3876	3314	NO WORK
10/15/2000	SU	3941	3877	NO WORK	NO WORK
10/16/2000	MO	3942	3878	3315	2752
10/17/2000	TU	3943	3879	3316	2753
10/18/2000	WE	3944	3880	3317	2754
10/19/2000	TH	3945	3881	3318	2755
10/20/2000	FR	3946	3882	3319	2756
10/21/2000	SA	3947	3883	3320	NO WORK
10/22/2000	SU	3948	3884	NO WORK	NO WORK
10/23/2000	MO	3949	3885	3321	2757
10/24/2000	TU	3950	3886	3322	2758
10/25/2000	WE	3951	3887	3323	2759
10/26/2000	TH	3952	3888	3324	2760
10/27/2000	FR	3953	3889	3325	2761
10/28/2000	SA	3954	3890	3326	NO WORK
10/29/2000	SU	3955	3891	NO WORK	NO WORK
10/30/2000	MO	3956	3892	3327	2762
10/31/2000	TU	3957	3893	3328	2763
11/01/2000	WE	3958	3894	3329	2764
11/02/2000	TH	3959	3895	3330	2765
11/03/2000	FR	3960	3896	3331	2766
11/04/2000	SA	3961	3897	3332	NO WORK
11/05/2000	SU	3962	3898	NO WORK	NO WORK
11/06/2000	MO	3963	3899	3333	2767
11/07/2000	TU	3964	3900	3334	2768
11/08/2000	WE	3965	3901	3335	2769
11/09/2000	TH	3966	3902	3336	2770
11/10/2000	FR	3967	3903	3337	2771
11/11/2000	SA	3968	3904	3338	NO WORK
11/12/2000	SU	3969	3905	NO WORK	NO WORK
11/13/2000	MO	3970	3906	3339	2772

DATE	DAY OF THE WEEK	CALENDAR DAY	7-DAY WORK WEEK W/HOLIDAYS	6-DAY WORK WEEK W/HOLIDAYS	5-DAY WORK WEEK W/HOLIDAYS
11/14/2000	TU	3971	3907	3340	2773
11/15/2000	WE	3972	3908	3341	2774
11/16/2000	TH	3973	3909	3342	2775
11/17/2000	FR	3974	3910	3343	2776
11/18/2000	SA	3975	3911	3344	NO WORK
11/19/2000	SU	3976	3912	NO WORK	NO WORK
11/20/2000	MO	3977	3913	3345	2777
11/21/2000	TU	3978	3914	3346	2778
11/22/2000	WE	3979	3915	3347	2779
11/23/2000	TH	3980	H	H	H
11/24/2000	FR	3981	3916	3348	2780
11/25/2000	SA	3982	3917	3349	NO WORK
11/26/2000	SU	3983	3918	NO WORK	NO WORK
11/27/2000	MO	3984	3919	3350	2781
11/28/2000	TU	3985	3920	3351	2782
11/29/2000	WE	3986	3921	3352	2783
11/30/2000	TH	3987	3922	3353	2784
12/01/2000	FR	3988	3923	3354	2785
12/02/2000	SA	3989	3924	3355	NO WORK
12/03/2000	SU	3990	3925	NO WORK	NO WORK
12/04/2000	MO	3991	3926	3356	2786
12/05/2000	TU	3992	3927	3357	2787
12/06/2000	WE	3993	3928	3358	2788
12/07/2000	TH	3994	3929	3359	2789
12/08/2000	FR	3995	3930	3360	2790
12/09/2000	SA	3996	3931	3361	NO WORK
12/10/2000	SU	3997	3932	NO WORK	NO WORK
12/11/2000	MO	3998	3933	3362	2791
12/12/2000	TU	3999	3934	3363	2792
12/13/2000	WE	4000	3935	3364	2793
12/14/2000	TH	4001	3936	3365	2794
12/15/2000	FR	4002	3937	3366	2795
12/16/2000	SA	4003	3938	3367	NO WORK
12/17/2000	SU	4004	3939	NO WORK	NO WORK
12/18/2000	MO	4005	3940	3368	2796
12/19/2000	TU	4006	3941	3369	2797
12/20/2000	WE	4007	3942	3370	2798
12/21/2000	TH	4008	3943	3371	2799
12/22/2000	FR	4009	3944	3372	2800
12/23/2000	SA	4010	3945	3373	NO WORK
12/24/2000	SU	4011	3946	NO WORK	NO WORK
12/25/2000	MO	4012	H	H	H
12/26/2000	TU	4013	3947	3374	2801
12/27/2000	WE	4014	3948	3375	2802
12/28/2000	TH	4015	3949	3376	2803
12/29/2000	FR	4016	3950	3377	2804
12/30/2000	SA	4017	3951	3378	NO WORK
12/31/2000	SU	4018	3952	NO WORK	NO WORK

431

INDEX